AWS Certified DevOps Engineer - Professional Certification and Beyond

Pass the DOP-C01 exam and prepare for the real world using case studies and real-life examples

Adam Book

BIRMINGHAM—MUMBAI

AWS Certified DevOps Engineer - Professional Certification and Beyond

Group Product Manager: Wilson Dsouza
Publishing Product Manager: Rahul Nair
Senior Editor: Arun Nadar
Content Development Editor: Rafiaa Khan
Technical Editor: Arjun Varma
Copy Editor: Safis Editing
Project Coordinator: Shagun Saini
Proofreader: Safis Editing
Indexer: Pratik Shirodkar
Production Designer: Joshua Misquitta

First published: February 2022
Production reference: 1031121

Published by Packt Publishing Ltd.
Livery Place
35 Livery Street
Birmingham
B3 2PB, UK.

ISBN 978-1-80107-445-2

www.packt.com

Writing a book is harder than I thought but more rewarding than I ever conceived. I'm eternally grateful for my wife Mariel, who not only told me this was an opportunity that I couldn't turn down, but for also keeping our two daughters, Sara and Emma, at bay so I had time to write and edit. I'd also like to thank the whole Packt team for presenting me with this opportunity. Finally, thank you to Presidio for all the support they have shown me when I said I was going to be taking this challenge on.

To all the people who take on the challenge of getting AWS certified and passing. – Adam Book

Contributors

About the author

Adam Book has been programming since the age of six and has been constantly tapped by founders and CEOs as one of the pillars to start their online or cloud businesses.

Adam has been developing applications and websites and has been involved in cloud computing and data center transformation professionally since 1996 and has made it a primary professional focus to bring the benefits of cloud computing to his clients. In addition, he has led technology teams in transformative changes such as the shift to programming in sprints, with Agile formats, and has also mentored organizations on the benefits of peer programming in local and remote formats along with test-driven development or business-driven development.

Adam is a passionate cloud evangelist with a proven track record of using the AWS platform from the very beginning to successfully migrate thousands of application workloads to the cloud and guiding businesses in understanding cloud economics to create business cases and identify operating model gaps.

Adam has passed the AWS Professional exams, both Architect and DevOps Engineer, multiple times along with the underlying associate exams. He has also achieved the Security Specialty Certification from AWS. In addition to this, Adam has helped multiple organizations achieve Premier Partner status, along with being instrumental in the attainment of numerous AWS Competencies.

About the reviewers

Sebastian Krueger started his AWS certification journey in 2014 and since then he has passed over 17 AWS Certification Exams (including re-certifications). He is the Founder of the Wellington (NZ) AWS User Group and has been an Organiser of the community group ever since. In 2019 he was a Founder of the AWS User Group Aotearoa and served as its first President.

After founding the successful AWS Consultancy 'API Talent' in New Zealand in 2014, his company was acquired by Deloitte in 2018. Sebastian then served as Partner in the Cloud Engineering practice.

Since 2021, he has been working at AWS in Wellington, New Zealand as a Principal Solution Architect.

Sebastian has also been a co-author of Java CAPS Basics (2008) and Tech Reviewer of Serverless Single Page Apps (2016) and Learn AWS Serverless Computing (2019).

Stoyan Zdravkov Yanev is a certified Solutions Architect and passionate cloud evangelist. His experience and specialization is on building modern event-driven architectures and creating scalable kubernetes infrastructures. Stoyan has broad knowledge in securing and scaling Enterprise projects, which he built during his stay in companies like NewsUK, VMware, MentorMate, and Schwarz IT. He is also the author of the book AWS Fundamentals, which will be published in 2022. Stoyan now works as a senior DevOps/Cloud consultant focusing on the middle market, and helping smaller organizations achieve the same benefits as large enterprises.

Acknowledgments

I would like to thank my wife Gergana for her continued support and encouragement with everything that I do. You have always pushed me towards new adventures, accomplishing my goals, and doing what is right. I appreciate what you have done for us and I love you. I'd also like to thank my son Boris for showing me what is actually important in life.

Stoyan Zdravkov Yanev

Table of Contents

2

Fundamental
AWS Services

3

Identity and Access Management and Working with Secrets
in AWS

4
Amazon S3 Blob Storage

5
Amazon DynamoDB

Section 2: Developing, Deploying, and Using Infrastructure as Code

6

Understanding CI/CD and the SDLC

7

Using CloudFormation Templates to Deploy Workloads

8

Creating Workloads with CodeCommit and CodeBuild

9

Deploying Workloads with CodeDeploy and CodePipeline

10

Using AWS OpsWorks to Manage and Deploy Your Stack

11
Using Elastic Beanstalk to Deploy Your Application

12
Lambda Deployments and Versioning

13

Blue/Green Deployments

Section 3: Monitoring and Logging Your Environment and Workloads

14
CloudWatch and X-Ray's Role in DevOps

15
CloudWatch Metrics and Amazon EventBridge

16

Various Logs Generated (VPC Flow Logs, Load Balancer Logs, CloudTrail Logs)

17

Advanced and Enterprise Logging Scenarios

Section 4: Enabling Highly Available Workloads, Fault Tolerance, and Implementing Standards and Policies

18

Auto Scaling and Lifecycle Hooks

19

Protecting Data in Flight and at Rest

22

Other Policy and Standards Services to Know

Section 5: Exam Tips and Tricks

23

Overview of the DevOps Professional Certification Exam

24

Practice Exam

Other Books You May Enjoy

Index

Preface

More and more companies are making the move to the cloud, and more specifically, the **Amazon Web Services** (**AWS**) cloud, every day. Once in the cloud, these companies and enterprises are looking to streamline their processes and **Software Development Life Cycles** (**SDLCs**) through the use of techniques found in DevOps practices. This includes automating build and release processes so that development teams can focus on writing the code and features that customers desire. This also includes making sure that logging and monitoring are in place, not just for the sake of checking a checkbox on a deployment list, but instead to empower teams to quickly find the root causes of their issues, be they performance, error, or security related.

The need for skilled, certified AWS individuals is at an all-time high and growing. Passing the AWS DevOps Engineer Professional exam allows you to be separated instantly from others, showing that you have taken the time and effort to not only learn these valuable skills but also pass the rigorous standard that is the AWS Professional certification.

Certifications offered by AWS, especially Professional certification exams, are not easy. Those that work and hire in this industry know this. Not only are the exams themselves time-consuming at around 3 hours to take, but they are also constantly being updated.

There is a lot of information to digest and understand on this exam. You not only need to know the topics that are covered in the blueprint provided by AWS, but you also need to have a solid foundation of AWS knowledge in order to use those services. It helps if you have real-world experience or at least hands-on practice with using these services. This is where the exercises that have been included in this book come in. They serve not as an ending point but hopefully as a starting point to build on and create other projects from so that you have the confidence when you press start on your test that you have the skills and the knowledge you need to pass the certification test and take your career to the next level.

This book is designed to help you gain a firm grasp of the services that are presented in the AWS DevOps Professional exam. This is accomplished in a variety of methods. There are sample architectures presented for many of the services so that you can visualize how different services work together. There are plenty of hands-on examples to try and therefore see how things work in the real world. There are also example cases of best use cases and anti-patterns for the different services. These are especially important to understand when evaluating the Professional exam questions, which are presented in a large scenario format. Understanding where one service is the best fit and where it is not a fit can help you decide on the correct answer.

Finally, this book is meant to be not only a study guide in passing the test but also a reference guide in performing your day-to-day activities once you have passed. This is where the *beyond* comes in. There is extra information presented that is not required for the exam, and this is included on purpose. I wanted to share some of the years of experience I have gained from working with all types of businesses on AWS, from small and medium-sized companies to large Fortune 100 enterprises.

Who this book is for

No matter where you are in your AWS journey, if you have the desire to become certified as a DevOps professional engineer, then this book will help you gain an understanding of the fundamental concepts and services, as well as examining the essential services that are covered by the exam blueprint.

With the opening chapter, we lay the foundation of what good looks like in the world of AWS, and although this seems like a lot of theory, it helps you work through many of the formidable scenarios that are presented in the exam questions. There are also plenty of hands-on exercises for using services that you may not be familiar with so that you have the confidence and experience both during the exam and afterward as well.

At the end of each chapter, there are knowledge assessments to help self-check to see whether you have a grasp of the information that is needed to not only pass this challenging exam but also succeed in the other tasks that lie ahead for you in the ever-changing world of cloud computing.

What this book covers

Chapter 1, Amazon Web Service Pillars, focuses on the foundational pillars that make up the Well-Architected Framework in AWS. By understanding these pillars, you will gain a better feel for the context of the questions being asked in the certification exam.

Chapter 2, Fundamental AWS Services, examines a large number of fundamental AWS services that are imperative to know going forward with future chapters. This may seem like a review for some that have already passed some of the lower associate exams. However, it can also serve as a quick refresher and provide a few tips that were previously unknown.

Chapter 3, Identity and Access Management and Working with Secrets in AWS, focuses on the fundamental security building blocks of AWS, which are identity and access management using the IAM service. After a quick look at the shared security model from AWS and the concepts of authorization versus authentication, we review creating users and groups. Providing access to other accounts via cross-account access is also covered in this chapter with a practical exercise. In this fundamental security chapter, we also talk about other essential security services that may appear in test questions, such as AWS Directory Service, Secrets Manager, and Systems Manager Parameter Store. There are comparisons on when to use and not to use the different versions of AWS Directory Service, along with which service would be better to store your secrets. Finally, we take a look at Amazon Cognito and how it can help with application authentication.

Chapter 4, Amazon S3 Blob Storage, focuses on one of the key services in AWS Simple Storage Service, or S3. Even though this service is easy to start using right away, it also has a number of features and functions available to it that you must be aware of if you are trying to become certified on AWS.

Chapter 5, Amazon DynamoDB, explains the native NoSQL database DynamoDB. It looks at not only some of the essential concepts of DynamoDB but also topics such as streams, understanding global tables, using DynamoDB Accelerator, and even using Web Federation to connect to your DynamoDB tables.

Chapter 6, Understanding CI/CD and the SDLC, focuses on many theoretical aspects of continuous integration, continuous development, and continuous deployment. We then look at the SDLC along with which services map to different stages of the SDLC.

Chapter 7, Using CloudFormation Templates to Deploy Workloads, teaches you about using Infrastructure as Code using the native CloudFormation service. First, we'll go over the basics of CloudFormation templates, but then we'll quickly ramp up to examples of creating a changeset for a basic template, and then move on to intrinsic functions and nested stacks. Using the knowledge of CloudFormation templates, we discuss how ServiceCatalog can be used to serve up templated designs for developers and non-developers in a quick and easy fashion. This chapter closes by going over the Cloud Development Kit, which can be programmed in your language of choice and then used to create CloudFormation templates.

Chapter 8, *Creating Workloads with CodeCommit and CodeBuild*, guides you through the initial steps of the SDLC using native AWS tooling. We start by creating a brand-new group and user, who is a developer, with a whole new set of permissions that are scoped to just this user's role. After creating an initial CodeCommit repository, we have our developer use Git to commit code onto a feature branch and then request a merge to the main branch. Next, we examine the CodeBuild service by having the service build a container using AWS CodeBuild.

Chapter 9, *Deploying Workloads with CodeDeploy and CodePipeline*, shows you how to create DevOps pipelines using the native AWS CodePipeline service. This is a chapter where many of the services that we have been talking about and practicing previously come into play. The pipeline example being used is crafted in a CloudFormation template. The developer user that we previously created also needs expanded access in order to view and run our pipeline, so there is an exercise to elaborate their IAM permissions. Also discussed in this chapter is how to deploy workloads using the AWS CodeDeploy service.

Chapter 10, *Using AWS OpsWorks to Manage and Deploy Your Application Stack*, focuses on how to create stacks and layers to deploy infrastructure and applications using the AWS OpsWorks service. There is a comparison of the different versions of OpsWorks available, along with an exercise to create a stack with layers and an application.

Chapter 11, *Using Elastic Beanstalk to Deploy Your Application*, walks through one of the key services on the DevOps Professional exam – Elastic Beanstalk. Creating and deploying an application in Elastic Beanstalk using the EB CLI not only lets you see things through the lens of the developer, but also allows you to think about how you would automate these types of tasks in the real world.

Chapter 12, *Lambda Deployments and Versioning*, explores the concepts of serverless and using the AWS Lambda platform for serverless computing. With the cost savings available from running compute needs on an on-demand, pay-per-usage basis, this is becoming a more and more desired state in organizations today. We talk about not only how to deploy and monitor Lambda functions but also how to implement versions and aliases. At the end of the chapter, we even go through orchestrating multiple functions in a step function.

Chapter 13, *Blue/Green Deployments*, focuses on blue/green deployment strategies and the different variations of those strategies, including which service can use the various strategies and how to implement the different strategies depending on the services that you are utilizing. There are specific strategies that you can employ when using EC2 instances and autoscaling groups, and there are others that are available when using a Lambda function. Ensuring that your end users and customers have a seamless experience, even if you have an issue during deployment, is what this chapter is truly about.

Chapter 14, CloudWatch and X-Ray's Role in DevOps, shows you the role that monitoring and logging play using the native CloudWatch and X-Ray services from AWS. Log streams and searching through logs can be tedious tasks and sometimes feel like looking for a needle in a haystack. The same can be said of performance problems. Adding the X-Ray service to your Lambda application can help you quickly pinpoint where the issues are and know where to remedy the issues.

Chapter 15, CloudWatch Metrics and Amazon EventBridge, shows you how to use the metrics from various services and then tie them in the Amazon EventBridge service to create automated alerts for your systems. We discuss which metrics are some of the most useful for different vital services to keep a watch over. We also walk through creating dashboards in the Amazon CloudWatch console.

Chapter 16, Various Logs Generated (VPC Flow Logs, Load Balancer Logs, and CloudTrail Logs), examines the other types of logs that can be generated by AWS services that are not CloudWatch Logs. These logs are all valuable when troubleshooting information and may need to be turned on some or all of the time. The ability to know where to retrieve these logs and how to search the logs can be a task that you are called upon to do as a DevOps professional.

Chapter 17, Advanced and Enterprise Logging Scenarios, shows you real-world scenarios and architectures for building and processing log files. This includes incorporating not only the CloudWatch and CloudTrail services but also services such as Elasticsearch, Kinesis, and Lambda for the real-time processing of multiple log streams. Understanding the concepts of how to gather and process massive amounts of log files is important both for real-world engagements and for potential scenarios that could appear on the DevOps Professional certification exam.

Chapter 18, Auto Scaling and Lifecycle Hooks, covers how autoscaling and autoscaling groups work in detail. This includes examining the autoscaling life cycle and life cycle hooks. There is an exercise that walks you through creating a launch template, which is the successor of the launch configuration. We also go through a practice of removing and terminating instances inside of an autoscaling group.

Chapter 19, Protecting Data in Flight and at Rest, illustrates how the use of services such as Key Management Service and Amazon Certificate Manager helps protect data that is both sitting at rest as well as in transit. If you are building systems using Infrastructure as Code, you need to incorporate these key pieces into your system so that your data is safe from the very start.

Chapter 20, Enforcing Standards and Compliance with System Manger's Role and AWS Config, focuses on how to use automation to keep your AWS environment in a compliant state. Using the AWS Config service, you can keep a constant check on what is being created in your AWS environment. Combine this with rules that flag violations for what is not allowed in your environment to either send alerts or do automated enforcement and remediation. Add to this the capabilities of System Manager, which can automatically install software on instances using runbooks for needed compliance items such as virus scanners or perform regular operating system upgrades; then, creating an audit trail of performed tasks becomes much easier for your organization.

Chapter 21, Using Amazon Inspector to Check your Environment, shows you how to add automated security scans to your DevOps life cycle using the Amazon Inspector service. We look at how to configure the Inspector service in both an automated and manual manner and then view and understand the different reports that Inspector generates.

Chapter 22, Other Policy and Standards Services to Know, covers some of the services that have the tendency to appear on the DevOps Professional exam but did not make it into other chapters. These include services such as AWS GuardDuty, Amazon Macie, and Server Migration Service. We also go over AWS Organizations once again with its incorporation with the Service Catalog service to make sure that you have a full understanding of how those services work hand in hand.

Chapter 23, Overview of the DevOps Professional Certification Test, explains the testing process itself. It also has a list of extra resources that you should use in conjunction with this book to read and study for the exam, as well as some tips for studying.

Chapter 24, Practice Exam 1, is primarily meant to be a readiness check for you. This chapter presents questions as you will be presented on the exam and then gives you the answers and an explanation of why you would choose the correct answers to help you.

To get the most out of this book

Prior knowledge in the following subjects will help you get the most understanding of this book:

- Knowledge of software architecture, programming languages, and application design
- Knowledge of relational and non-relational databases
- Knowledge of AWS regions and geographical locations
- Understanding of JSON and YAML file formats

- Basic knowledge of operating systems and system commands
- Knowledge of logging and monitoring for applications and system availability

Software/hardware covered in the book	Operating system requirements
Python 3.x	Windows, macOS, or Linux
Amazon Command-Line Interface (CLI)	Windows, macOS, or Linux
Git	Windows, macOS, or Linux

Setting up the AWS CLI will be necessary to complete many of the exercises that have been presented throughout the book. A step-by-step walkthrough of installing the CLI is given in Chapter 2, Fundamental AWS Services, *if you do not already have the CLI installed on your computer.*

If you are using the digital version of this book, we advise you to type the code yourself or access the code from the book's GitHub repository (a link is available in the next section). Doing so will help you avoid any potential errors related to the copying and pasting of code.

Download the example code files

You can download the example code files for this book from GitHub at `https://github.com/PacktPublishing/AWS-Certified-DevOps-Engineer-Professional-Certification-and-Beyond`. If there's an update to the code, it will be updated in the GitHub repository.

We also have other code bundles from our rich catalog of books and videos available at `https://github.com/PacktPublishing/`. Check them out!

Download the color images

We also provide a PDF file that has color images of the screenshots and diagrams used in this book. You can download it here: `https://static.packt-cdn.com/downloads/9781801074452_ColorImages.pdf`.

Conventions used

There are a number of text conventions used throughout this book.

`Code in text`: Indicates code words in text, database table names, folder names, filenames, file extensions, pathnames, dummy URLs, user input, and Twitter handles. Here is an example: "You will see both the username and password returned unencrypted in the `SecretString` field for you to use."

A block of code is set as follows:

```
{
    "Project_ID": {"N": "0100"},
    "Dept": {"S": "Test Team"},
    "Dept_ID": {"N": "0001"},
    "Project_Name": {"S": "Serverless Forms"},
    "Owner": {"S": "Jerry Imoto"},
    "Builds": {"NS": ["2212121"] },
    "Language": {"S": "python" },
    "Contact": {"S": "test_team@testcompany.com" }
}
```

When we wish to draw your attention to a particular part of a code block, the relevant lines or items are set in bold:

```
[default]
exten => s,1,Dial(Zap/1|30)
exten => s,2,Voicemail(u100)
exten => s,102,Voicemail(b100)
exten => i,1,Voicemail(s0)
```

Any command-line input or output is written as follows:

```
$ aws iam list-groups --output text
$ aws iam create-group --group-name Admins
```

Bold: Indicates a new term, an important word, or words that you see onscreen. For instance, words in menus or dialog boxes appear in **bold**. Here is an example: "You have the option to encrypt the reports either with SSE-S3 or with a **Key Management Service (KMS)** key of your choosing."

> **Tips or important notes**
> Appear like this.

Get in touch

Feedback from our readers is always welcome.

General feedback: If you have questions about any aspect of this book, email us at customercare@packtpub.com and mention the book title in the subject of your message.

Errata: Although we have taken every care to ensure the accuracy of our content, mistakes do happen. If you have found a mistake in this book, we would be grateful if you would report this to us. Please visit www.packtpub.com/support/errata and fill in the form.

Piracy: If you come across any illegal copies of our works in any form on the internet, we would be grateful if you would provide us with the location address or website name. Please contact us at copyright@packt.com with a link to the material.

If you are interested in becoming an author: If there is a topic that you have expertise in and you are interested in either writing or contributing to a book, please visit authors.packtpub.com.

Share your thoughts

Once you've read *AWS Certified DevOps Engineer - Professional Certification and Beyond*, we'd love to hear your thoughts! Scan the QR code below to go straight to the Amazon review page for this book and share your feedback.

https://packt.link/r/1801074453

Your review is important to us and the tech community and will help us make sure we're delivering excellent quality content.

Section 1: Establishing the Fundamentals

In this part, we will look at the fundamentals of the AWS cloud, including the basis of the Well-Architected Framework, security, and storage.

This part of the book comprises the following chapters:

- *Chapter 1, Amazon Web Service Pillars*
- *Chapter 2, Fundamental AWS Services*
- *Chapter 3, Identity and Access Management and Working with Secrets in AWS*
- *Chapter 4, Amazon S3 Blob Storage*
- *Chapter 5, Amazon DynamoDB*

1
Amazon Web Service Pillars

DevOps is, at its heart, a combination of the skills of development and operations and breaking down the walls between these two different teams. DevOps includes enabling developers to perform operational tasks easily. DevOps also involves empowering operational team members to create their Infrastructure as Code and use other coding techniques, such as continuous integration pipelines, to spin up the same infrastructure in multiple regions quickly.

In this book, we will go through the services and concepts that are part of the DevOps professional exam so that you have a solid understanding from a practical standpoint, in terms of both explanations and hands-on exercises.

Becoming Amazon Web Services (AWS) Certified not only gives you instant validation of the technical skills that you hold and maintain – it also strengthens you as a technical professional. The AWS DevOps Engineer Professional Certification is a cumulative test that incorporates the base knowledge of fundamental AWS services, including system operations capabilities for running, managing, and monitoring workloads in AWS. This is in addition to developing and deploying code to functions, containers, and instances.

We go look at the test itself in more depth in *Chapter 23*, *Overview of the DevOps Professional Certification Test*, as well as provide tips for taking the exam.

The AWS pillars are the five guiding principles that guide architects and developers in generally accepted cloud architecture and design. They are subtly referenced in the DevOps Professional exam, but the pillars and their guidelines are tenets of best practices for working with any cloud service provider – especially Amazon Web Services. These are all guiding principles in DevOps practices and pipelines, and having a sound understanding of these five items will not only help you come exam time, but serve you throughout your DevOps career journey.

In this chapter, we're going to cover the following main topics:

- Operational excellence
- Security
- Reliability
- Performance efficiency
- Cost optimization

Service pillars overview

At first glance, you may be wondering why we aren't just jumping right into AWS, **continuous integration/continuous delivery** (**CI/CD**), and other DevOps topics. The main reason is that these five pillars are the foundational fabric of the exams. In addition, they will help you provide the most effective, dependable, and efficient environment for your company or clients. These design principles are not only important when architecting for success on Amazon Web Services, or any cloud provider for that matter, but in guiding the practices that you use throughout your day-to-day endeavors.

Once you become familiar with these pillars, you will see them and their themes in the testing questions as you go down your path for certification. This is especially true when working to obtain the DevOps Professional Certification as there are specific sections for Operations, Security, and Reliability.

The following are the five pillars of a well-architected framework:

- Operational excellence
- Security
- Reliability
- Performance efficiency
- Cost optimization

Use these pillars as the guiding principles, not only for designing your workloads in AWS but also for improving and refactoring current workloads. Every organization should strive to achieve well-architected applications and systems. Therefore, improving any AWS applications you are working on will make you a valuable asset. Now, let's look at each of these pillars in detail.

Operational excellence

As we look at the *operational excellence* pillar, especially in the context of DevOps, this is one – if not the most – important service pillar for your day-to-day responsibilities. We will start by thinking about how our teams are organized; after all, the DevOps movement came about from breaking down silos between Development and Operations teams.

> **Question – How does your team determine what its priorities are?**
>
> * Does it talk to customers (whether they're internal or external)?
>
> * Does it get its direction from product owners who have drawn out a roadmap?

Amazon outlines five design principles that incorporate operational excellence in the cloud:

- Performing Operations as Code
- Refining operations frequently
- Making small, frequent, and reversible changes
- Anticipating failure
- Learning from all operational failures

Let's take a look at each of these operational design principals in detail to see how they relate to your world as a DevOps engineer. As you go through the design principles of not only this pillar but all the service pillars, you will find that the best practices are spelled out, along with different services, to help you complete the objective.

Performing Operations as Code

With the contrivance of Infrastructure as Code, the cloud allows teams to create their applications using code alone, without the need to interact with a graphical interface. Moreover, it allows any the underlying networking, services, datastores, and more that's required to run your applications and workloads. Moving most, if not all, the operations to code does quite a few things for a team:

- Distributes knowledge quickly and prevents only one person on the team from being able to perform an operation

- Allows for a peer review of the environment to be conducted, along with quick iterations

- Allows changes and improvements to be tested quickly, without the production environment being disrupted

In AWS, you can perform Operations as Code using a few different services, such as **CloudFormation**, the **Cloud Development Kit (CDK)**, language-specific **software development kits (SDK)**, or by using the **command-line interface (CLI)**.

Refining operations frequently

As you run your workload in the cloud, you should be in a continual improvement process for not only your application and infrastructure but also your methods of operation. Teams that run in an agile process are familiar with having a retrospective meeting after each sprint to ask three questions: what went well, what didn't go well, and what has room for improvement?

Operating a workload in the cloud presents the same opportunities for retrospection and to ask those same three questions. It doesn't have to be after a sprint, but it should occur after events such as the following:

- Automated, manual, or hybrid deployments

- Automated, manual, or hybrid testing

- After a production issue

- Running a game day simulation

After each of these situations, you should be able to look at your current operational setup and see what could be better. If you have step-by-step runbooks that have been created for incidents or deployments, ask yourself and your team whether there were any missing steps or steps that are no longer needed. If you had a production issue, did you have the correct monitoring in place to troubleshoot that issue?

Making small, frequent, and reversible changes

As we build and move workloads into the cloud, instead of placing multiple systems on a single server, the best design practices are to break any large monolith designs into smaller, decoupled pieces. With the pieces being smaller, decoupled, and more manageable, you can work with smaller changes that are more reversible, should a problem arise.

The ability to reverse changes can also come in the form of good coding practices. AWS CodeCommit allows Git tags in code repositories. By tagging each release once it has been deployed, you can quickly redeploy a previous version of your working code, should a problem arise in the code base. Lambda has a similar feature called versions.

Anticipating failure

Don't expect that just because you are moving to the cloud and the service that your application is relying on is labeled as a managed service, that you no longer need to worry about failures. Failures happen, maybe not often; however, when running a business, any sort of downtime can translate into lost revenue. Having a plan to mitigate risks (and also test that plan) can genuinely mean the difference in keeping your **service-level agreement** (**SLA**) or having to apologize or, even worse, having to give customers credits or refunds.

Learning from failure

Things fail from time to time, but when they do, it's important not to dwell on the failures. Instead, perform post-mortem analysis and find the lessons that can make the team and the workloads stronger and more resilient for the future. Sharing learning across teams helps bring everyone's perspective into focus. One of the main questions that should be asked and answered after failure is, *Could the issue be resolved with automatic remediation?*

One of the significant issues in larger organizations today is that in their quest of trying to be great, they stop being good. Sometimes, you need to be good at the things you do, especially on a daily basis. It can be a steppingstone to greatness. However, the eternal quest for excellence without the retrospective of what is preventing you from becoming good can sometimes be an exercise in spinning your wheels, and not gaining traction.

Example – operational excellence

Let's take a look at the following relevant example, which shows the implementation of automated patching for the instances in an environment:

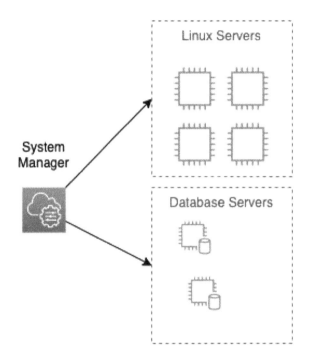

Figure 1.1 – Operational excellence – automated patching groups

If you have instances in your environment that you are self-managing and need to be updated with patch updates, then you can use **System Manager – Patch Manager** to help automate the task of keeping your operating systems up to date. This can be done on a regular basis using a Systems Manager Maintenance Task.

The initial step would be to make sure that the **SSM** agent (formally known as **Simple Systems Manager**) is installed on the machines that you want to stay up to date with patching.

Next, you would create a patching baseline, which includes rules for auto-approving patches within days of their release, as well as a list of both approved and rejected patches.

After that, you may need to modify the IAM role on the instance to make sure that the SSM service has the correct permissions.

Optionally, you can set up patch management groups. In the preceding diagram, we can see that we have two different types of servers, and they are both running on the same operating system. However, they are running different functions, so we would want to set up one patching group for the Linux servers and one group for the Database servers. The Database servers may only get critical patches, whereas the Linux servers may get the critical patches as well as the update patches.

Security

Next is the Security pillar of the AWS Well-Architected Framework. Today, security is at the forefront of everyone's minds. Bad actors are consistently trying to find vulnerabilities in any code and infrastructure (both on-premises and in the cloud). When looking back at the lessons learned from the first 10 years of AWS, CTO Werner Vogels said *Protecting your customers should always be your number one priority… And it certainly has been for AWS.* (Vogels, 2016)

 It is everyone's job these days to have secure practices across all cloud systems. This (protection) includes the infrastructure and networking components that serve the application and using secure coding practices and data protection, ultimately ensuring that the customer has a secure experience.

When you think about security, there are four main areas that the security pillar focuses on. They are shown in the following diagram:

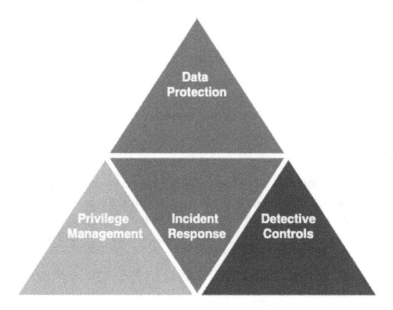

Figure 1.2 – The four main areas of security in the security pillar

The security pillar is constructed of seven principles:

- Implementing a strong identity foundation
- Enabling traceability
- Applying security at all layers
- Automating security best practices
- Protecting data in transit and at rest
- Keeping people away from data
- Preparing for security events

As we move through this book, you will find practical answers and solutions to some of the security principles introduced here in the security pillar. This will help you develop the muscle memory needed to instill security in everything you build, rather than putting your piece out there and letting the *security team* worry about it. Remember, security is everyone's responsibility. Initially, we will look at these security principles in a bit more detail.

Implementing a strong identity foundation

When building a strong identity foundation, it all starts with actualizing the principle of least privilege. No user or role should have more or less permissions than it actually needs to perform its job or duties. Taking this a step further, if you are using IAM to manage your users, then ensure that a password policy is in place to confirm that passwords are being rotated on a regular basis, and that they don't become too stale. It is also a good idea to check that the IAM password policy is in sync with your corporate password policy.

Also, as your organization grows and managing users and permissions starts to become a more complex task, you should look to establish central identity management either with Amazon Single Sign-on or by connecting a corporate Active Directory server.

Enabling traceability

Security events can leave you in a reactive state; however, your ability to react can rely on the amount of information you can gather about the event. Putting proper monitoring, logging, alerting, and the ability to audit your environments and systems in place before an event happens is crucial to being able to perform the correct assessments and steps, when the need arises.

Capturing enough logs from a multitude of sources can be done with AWS services such as CloudWatch Logs, VPC Flow Logs, CloudTrail, and others. We will look at logging and monitoring extensively in Part 3 of this book as it is important to the DevOps Professional exam.

Think about the following scenario:

Someone has gained access to a server via a weak password and compromised some data. You feel that you are currently capturing many logs; however, would you be able to figure out the following?

- The username used to access the system

- The IP address that was used where the access originated

- The time access was started

- The records that were changed, modified, or deleted

- How many systems were affected

Applying security at all layers

Securing all the levels of your environment helps protect you by giving your actions an extra expansiveness throughout your environment. To address network-level security, different VPCs can be secured using simple techniques such as Security Groups and Network ACLs. Seasoned AWS professionals know that additional security layers add an expansive security footprint – for example, at the edge (network access points to the AWS cloud), at the operating system level, and even making a *shift left* to secure the application code itself.

Automating security best practices

As you and your team get more educated about secure practices in the cloud, repetitive tasks should become automated. This allows you to react quicker to events that are happening and even react when you don't realize when things are happening.

This should be a topic when you start to dive in headfirst. As a DevOps specialist, you are used to taking repetitive manual processes and making them more efficient with automation. Automation can take the form of automatically analyzing logs, removing or remediating resources that don't comply with your organization's security posture, and intelligently detecting threats.

Amazon Web Services has come out with tools to help with this process, including GuardDuty, CloudWatch, EventHub, and AWS Config.

Protecting data in transit and at rest

Bad actors are all around, constantly looking for exploitable data that is traveling across the internet unprotected. You definitely can't rely on end users to use best practices such as secure communications over VPN, so it is up to you and your team to put the best practices in place on the server side. Basic items such as implementing certificates on your load balancers, on your CloudFront distribution, or even at the server level allows transmissions to be encrypted while going from point to point.

On the same token, figuratively speaking, making sure that you authenticate network communications either by enabling **Transport Layer Security** (**TLS**) or IPsec at the protocol layer helps ensure that network communications are authenticated.

There are AWS services to help protect your data, both in transit and at rest, such as AWS Certificate Manager, AWS Shield, AWS **Web Application Firewall** (the other **WAF**), and Amazon CloudFront. The **Key Management Service** (**KMS**) can also help protect your data at rest by allowing you to create, use, and rotate cryptographic keys easily.

For a deeper look at protecting data in transit and at rest, see *Chapter 19, Protecting Data in Flight and at Rest*.

Using mechanisms to keep people away from data

There are ways to automate how data is accessed, rather than allowing individuals to directly access the data. It is a better idea to have items that can be validated through a change control process. These would be items, such as System Manager runbooks or Lambda Functions, that would access the data. The opposite of this would be allowing direct access to data sources through either a bastion host or Elastic IP address/CNAME.

Providing this direct access can either lead to human mistakes or having a username and password compromised, which will ultimately lead to data loss or leakage.

Preparing for security events

Even if you enact all the security principles described previously, there is no guarantee that a security event won't be coming in the future. You are much better off practicing and having a prepared set of steps to enact quickly in case the need ever arises.

You may need to create one or more runbooks or playbooks that outline the steps of how to do things such as snapshotting an AMI for forensic analysis and moving it to a secured account (if available). If the time comes when these steps are necessary, there will be questions coming from many different places. The answers will have a timeline aspect to them. If the team whose responsibility is to perform these duties has never even practiced any of these tasks, nor has a guide been established to help them through the process, then valuable cycles will be wasted, just trying to get organized.

The following is the *Shared Responsibility Model* between AWS and you, the customer:

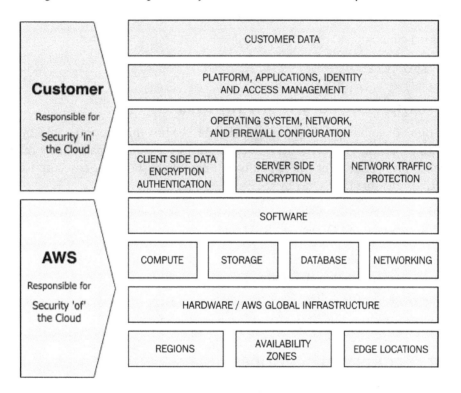

Figure 1.3 – The AWS shared responsibility model

Questions to ask

* How do you protect your root account?

 - Is there a **Multi-Factor Authentication** (**MFA**) device on the root account?

 - Is there no use of the root account?

* How do you assign IAM users and groups?

* How do you delegate API/CLI access?

Next, let's learn about the five design principles for reliability in the cloud.

Reliability

There are five design principles for reliability in the cloud:

- Automating recover from failure

- Testing recovery procedures

- Scaling horizontally to increase aggregate workload availability

- Stopping guessing capacity

- Managing changes in automation

Automating recovery from failure

When you think of automating recovery from failure, the first thing most people think of is a technology solution. However, this is not necessarily the context that is being referred to in the reliability service pillar. These points of failure really should be based on **Key Performance Indicators** (**KPIs**) set by the business.

As part of the recovery process, it's important to know both the **Recovery Time Objective** (**RTO**) and **Recovery Point Objective** (**RPO**) of the organization or workload:

- **RTO** (**Recovery Time Objective**): The maximum acceptable delay between the service being interrupted and restored

- **RPO** (**Recovery Point Objective**): The maximum acceptable amount of time since the last data recovery point (backup) (Amazon Web Services, 2021)

Testing recovery procedures

In your cloud environment, you should not only test your workload functions properly, but also that they can recover from single or multiple component failures if they happen on a service, Availability Zone, or regional level.

Using practices such as Infrastructure as Code, CD pipelines, and regional backups, you can quickly spin up an entirely new environment. This could include your application and infrastructure layers, which will give you the ability to test that things work the same as in the current production environment and that data is restored correctly. You can also time how long the restoration takes and work to improve it by automating the recovery time.

Taking the proactive measure of documenting each of the necessary steps in a runbook or playbook allows for knowledge sharing, as well as fewer dependencies on specific team members who built the systems and processes.

Scaling horizontally to increase workload availability

When coming from a data center environment, planning for peak capacity means finding a machine that can run all the different components of your application. Once you hit the maximum resources for that machine, you need to move to a bigger machine.

As you move from development to production or as your product or service grows in popularity, you will need to scale out your resources. There are two main methods for achieving this: scaling vertically or scaling horizontally:

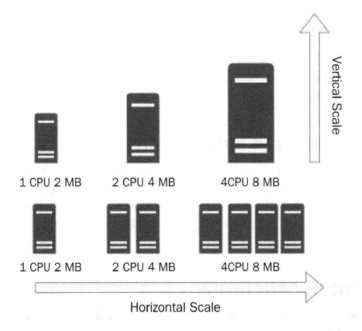

Figure 1.4 – Horizontal versus vertical scaling

One of the main issues with scaling vertically is that you will hit the ceiling at some point in time, moving to larger and larger instances. At some point, you will find that there is no longer a bigger instance to move up to, or that the larger instance will be too cost-prohibitive to run.

Scaling horizontally, on the other hand, allows you to gain the capacity that you need at the time in a cost-effective manner.

Moving to a cloud mindset means decoupling your application components, placing multiple groupings of the same servers behind a load balancer, or pulling from a queue and optimally scaling up and down based on the current demand.

Stop guessing capacity

If resources become overwhelmed, then they have a tendency to fail, especially on-premises, as demands spike and those resources don't have the ability to scale up or out to meet demand.

There are service limits to be aware of, though many of them are called *soft limits*. These can be raised with a simple request or phone call to support. There are others called *hard limits*. They are set a specified number for every account, and there is no raising them.

> **Note**
> Although there isn't a necessity to memorize all these limitations, it is a good idea to become familiar with them and know about some of them since they do show up in some of the test questions – not as pure questions, but as context for the scenarios.

Managing changes in automation

Although it may seem easier and sometimes quicker to make a change to the infrastructure (or application) by hand, this can lead to infrastructure drift and is not a repeatable process. A best practice is to automate all changes using Infrastructure as Code, a code versioning system, and a deployment pipeline. This way, the changes can be tracked and reviewed.

Performance efficiency

If you and your architectural design team are coming from a data center infrastructure and a provisioning process takes weeks or months to get the system you need, then the quickness and availability of cloud resources is certainly a breath of fresh air. There is a need to understand how to select either the correct instance type or compute option (that is, server-based, containerized, or function-based compute) based on the workload requirements.

Once you have made an initial selection, a benchmarking process should be undertaken so that you can see if you are utilizing all the CPU and memory resources that you have allocated, as well as to confirm that the workload can handle the duty that it is required to handle. As you select your instance types, don't forget to factor in costs and make a note of the cost differences that could either save you money or cost you more as you perform your baseline testing.

AWS provides native tools to create, deploy, and monitor benchmark tests, as shown in the following diagram:

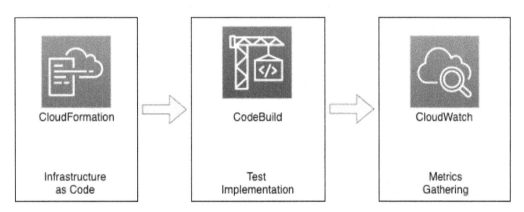

Figure 1.5 – Baseline testing with AWS tooling

Using the tools provided by AWS, you can quickly spin up an environment for right-sizing, benchmarking, and load testing the initial value that you chose for your compute instance. You can also easily swap out other instance types to see how performant they are with the same test. Using CloudFormation to build the infrastructure, you can, in a quick and repeated fashion, run the tests using CodeBuild, all while gathering the metrics with CloudWatch to compare the results to make sure that you have made the best decision – with data to back up that decision. We will go into much more detail on how to use CodeBuild in *Chapter 7, Using CloudFormation Templates to Deploy Workloads*.

The *performance efficiency* pillar includes five design principles to help you maintain efficient workloads in the cloud:

- Making advanced technologies easier for your team to implement
- Being able to go global in minutes
- Using serverless architectures
- Allowing your teams to experiment
- Using technology that aligns with your goals

Making advanced technologies easier for your team to implement

Having the ability to use advanced technologies has become simplified in the cloud with the advent of managed services. No longer do you need full-time DBAs on staff who specialize in each different flavor of database, to test whether Postgres or MariaDB will perform in a more optimal fashion. In the same way, if you need replication for that database, you simply check a box, and you instantly have a Highly Available setup.

Time that would otherwise be spent pouring over documentation, trying to figure out how to install and configure particular systems, is now spent on the things that matter the most to your customers and your business.

Being able to go global in minutes

Depending on the application or service you are running, your customers may be centralized into one regional area, or they may be spread out globally. Once you have converted your infrastructure into code, there are built-in capabilities, either through constructs in CloudFormation templates or the CDK, that allow you to use regional parameters to quickly reuse a previously built pattern or architecture and deploy it to a new region of the world.

Even without deploying your full set of applications and architecture, there are still capabilities that allow you to serve a global audience using the **Content Delivery Network** (**CDN**) known as CloudFront. Here, you can create a secure global presence using the application or deploy content in the primary region, which is the origin.

Using serverless architectures

First and foremost, moving to serverless architectures means servers are off your to-do list. This means no more configuring servers with packages at startup, no more right-sizing servers, and no more patching servers.

Serverless architectures also mean that you have decoupled your application. Whether you are using functions, events, or microservices, each of these should be doing a specific task. And with each component doing only their distinct task, it allows you to fine-tune memory and utilize CPU at the task level, as well as scale out at a particular task level.

This is not the best option for every workload, but don't allow a workload to be disqualified just because it would need a little refactoring. When an application can be moved to a serverless architecture, it can make life easier, the application itself more efficient, and there are usually cost savings to reap as a result – especially in the long run.

Allowing your teams to experiment

Once you move to the cloud, you can quickly and constantly refactor your workload to improve it for both performance and cost. If you have built your Infrastructure as Code, creating a new temporary environment just for testing can be a quick and cost-efficient way to try new modular pieces of your application, without having to worry about disrupting any customers or other parts of the organization.

Many of the experiments may not work, but that is the nature of experimentation. Business is extremely competitive in this day and age, and finding an item that does work and makes your service faster, cheaper, and better can be a real game changer.

Using technology that aligns with your workload's goals

List your business goals and let the product owner help drive some of the product and service selections based on those goals. If a development team has previous familiarity with certain technologies, they may be inclined to sway toward those technologies that they already feel confident using.

On the other hand, there are other teams that strive to use the latest and greatest technologies – but not necessarily because the technology solves a problem that has been identified. Rather, they are interested in constantly resume-building and making sure that they have both exposure to and experience with cutting-edge services as soon as they become available.

Cost optimization

Many have the misconception that moving to the cloud will become an instant cost saver for their company or organization. A stark reality is faced once more, and more teams find out that provisioning new resources is as easy as clicking a button. Once the bills start appearing from an environment that doesn't have strict guardrails or the ability to chargeback the workloads to the corresponding teams, most of the time, there comes a cost reduction movement from the top down.

As you look to optimize your costs, understand that cloud services that have been proven to be managed services come at a high cost per minute; however, the human resources cost is much lower. There is no need to *care and feed* the underlying servers, nor worry about updating the underlying operating systems. Many of the services allow you to scale to user demands, and this is taken care of for you automatically.

The ability to monitor cost and usage is also a key element in a cost optimization strategy. Having a sound strategy for resource tagging allows those who are responsible for financially managing the AWS account to perform chargebacks for the correct department.

There are five design principles for cost optimization in the cloud:

- Implementing cloud financial management
- Adopting a consumption model
- Measuring overall efficiency
- Stop spending money on undifferentiated heavy lifting
- Analyzing and attributing expenditure

Implementing cloud financial management

Cloud financial management is something that is starting to grow across organizations, both big and small, rapidly. It takes a dedicated team (or a group of team members that has been partially allocated this responsibility) to build out the ability to see where the cloud spend is going. This part of the organization will be looking at the cost usage reports, setting the budget alarms, tracking the spend, and hopefully enforcing a costing tag that can show the chargebacks for each department, cost center, or project.

> **What is a chargeback?**
>
> An IT chargeback is a process that allows units to associate costs with specific departments, offices, or projects to try and accurately track the spend of IT. We are specifically referring to cloud spend in this example, but IT chargebacks are used in other areas of IT as well, such as for accounting purposes.

Adopting a consumption model

Using the cloud doesn't require a sophisticated forecasting model to keep costs under control, especially when you have multiple environments. Development and testing environments should have the ability to spin down or be suspended when not in use, hence saving on charges when they would otherwise be sitting idle. This is the beauty of on-demand pricing in the cloud. If developers gripe over the loss of data, then educate them on the use of snapshotting database instances before shutting them down; then, they can start their development from where they left off.

An even better strategy is to automate the process of shutting down the development and test environments when the workday is finished and require a specialized tag, which would prevent an instance from being shut down after hours or on the weekend. You can also automate the process of restarting instances 30 to 60 minutes before the workday begins so that there is ample time for operating systems to become functional, allowing the team to think that they had never been turned off in the first place. Just be sure to watch out for any EC2 instances running on the instance store that may lose data.

Measuring overall efficiency

One of the most evident ways that organizations lose efficiency when it comes to their cloud budgets is neglecting to decommission unused assets. Although it is easier to spin up new services and create backups in the cloud, not having a plan to retire depreciated data, volume backups, machine images, log files, and other items adds to the bottom line of the monthly bill. This should be done with a scalpel and not a machete. Data, once deleted, is gone and irretrievable; however, there is no need to keep everything forever. Even with compliance, there is a fade-out period, and data can be stored in cold storage at a much more reasonable rate.

A perfect example is EBS snapshots. A customer who is trying to be proactive about data protection may be both snapshotting volumes multiple times per day as well as copying those snapshots to a Disaster Recovery region. If there is no way to depreciate the old snapshots after 30, 60, or 90 days, then this cost center item can become an issue rather quickly.

Stop spending money on undifferentiated heavy lifting

When we talk about heavy lifting, we're talking about racking, stacking, and cooling servers in a data center. Running a data center is a 24/7/365 endeavor, and you can't easily turn off machines and storage when you're not using them. Moving workloads to the cloud takes the onus of running those data centers off your team members and allows more time and effort to go into focusing on customer needs and features, rather than caring for and feeding servers and hardware.

Analyzing and attributing expenditure

The cloud – and the AWS cloud, in particular – has tools available to help you analyze and reference where the charges for your account(s) are coming from. The first tool in your toolbox is tags and a tagging strategy. Once you have decided on a solid set of base tags, including things such as cost center, department, and application, then you have a foundation for the rest of the tools that are available to you.

Breaking out from a single account structure into multiple accounts and organizational units using AWS Organizations can automatically categorize spend, even without the use of tags at the account level.

AWS Cost Explorer allows your financial management team to dig into the services and regions where spend is occurring, as well as create automatic dashboards to visualize the spend quickly. Amazon Web Services also has pre-set service quotas in place, some of which are hard quotas that cannot be changed, but many of which are soft quotas that allow you to raise the number of particular services (in a region) in a simple request.

Overarching service pillar principals

The Well-Architected Framework identifies a set of general design principles to facilitate good design in the cloud:

- Test systems at full production scale.

- Automate as many components as possible to make experimentation as easy as possible.

- Drive architectures using data.

- Stop guessing capacity needs.

- Allow architectures to evolve with new technologies and services.

- Use game days to drive team improvement.

As you are thinking about these service principals and how to put them into practice, realize that sometimes, the principles can feel like they are contradicting each other. The most obvious case is with the *cost optimization* pillar. If this is the pillar that the organization you are working for is trying to give the most attention, the other pillars can get in the way of pure cost savings. Strengthening weaknesses that you have found in the *reliability* pillar, most times, means more assets, and assets mean money. However, you can still strive to make those assets as cost-effective as possible so that you comply with all the pillars.

Summary

In this chapter, we learned about the five service principals that guide architects and developers to be well architected. We talked about how these are the underlying themes that run through the test questions in the DevOps pro exam, and how having this foundational knowledge can help when you're trying to determine the correct answer to a question. As we discussed each service pillar, we also talked about their underlying design principals.

We briefly mentioned several different AWS services and which service pillar or design principals where specific services come into play. In the next chapter, we will learn about the fundamental AWS services that are used throughout the environments and accounts you will be working in.

Review questions

1. What are the five pillars of the Well-Architected Framework?
2. What are the five main areas that security in the cloud consists of?
3. What are the four areas that the performance efficiency pillar consists of?
4. What are the three areas that the reliability pillar consists of?
5. What is the definition of RTO?
6. What is the definition of RPO?

Review answers

1. **C**ost Optimization, **R**eliability, **O**perational Excellence, **P**erformance Efficiency, and **S**ecurity. (Use the Mnemonic **CROPS** to help remember the five pillars using the first letter of each pillar.)
2. Data protection, infrastructure protection, privilege management, incident response, and detective controls.
3. Compute, storage, database, and network.
4. Foundations, change management, and failure management.
5. **Recover Time Objective** – The maximum acceptable delay between interrupting the service and restoring the service.
6. **Recovery Point Objective** – The maximum acceptable amount of time since the last data recovery point (backup).

Further reading

The following is a list of recommended papers and resources for further reading:

- AWS Well-Architected Whitepaper: `https://d1.awsstatic.com/whitepapers/architecture/AWS_Well-Architected_Framework.pdf`

- Cost Optimization Whitepaper: `https://docs.aws.amazon.com/wellarchitected/latest/cost-optimization-pillar/wellarchitected-cost-optimization-pillar.pdf`

- Reliability Whitepaper: `https://docs.aws.amazon.com/wellarchitected/latest/reliability-pillar/wellarchitected-reliability-pillar.pdf#welcome`

- Operational Excellence Whitepaper: `https://docs.aws.amazon.com/wellarchitected/latest/operational-excellence-pillar/wellarchitected-operational-excellence-pillar.pdf`

- Performance Efficiency Whitepaper: `https://docs.aws.amazon.com/wellarchitected/latest/performance-efficiency-pillar/wellarchitected-performance-efficiency-pillar.pdf#welcome`

- Security Whitepaper: `https://docs.aws.amazon.com/wellarchitected/latest/security-pillar/wellarchitected-security-pillar.pdf`

- Calculating Total System Availability:

  `https://www.delaat.net/rp/2013-2014/p17/report.pdf`

- Well-Architected Labs:

  `https://www.wellarchitectedlabs.com/`

- AWS Well-Architected Framework pillars:

  `https://wa.aws.amazon.com/wat.pillars.wa-pillars.en.html`

- 10 Lessons from 10 Years of AWS: `https://www.allthingsdistributed.com/2016/03/10-lessons-from-10-years-of-aws.html`

2
Fundamental AWS Services

Now that we have an understanding of the service principles and pillars that make up the best practices when using **Amazon Web Services** (**AWS**), it's time to look at some of the fundamental services that are used throughout the environments and accounts you will be working in. The fundamental services that we are referring to are compute services such as **Elastic Cloud Compute** (**EC2**), the global **Domain Name System** (**DNS**) service of **Route 53**, database services such as **RDS** and **Aurora**, and the advisory service of Trusted Advisor. This may seem like a review of services you are already familiar with if you have taken either the cloud practitioner, SysOps, or developer exams. However, since there is no longer a requirement to pass any of the lower associate exams before attempting (and passing) the DevOps professional test, it's not a bad idea to level-set on some of the basic services.

This is not meant to be an exhaustive look at these services. The services mentioned will be brought into the context of the DevOps exam, so skipping over this chapter is not advised. However, if you feel that you have a strong grasp of the topics mentioned, then verify this by checking the review questions, along with the reference material. Likewise, you can review any topics that you feel that you already have a strong grasp of.

In this chapter, we're going to cover the following main topics:

- Setting up and accessing your AWS account
- Virtual Private Cloud networking and Route 53 networking
- Cloud databases
- Message and queueing systems
- Trusted Advisor

Technical requirements

You will need an AWS account to access the Management Console and CLI, which will be mentioned in the initial part of this chapter. If you need assistance with creating your account, then the `https://aws.amazon.com/premiumsupport/knowledge-center/create-and-activate-aws-account/` page will walk you through the steps of creating a new account. Basic knowledge of how to use a terminal to complete the shell commands is also required.

> **Important Note**
> In this book, we will not be going over the geography, regions, Availability Zones, or edge locations of AWS. These are basic concepts that you should have a firm grasp of before attempting the DevOps exam.

Setting up and accessing your AWS account

At this point, you most likely have an AWS account to work with; nevertheless, you may only have access through your workplace, where some of the permissions are restricted, so you would not be able to practice all of the skills that you need to feel confident in passing the DevOps professional exam.

If you don't already have a personal account for testing things, then this would be the perfect time to set one up. Even if you do have an account, you may want to take the time to set up a new account just for the exam to ensure that you are allowed to take advantage of the free tier (allocated for your first year in AWS) on a number of the provided services.

If you do already have an account, making the switch to **AWS Organizations**, especially using Control Tower, is an excellent exercise if you wish to create service control policies, organizational units, **Single Sign-On** (**SSO**), and cross-account roles.

> **Important Note**
> Control Tower takes a minimum of three distinct email accounts to get started
> for the three separate accounts that are created.

Once you have created your master account, you can send an invitation to the previous
account that you created and allow it to join the organization.

Accessing the AWS Management Console

The AWS Management Console is the front door (GUI) to accessing your AWS account.

Start by opening any web browser and going to `https://aws.amazon.com/`.

On the start page, look for the **My Account** menu item. Hovering over this menu should
give you the option of choosing **AWS Management Console**:

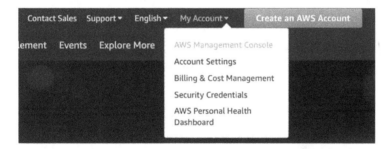

Figure 2.1 – Accessing the AWS Management Console

When following the link to log into AWS, so long as you are not going through an SSO
portal, you will be presented with three fields: **Account ID or account alias**, **IAM user
name**, and **Password**. Initially, when creating your account, you will only have the long
account number to use, and this only has context to AWS. It is a good practice to create an
account alias, enabling you to quickly associate the account ID with a logical name for the
account. This becomes especially handy when adding multi-factor authentication to your
accounts and seeing that logical names bring much more context than multiple rows of
numbers.

Once you have logged in, you can search for services, change regions, designate favorite
services for quick access, and open support requests. If you have the correct permissions,
you can also access the billing console.

As an initial step, to perform the exercises, we need to create a user, associate that user
with a role containing permissions, and then create a set of keys that we will download
and use later when we configure the CLI.

Once you have logged into the AWS Management Console, follow these steps to create your user and a set of access keys:

1. In the top search bar, type IAM.

2. Click on the **IAM** service; you will be taken to the IAM dashboard.

3. In the left-hand navigation column, click on **Users**.

4. Once you are in the user menu, click on the blue button at the top of the main screen, which is labeled **Add user**.

5. For the username, you can choose anything, but for our user, we are going to name them devops.

6. We are going to enter a custom password in the field that meets our password requirements and leave the box that is labeled **Require password reset** unchecked:

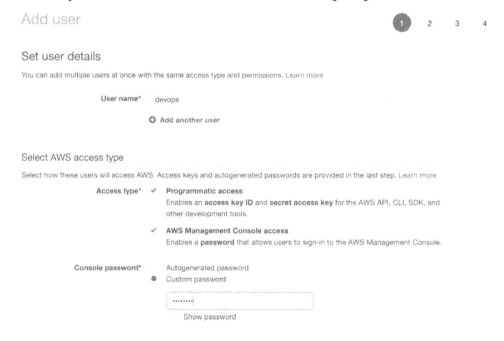

Figure 2.2 – Creating our user

7. We can then click **Next** to move on to the permissions.

8. Under **Set permissions**, we want to choose **Attach existing policies directly**.

9. For this initial user, we are going to use the **AdministratorAccess** job function policy. Select this policy so that the box on the left is checked, and then click the button at the bottom that says **Next: Tags**:

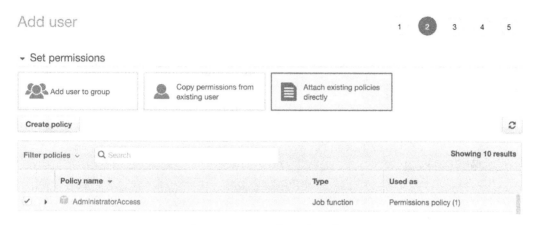

Figure 2.3 – Attaching the policy

10. Just click the button at the bottom that says **Next: Review**.

11. If everything looks correct, click the blue button that is labeled **Create User**.

12. Because we said that we wanted programmatic access, once we have created our user, we will be given the opportunity to both see the secret access key and download the access key and secret key pair in a CSV file. Take note of your secret key for this user or download the file as this will be the only time that the secret access key will be available.

With that, we have set up our first user, along with their access policy, password, access key, and secret key. We will be referencing this user throughout other exercises in this book. You are free to set up any username you are comfortable with but for an administrative user account that is not using the root account, we will be using the *devops* user.

Setting up and using the AWS CLI v2

You can indeed perform most tasks using the graphical web-based interface of the Management Console. As a DevOps engineer, you will want to automate items in your environment, and the CLI gives you the power, via a set of scripting abilities, to do this. The CLI is one of the favorite tools of many for both its speed and powerful capabilities.

> **Note for Previous CLI v1 Users**
>
> If you have previously installed the AWS CLI v1, then it is strongly recommended that you uninstall v1 *before* installing CLI v2. Both CLI commands use the same command name of aws. If you do not want to uninstall CLI v1, then you can put an alias in your path.

Mac setup

So long as you have sudo privileges on your machine, you can easily install AWS CLI v2 on your Mac using the bundled installer:

1. Open a Terminal window.
2. Run the following two commands to install the AWS CLI:

    ```
    $ curl "https://awscli.amazonaws.com/AWSCLIV2.pkg" -o
    "AWSCLIV2.pkg"
    $ sudo installer -pkg AWSCLIV2.pkg -target /
    ```

3. Check that the correct version has been installed by running the following command:

    ```
    $ aws --version
    ```

    The output should be similar to the following:

    ```
    aws-cli/2.1.29 Python/3.8.8 Darwin/18.7.0 exe/x_86_64
    ```

So long as the number after `cli` starts with a 2, you have successfully installed AWS CLI v2.

You can now skip to the Configuring the CLI section.

PC setup

> **Note for PC Users**
>
> To run CLI v2, you need to be running at least a 64-bit version of Windows XP or a later version of the Windows operating system.

If you have administrative rights to install software on your machine, then you can follow these instructions to install AWS CLI v2:

1. Download the AWS CLI MSI installer for Windows: `https://awscli.amazonaws.com/AWSCLIV2.msi`.
2. Run the downloaded MSI installer and follow the onscreen instructions.
3. Confirm that the correct version has been installed by running the following command:

    ```
    C:\> aws --version
    ```

Linux setup

Before installing CLI v2 on a Linux machine, there are a few prerequisites that must be taken care of first:

- You need the ability to unzip a package, either with the system `unzip` command or some other installed package.
- To have AWS CLI v2 run correctly, you need to make sure that the `glibc`, `groff`, and `less` packages are installed on your distribution. Most major distributions already have these packages installed by default.
- AWS supports AWS CLI v2 on newer 64-bit versions of CentOS, Fedora, Ubuntu, Amazon Linux 1, and Amazon Linux 2.

Now that the prerequisites have been met, you can use the following instructions to install the CLI:

1. Run the following `curl` command to download the AWS CLI v2 ZIP file:

   ```
   $ curl "https://awscli.amazonaws.com/awscli-exe-
   linux-x86_64.zip" -o "awscliv2.zip"
   $ sudo installer -pkg AWSCLIV2.pkg -target /
   ```

2. Now that the ZIP file has been downloaded, we can unzip it:

   ```
   $ unzip awscliv2.zip
   ```

3. Once unzipped, we can run the `install` program:

   ```
   $ sudo ./aws/install
   ```

4. Check that the correct version has been installed by running the following command:

   ```
   $ aws --version
   ```

5. The output should be similar to the following:

   ```
   aws-cli/2.1.29 Python/3.8.8 Linux/4.14.133-133.105.amzn2.
   x86_64 botocore/2.0.0
   ```

A reference to setting up the CLI from AWS can be found in their documentation at `https://docs.aws.amazon.com/cli/latest/userguide/cli-chap-install.html`.

Configuring the CLI

Now that you have installed the CLI, it is a good idea to configure it. If you had the CLI v1 installed and had used profiles in the past, you should be able to use those same profiles going forward. To quickly configure your CLI, you can use the `aws configure` command; however, as a prerequisite, it's a good idea to have already created and downloaded a key and a secret key for your user from IAM. You will need this pair of credentials if you are configuring either your default profile or a secondary profile to use with the examples shown.

If you haven't already done so, log back into your AWS account via the Management Console and then navigate to the IAM service so that you can create yourself a user and role. Using this user, you can allocate an **access key ID** and **secret key ID** to input into the CLI that you can use for the tutorials in this book.

Once you have your key pair, follow these steps to configure your CLI:

1. Run the `aws configure` command:

    ```
    $ aws configure
    ```

2. When prompted, cut and paste the access key ID.

3. When prompted, cut and paste the secret key ID.

4. When prompted, set your default region (we will be using `us-east-2`) for the examples.

5. You can just hit *Enter* to exit here and use the default JSON output; however, I find setting the output as a table is a lot more user-readable.

For many of the examples you will see in this book, a *profile* will be added to the CLI commands.

Cloud compute in AWS

When we talk about compute in AWS, we are talking about several services, including Amazon EC2, Elastic Load Balancing, AWS Batch, Elastic Container Service, and Elastic Kubernetes Service, along with AWS Fargate, the managed service that allows you to run your containers with minimal overhead. It even includes Lightsail, which is one of the quickest ways to get up and running on the cloud for developers, with no need to configure software or networking:

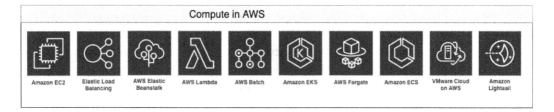

Figure 2.4 – Compute services in AWS

Although many services fall under the compute umbrella in AWS, the most foundational service is EC2. This is your virtualized instance in the Amazon cloud. While other services, such as Elastic Container Service, Elastic Kubernetes Service, and even Elastic Beanstalk, can allow you to run containers in AWS, at their core, they are running on EC2 instances. Thus, knowing the foundational pieces of the EC2 service, such as how to select the correct instance type, how to use the optimal load balancer (as there are three to choose from), and how to add volumes to an instance, all become relative information, both when processing questions for the DevOps professional exam and in your day-to-day duties as a professional:

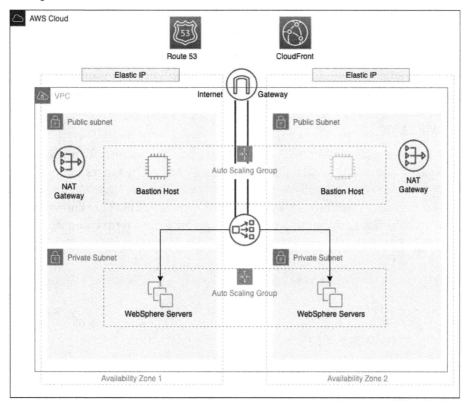

Figure 2.5 – EC2 in a real-world architecture

In the preceding diagram, we can see a real-life scenario where EC2 instances are being used in an Auto Scaling group to service the WebSphere platform. There are multiple EC2 instances in a private subnet that can only be accessed by an application load balancer that outside users see from a DNS entry in Route 53. If internal users need to access any of the WebSphere servers, there is a bastion host that allows the SSH protocol in the public subnet. The bastion host is in an Auto Scaling group that spans both Availability Zones, but only one host is up at a time.

We will now take a closer look at some of these services, especially the EC2 service.

Amazon Elastic Cloud Compute (EC2)

Amazon EC2 allows you to create a virtual server to perform any number of tasks in the cloud. EC2 allows for a whole array of customization. There is a multitude of operating systems that you can use to meet your application needs. Appropriating the correct amount of memory and processing power is simply a matter of choosing the correct instance type based on the needs of your workload.

EC2 also has three different pricing models. Each allows you the flexibility or discounting to fit your needs. There are On-Demand Instances, Reserved Instances, and Spot Instances.

On-Demand Instances are the default instance type and offer no long-term commitments when requesting an EC2 compute instance. You determine when you want to start, stop, hibernate, launch, or terminate the instance without any repercussions. From a pricing standpoint, you are only paying for On-Demand Instances per second while the instance is in the *running* state.

If you have known workloads with EC2 that will be running constantly for a year or need a desired capacity in a certain Availability Zone, then **Reserved Instances** can provide cost savings, along with reserved capacity in that particular Availability Zone. There are two different term commitments available with Reserved Instances. These are known as 1 year and 3 years, with the latter establishing greater savings for longer-term commitments.

When AWS has extra capacity that isn't being utilized, then those instances are made available as **Spot Instances** at precipitous discounts. The supply and demand of different types of instances that are available makes the price fluctuate, sometimes rather rapidly. These savings can reach up to 80% off of the normal on-demand pricing; however, there are a few caveats that come along with it. First, you have to launch the instance immediately and you cannot stop or hibernate Spot Instances. When launching a Spot Instance, you will set a maximum price for what you are looking to spend, such as the current on-demand price, and if the price rises above the maximum price that you set, then AWS will send a signal and you will be given 2 minutes to save your work before the instance is terminated.

EC2 instance types

At the time of writing, there are over 170 instance types that allow you to customize your compute needs for any type of workload that you would want to run into in the cloud. It's not important to try and memorize all the different types and sizes of the instances, along with their compute and memory specifications. However, it is a good idea to know the different categories that EC2 breaks down into for workload specificity and which EC2 families belong to those categories.

> **Note**
> Only some EC2 instances allow enhanced networking and this can be a deciding factor when trying to choose the correct instance type. The enhanced networking feature can be especially important when dealing with troubleshooting or workloads as it supports higher bandwidth and higher packets per second.

Since there are many types of workloads moving to the cloud, each with its own specific needs, AWS has created several different EC2 instance families. Each of these families contains one or more instance types and has a preferred application profile.

These instance families can be broken down into groupings, as follows:

- General-purpose instances
- Compute-optimized instances
- Memory-optimized instances
- Accelerated computing instances
- Storage-optimized instances

Let's take a look at these in more detail.

General-purpose instances

General-purpose instances balance memory, compute, and network resources and are a good choice for an assortment of workloads. This includes the T class of instances, which have burstable credits that build up over time. General-purpose instances are a good starting point if you don't have a full classification of your workload.

Use cases: Web servers, development and test servers, code repositories, small to mid-sized databases, and backend servers for SAP.

Compute-optimized instances

Compute-optimized instances are tailored for workloads that benefit from high-performance processors. The instances in this family also have the ability for enhanced networking pre-built in, along with being **Elastic Block Store** (**EBS**)-optimized by default.

Use cases: Batch processing, video encoding, ad serving, distributed analytics, CPU-based machine learning, gaming, scientific modeling, and high-performance science and engineering applications such as genome analysis or computational fluid dynamics.

Memory-optimized instances

This family of instances is, as its name implies, designed for memory-intensive applications. They are designed to deliver fast performance capabilities on jobs that need lots of memory.

Use cases: Open source databases, in-memory caches, and real-time analytics.

Accelerated computing instances

Accelerated computing instances contain co-processors or hardware accelerators that perform special functions much more efficiently than other processors. These functions can include data pattern matching, floating-point number calculations, and graphics processing.

Use cases: Speech recognition, high-performance computing, drug discovery, computational finance, and autonomous vehicles.

Storage-optimized instances

Storage-optimized instances provide directly attached storage options that allow for specialized storage needs. This means that the instance storage can be optimized for very customized **input and output** (**I/O**) performance in the case of H1 instances or very high storage density in the case of **high storage** (**HS**) instances.

Use cases: NoSQL databases, data warehousing, Elasticsearch, in-memory databases, traditional databases, and analytics workloads.

Instances are backed by two types of storage: **instance store** and EBS. When selecting an instance type, it will show whether the instance is backed by EBS or an instance store. There are some major differences between these two backing types, especially when it comes to persistence. One of the major advantages of an instance store is that it has high I/O and throughput. This comes from it being directly attached to the instance. The disadvantage of instance store-based volumes comes from persistence. If you restart an EC2 instance backed by an instance store, then you will lose all ephemeral data, such as logs and temporary files. This is not the case with EBS-based instances since the storage is not directly attached to the instance.

Understanding Amazon Machine Images (AMIs)

Whenever you launch an EC2 instance, it must start from an **Amazon Machine Image (AMI)** so that it contains the required information to launch. These can be the base operating system images that are, for all intents and purposes, clean slates. Or they can be AMIs that you or some other entity has created as a valid checkpoint for a working system or systems running on a single instance.

AMIs can be provided by Amazon itself, your own user account, shared privately with your other accounts or from a partner account, created by a member from the community, or even available for free or for a per-hour cost on the AWS Marketplace.

Use cases for AMIs

You can create your own AMIs to use in Auto Scaling groups or to speed up the launch of complex instances that require multiple steps to download, install, and configure the software.

There is also a case for a base image in an organization that has a pre-approved operating system, as well as security settings pre-installed for all users to conform to.

There are community AMIs available; however, you run these at your own risk since there may be unknown packages installed on them.

Another option is to use marketplace AMIs provided by vendors and partners that have vetted and known software already preconfigured. These AMIs often have an extra price per hour when running the instance.

Backing up Amazon EC2 instances

If you want to back up your instance either for point-in-time recovery purposes or to use it in a launch configuration with autoscaling, then you need to create an AMI:

1. First, we must find the latest version of the Amazon Linux2 AMI with a little help from the Systems Manager service and throw this into a variable for the next step:

```
$IMAGE='aws ssm get-parameters --names /aws/service/
ami-amazon-linux-latest/amzn2-ami-hvm-x86_64-gp2 --query
'Parameters[0].[Value]' --output text --region us-east-2'
```

2. Launch an EC2 instance. Now that we know the latest version of our base AMI that we are going to use, we need to have an instance to back up and create the custom AMI:

```
$ aws ec2 run-instances \
--image-id $IMAGE \
--instance-type t2.micro \
--region us-east-2
```

Once it launches, it should return a JSON statement that contains the instance ID. We will need this instance ID for our next step to create the AMI. In my case, the instance ID that was returned was i-02666d4b05eee61c1.

Create the AMI as a backup:

```
$aws ec2 create-image \
--instance-id i-02666d4b05eee61c1 \
--name "DevOps_chapter2" \
--no-reboot
```

If the image is created successfully, then some JSON should be returned with the ImageId value:

```
{
    "ImageId": "ami-0aa7408c42669e27b"
}
```

3. Verify the image:

```
$aws ec2 describe-images \
--region us-east-2 \
--image-ids ami-0aa7408c42669e27b
```

This should return a block of JSON with multiple values. However, the main value we are looking for is the line that says `State`. If the value says `available`, then you have successfully backed up your EC2 instance and are ready to test.

4.   Let's test our backup.

Here, we are going back to our original command that we used in *Step 2* to create the instance, but now, we are going to substitute the original image ID for the AMI that we have just created:

```
$aws ec2 run-instances \
--image-id ami-0aa7408c42669e27b \
--instance-type t2.micro \
--region us-east-2
```

As you can see, with just a few simple commands, we have taken our running EC2 instance and not only backed it up but also created a launchable AMI from it. We could take our backup a step further and copy our AMI to another region; so long as there are no hardcoded region-specific items on the image, it should launch and run without issue.

Once you have finished launching your EC2 instance and created your AMI, it is a good idea to terminate the instance and remove the AMI so that you don't get charged extra on your AWS account.

Using user data scripts to configure EC2 instances at launch

Although we can launch our EC2 instances and then configure the software and packages that we need on them by hand, this is not the most efficient approach. Apart from following manual steps leading to human mistakes, this approach is much more time-consuming than using an automated process.

Example user data script

Next, we will look at an example of a user data script that can configure an EC2 instance without user interaction by pre-forming items such as creating files from scratch, updating previously installed packages, updating software repositories, and even running commands:

```
#cloud-config
package_upgrade: true
repo_update: true
repo_upgrade: all
```

```
packages:
  - boto3
  - aws-cfn-bootstrap

write_files:
  - path: /test.txt
    content: |
      This is my file.
      There are many like it but this one is mine.

runcmd:
  - [ sed, -i, -e, 's/is/was/g', /test.txt]
  - echo "modified my file"
  - [cat, /test.txt]
```

With that, we have learned how to automatically configure EC2 instances at launch using user data scripts, which can do things such as install and upgrade packages, as well as run scripts and commands. Next, we will look at networking interfaces for EC2 instances.

Elastic Networking Interfaces (ENIs)

Elastic Networking Interfaces (**ENIs**) work like virtual networking cards, so they allow the instance to have an IP address and be connected to a specific subnet. EC2 instances allow multiple ENIs to be attached, and each of those network interfaces can be on the same subnet or can traverse different subnets for specific reasons:

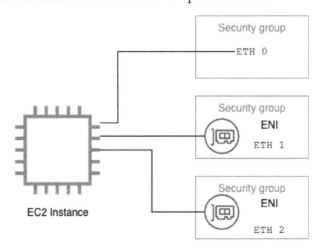

Figure 2.6 – ENIs in separate security groups

Since security groups are attached at the network interface level and not at the instance level, adding additional ENIs to your instances allows you to have your instance join more than one security group for specialized purposes. If you have a web server that needs to access the public internet, you can attach one interface to the security group that serves this purpose. Also, in the same instance, you may need to SSH into the machine so that a team member can check the logs or processes running on the server. Locking the security group down that is attached to a particular ENI that allows access to the SSH port (port 22) can be done in this manner.

Elastic Block Store (EBS)

Although EBS and EC2 are closely tied together, it is pertinent to remember that they are both separate services. EBS is a storage service that provides network-based storage that is allocated in the same Availability Zone as the instance and then mounted for usage. The amount of instance storage that's allocated to an instance varies by the instance type, and not all types of EC2 instances contain an instance store volume.

EBS is different from an **instance store** based on some key attributes. The instance store volume is the storage that is physically attached to the EC2 instance. It is best used for temporary storage since the instance store does not persist through instance stops, terminations, or instance failures. In contrast, data stored on an EBS volume will persist:

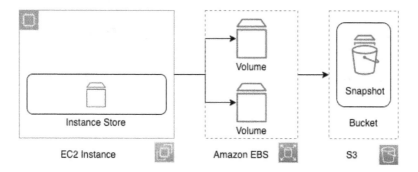

Figure 2.7 – EBS overview in AWS

EBS volumes can either be allocated at the time of instance creation or created once the instance has been placed in a service as additional storage.

One of the key features to remember from a DevOps perspective, when it comes to allocating and restoring EBS volumes, is that a volume must stay in the same Availability Zone where it was created.

One of the key terms to understand when trying to decide which of the different types of EBS volumes to use is **input/output operations per second** (**IOPS**) or **Provisioned IOPS** (**PIOPs**). IOPS is the measure of I/O, measured in kilobytes, that a volume can perform in a second. Using CloudWatch Metrics, you can monitor the performance of a particular volume using volume metrics such as the following:

- `VolumeReadOps`
- `VolumeReadBytes`
- `VolumeWriteOps`
- `VolumeWriteBytes`

Now that we know the basics of how EBS volume performance is measured, let's look at some of the different types of EBS volumes available.

Types of EBS volumes

There are three main types of EBS volumes, all of which differ in terms of performance, optimal use cases, and cost:

- **Solid State Drives** (**SSD**): This is a type of drive that's optimized for heavy read and write operations and where a higher number of IOPS is needed. There are two types of SSD EBS volumes available for provisioning:

  a. **General-purpose SSD**: Gives a balance of cost and performance, best used for development and test environments.

  b. **Provisioned IOPS SSD**: Used for mission-critical workloads where performance is key, such as databases or caches.

- **Hard Disk Drives** (**HDD**): This is a type of drive that is optimized for streaming workloads where performance is a necessity due to constant reads and/or writes. There are two types of HDD EBS volumes available for provisioning:

  a. **Throughput-optimized HDD**: This type of EBS volume is best for data warehouses, log servers, or big data workloads.

  b. **Cold HDD**: This is a low-cost HDD that is best used for infrequently accessed data.

- **Previous generation**: This is a type of drive that is best used for smaller datasets that are not of critical importance. Rather than being on SSDs, these EBS volumes are on magnetic disks, which means they are not as performant as the other two types of EBS volumes. There is only one type of previous-generation EBS drive available and it's best used for data that is infrequently accessed.

Looking at the EC2 service, which also covers AMIs and EBS volumes, there are many options to choose from, which allows you to choose the right size for your instance. This spans much deeper than the operating system that you choose, but also how fast you provision your instances and how performant you need your storage to be.

Next, we will look at AWS Batch, a service that allows us to perform large operations with ease either on-demand or on a schedule.

AWS Batch

Sometimes, you need a large amount of compute resources.

An example to think about in this instance is tabulating voting results. Once voting has ended, multiple people and machines start the count at different regional centers. Each of the regional centers counts the votes and submits that sub-total to get the final results.

This process of counting the votes is a batch process. If the votes were all tallied in different **Comma Separated Value (CSV)** files, then they could all be uploaded and processed by the AWS Batch service.

AWS Batch can run its jobs as either shell scripts, Linux executables, or Docker containers.

There are five main components in AWS Batch:

- Jobs
- Job definitions
- Job queue
- Job scheduler
- Compute environments

Let's look at these in more detail.

Jobs

Jobs in AWS Batch are nothing more than units of work. These units can take the form of a shell script, a Docker container image, or a Linux executable, but regardless of how the job is submitted, it will be run in a containerized environment.

Job definitions

Telling the job how it should be run is the function of the AWS Batch job definition. This is also the place where you give extra detail to your jobs, such as which IAM role to use if the job needs to access other AWS resources, as well as defining how much memory and compute power the job will have.

Job queue

Until a compute environment is available to run the job, the job will sit in a job queue. You are not limited to a single job queue per AWS account as queues can be associated with different types of compute environments.

High-priority jobs may be placed in a queue where the compute environment is made up of On-Demand Instances. Lower-priority jobs may wait in a queue that is made up of Spot Instances and can run once the spot becomes available.

Job scheduler

Once an AWS Batch job has been submitted to a queue, the scheduler assesses the current compute environments to see if they can run the job. Since jobs may be dependent on other jobs running and/or finishing before they can start, the scheduler also considers this. It is for this fact that jobs are run in a reasonable order from which they were previously submitted to the queue, but an exact order is not always possible.

Compute environments

Once you have set up your compute environments where the job queues will perform the tasks, you have several choices. You can choose managed or unmanaged compute and specify a particular instance type, or even only run on the newest types of instances available.

If one of the reasons you're using Batch is that that you are interested in cost savings, you can also determine the spot price for your compute instances. If the reliability of the compute environment is a necessity, then you are better off setting up your environment with On-Demand Instances.

Virtual Private Cloud networking and Route 53 networking

The **Virtual Private Cloud** (**VPC**) service from AWS allows you to create a virtual network in the cloud. It allows your compute and database instances to either allow internet connectivity or segment it off from the internet. Security can be accomplished through either stateful or stateless virtual firewall rules, which provide the amount of network connectivity that you see fit:

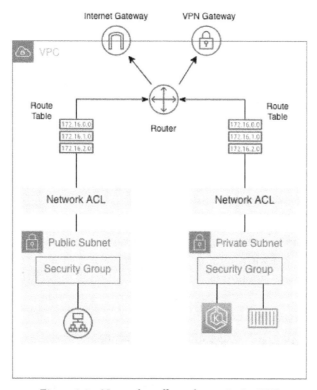

Figure 2.8 – Network traffic and security in AWS

The VPC service comprises multiple components that allow you to route and secure traffic from your AWS services and, optionally, the internet and/or your on-premises network.

VPC

Although the solution architect (or possibly the network architect) will often determine the CIDR address range that's used for the VPC, many times, it falls on the DevOps engineer to implement the VPC with **Infrastructure as Code (IaC)**.

There are quite a few components that can help make up a VPC. While we will not be covering VPC extensively, we will look at some items that you should know about in case they come up in any exam questions.

Subnets

A **subnet** defines a range of IP addresses in a VPC. There are both public and private subnets. A public subnet should be used for resources that will be accessed by the internet. A private subnet should be used for resources that won't be accessible from the internet. Each subnet can only inhabit one Availability Zone and cannot traverse multiple Availability Zones.

Security groups

Security groups act as virtual firewalls in Amazon VPC. You can have up to five security groups per EC2 instance, and security groups are enforced at the instance level (not at the subnet level).

Security groups allow for stateful traffic, meaning that if the traffic is allowed in via the rules, then it's returned, notwithstanding any rules in place. You can specify inbound traffic rules based on a combination of port, IP range, or another security group.

Network Access Control Lists (NACLs)

Network Access Control Lists (**NACLS**) work at the subnet level (unlike security groups, which work at the instance level).

Whereas security groups are stateful, NACLs are stateless, and any traffic that needs to return through an NACL needs to have the port and IP range opened. NACL rules are evaluated in order, with the lowest rule being processed first.

Internet gateways

Unless you are running your VPC as a private extension of your data center, then you will need internet connectivity. **Internet gateways** provide the connection to the internet for VPCs in a highly available, scalable, and redundant manner. There are two main functions provided by the internet gateway: the first is to provide internet access to the subnets designated in the route table. The second is to provide network address translation for instances that have been assigned an IPv4 public address.

Egress-only internet gateways

If you are using IPv6 for your VPC and instances, then you need an **egress-only internet gateway** rather than a regular internet gateway. This type of internet gateway prevents the internet from initiating connections to your instances, but still offers the same scalability, redundancy, and high availability as the other internet gateway.

Network Address Translator (NAT)

When you have instances in a private subnet, by default, they cannot talk to the internet. Even if you have set up an internet gateway, you want to separate your private instances from direct internet access. This is where either a NAT instance or a NAT gateway comes into play.

A **NAT** device forwards traffic from the instances in the private subnet to the internet or other AWS services. Since the advent of **VPC Endpoints**, using a NAT to talk to other AWS services in your account is considered a non-secure practice and should never be done in production environments.

VPC endpoints

If you need to talk to your AWS services securely from your VPC, then you can create VPC endpoints. A VPC endpoint allows EC2 instances and other AWS compute services (such as Lambda and Fargate) to communicate with supported AWS services such as S3 and DynamoDB, without the need to create an internet gateway or even the need to have a public DNS name.

This is especially useful for EC2 instances that neither need to nor should be connecting to the internet. Using a VPC endpoint allows the data connections to travel on the internal AWS network and does not require an internet gateway, a virtual private gateway, a NAT device, a VPN connection, or an AWS Direct Connect connection. VPC endpoints are virtual devices that can scale to meet demand, along with being redundant and highly available.

DHCP option sets

When you create a VPC, Amazon automatically creates a set of **Dynamic Host Configuration Protocol** (**DHCP**) options for you by default. You can, however, customize some of the available settings by creating a new **DHCP option set** and then attaching that DHCP option set to your VPC (and therefore removing the default DHCP option set).

Some of the allowed options include the following:

- `domain-name-servers`: To use your own DNS instead of the AWS-provided DNS.

- `domain-name`: The default domain name for unqualified servers.

- `ntp-servers`: You can specify up to four Network Time Protocol servers.

- `netbios-name-servers`: You can specify up to four NetBIOS name servers.

- `netbios-node-type`: The NetBIOS node type (1, 2, 4, or 8); this setting is blank by default and Amazon does not support broadcast or multicast.

> **Note**
> Knowing the DHCP options is useful for configuring your VPCs in the real world, but memorizing the available options is not necessary for the DevOps professional exam.

Using a custom DNS with your VPC

Amazon provides a default DNS server (Route 53 Resolver); however, you can use your own DNS server if you like. Some companies either run the DNS from their data centers in a hybrid environment, mirror their current DNS server to the cloud so that they do not have to create Route 53-hosted zones with all of the required information, or just choose to manage the DNS themselves on the platform that they feel the most comfortable with.

If this is the case and you want to specify the DNS versus using the default DNS server, then you would need to create a DHCP option set and complete the value for your DNS server. Once you've done this, you can attach the DCHP option set to your VPC, and the instances in that VPC will start using the designated DNS server that you have specified.

Ways to connect multiple networks

In the world of AWS networking, multiple tools have been provided to help you connect your different networks.

VPN connections

You can network AWS with your own **Virtual Private Network (VPN)** using a variety of options:

- **AWS Site-to-Site VPN**: This option creates an IPSec connection between your VPN and the remote VPC.

- **AWS Client VPN**: This option uses a managed client that allows your users to connect from almost any location using an OpenVPN-based client.

- **AWS VPN CloudHub**: This option allows you to create a Site-to-Site VPN connection to multiple VPCs with your virtual private gateway. (The other options mentioned only create a VPN connection with a single VPC.)

- **Third-party VPN appliance**: This option allows you to create a VPN connection using third-party software running on an EC2 instance. However, it's important to note that when choosing this option, you are responsible for the maintenance and upkeep of the EC2 instance.

AWS Transit Gateway

AWS Transit Gateway allows you to connect both on-premises networks and multiple VPCs through a central networking hub. Transit Gateway acts as a router for your cloud connections, allowing you to connect each network to the Transit Gateway only once. Once connected, those other networks can then talk to each other, using defined network rules, through the AWS global private network and not the public internet.

AWS Direct Connect

If you or your company have consistent data transfers both to and from AWS, then the **Direct Connect** service can provide a private network connection between your premise and AWS. These dedicated connections are available in 1 Gbps, 10 Gbps, or 100 Gbps denominations through direct connect providers.

Unlike connecting directly over the public internet, using an AWS Direct Connect connection gives you consistent network performance by transferring data from your data center or office to an from AWS over dedicated channels rather than going over the public internet.

Route 53

The global DNS service that AWS provides is **Route 53**. This is one of the few services in AWS that is not tied to any specific region. The Route 53 service also has one of the strongest commitments, stating that it *will use commercially reasonable efforts to make Amazon Route 53 100% available* (Amazon Web Services, 2021).

There are three main components of Route 53 that are of foundational importance:

- The ability to register (and manage) domain names
- The DNS service
- The ability to perform health checks (and subsequently route traffic) on your web application based on the fact that it's functional, available, and reachable

In this section, we will cover some of the basic information about the Route 53 service, especially those topics that would be relevant to know for the DevOps exam.

Understanding the different types of records available in Route 53

There are many different DNS record types and at the time of writing, Route 53 supports the following types of DNS records:

- **Address (A)** record
- AAAA (IPV6 address record)
- **Canonical name (CNAME)** record
- **Certification Authority Authorization (CAA)**
- **Mail exchange (MX)** record
- **Name authority pointer record (NAPTR)**
- **Name server (NS)** record
- **Pointer (PTR)** record
- **Start of Authority (SOA)** record
- **Sender policy framework (SPF)**
- **Service (SRV)** location
- **Text (TXT)** record
- Alias records (which are a Route53-specific extension of DNS)

Now that we know what types of records Route 53 supports, let's take a look at the difference between a domain and a hosted zone.

Knowing the difference between a domain and a hosted zone

One of the first things to understand about domains and a hosted zones is that, first, a domain is an internet construct of domain name servers that associates the unique name of a person or organization with a numerically addressed internet resource.

Domains have zone files that are text mappings of the different resources and their associated names, addresses, and type of record that the asset is currently mapped in. A hosted zone, on the other hand, is something only found in Route 53. It is similar to a DNS zone file (and you can import DNS zone files into your Route 53 hosted zones) in terms of its structure and mapping. One major difference, however, is that it can be managed and modified using the Route 53 interface, CLI, or API.

Route 53 health checks

Route 53 allows you to check the health of your applications and then reroute traffic to other servers or resources based on the rules that you provide. You can even see the recent status of your health checks in the Route 53 web console.

Checking the health of a specific endpoint

In this case, you will be creating a check from Route 53 that performs checks at regular intervals specified by you. Your health checks are monitoring an endpoint that is either an IP address or a domain name. Route 53 then goes out at the interval and checks whether the server, the application, or another resource is available and operational. You can also request a specific web page or URL that would mirror most of the actions of your users, rather than just a simple health check page placed on the server that returns a simple 200 code if the system is up and running.

Calculated health checks (health checks that monitor other health checks)

If you have multiple resources that all perform the same function, then you may be wondering if a minimum number of resources are healthy. This is where calculated health checks come into play:

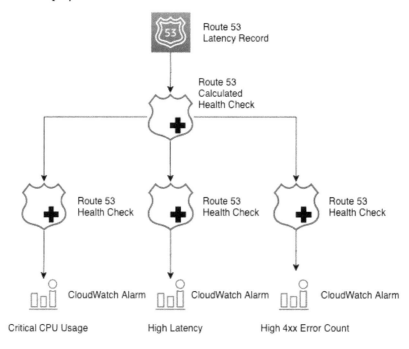

Figure 2.9 – An example of a calculated health check

The calculated health check acts as a root health check where descendant checks can fail before the origin is considered unhealthy.

This type of health check is designed to fail if any of the alarms are set off.

Checking the status of a CloudWatch alarm

Route 53 can utilize the power of CloudWatch metrics and alarms. Once a CloudWatch alarm has been created, you can create a health check in Route 53 that observes the same data stream as the CloudWatch alarm.

To help improve both the availability and flexibility of the CloudWatch alarm health check, Route 53 looks at the data coming from CloudWatch, which then determines whether that route will be healthy or unhealthy. It does not wait for an ALARM state to be reached before setting the route to unhealthy.

Route 53 routing policies

Routing policies in Route 53 tell the service how to handle queries. These policies can be simple to complex, as we will see next.

Simple routing

This is where the DNS is configured with no special configurations. You are simply setting the single record that the DNS file should point to. There is no weighting and no failover – just keep it simple.

Failover routing

With failover routing, you will have an alternative record in case the primary resource that is listed in the initial record set is unavailable or fails the defined health check(s).

This could be another server in a different region or a backup website being served by an S3 bucket.

Geolocation routing

If you have an audience or userbase that spans either a country, a continent, or the world, and depending on their location would like them to have a custom message or website (without needing to use complex cookies and dynamic content), then you can use the **geolocation routing** feature of Route 53 to send them to the server or content origin that would be most relevant for their originating location.

You can specify the geographic locations by continent, country, or state in the United States.

Geoproximity routing

You may have a geographically diverse userbase but want the ability to push more (or less) traffic to a certain region or set of resources. This is where **geoproximity routing** differs from geolocation routing. It can create a bias for each specific resource that routes more or less traffic to that resource based on the bias number and the original location of the requesting user. A bias expands or shrinks the size of the geographic region in which traffic is routed to a resource.

Latency-based routing

When you have an audience or userbase that is geographically spread out and you have resources running in multiple regions, then you can use latency-based routing to direct each user to the resource or content that will give them the quickest possible response time.

> **Question**
>
> What is the difference between geolocation routing and latency-based routing?
>
> Although in both cases you are dealing with a geographically expanded user base, **latency-based routing** is based on latency records that Route 53 creates and stores on points of origin and IP or DNS points of record. These latency records can change from time to time, so it's best to have a uniform experience or have a customized experience based on customer settings.
>
> **Geolocation routing**, on the other hand, is matching the user request with the resource that has been geographically tied to that originating IP address, so you can have local content customized for the end user experience.

Multi-answer routing

Although multi-answer routing doesn't take the place of a load balancer, it does allow Route 53 to respond with up to eight healthy values chosen at random.

Some scenarios where you would want to use multi-answer routing would be as follows:

- Creating more than one record of the same name type

- Routing traffic to multiple sources

- Associating a Route 53 health check with records

Weighted routing

If you have multiple resources that you would like the requests to be sent to while distributing the traffic unevenly, you can set up weighted routing from Route 53. This has a few applications, such as launching a canary release with major or minor updates and redirecting only a portion of the traffic to the new environment. You can then gauge key performance indicators, such as the number of abandoned carts.

Implementing weighted routing starts with having more than one resource that you want the traffic to be balanced between. You can then create records inside your Route 53-hosted zone to reflect these resources. These have to be records of the same type within a domain or subdomain; that is, both need to be A records.

Once the records have been set, you can move on to configuring the routing by assigning a weight to the specific resource. The weighted value can be any number from 0 through 255 (if you specify 0, then that record will no longer receive traffic).

The traffic is balanced using a calculation that takes the weight of the current record. That is then divided by the sum of all the records.

Example: If you had two servers running and the record for server A had a weight of 75 and the weight for server B had a weight of 150, then 75 + 150 = 225 total weight, and server A would get 33% of the traffic using the formula 75 / 225 = 0.3333333:

Quick create record   Info   Switch to wizard   Add another record

▼ Record 1   Delete

Record name Info   Record type Info   Value Info   ⬤ Alias

weighted   .devopsandbeyond.com   A – Routes traffic to an IPv4 address and so... ▼   172.31.6.128

Valid characters: a-z, 0-9, ! " # $ % & ' ( ) * + , - / : ; < = > ? @ [
\ ] ^ _ ` { | } . ~
                                                                            Enter multiple values on separate lines.

TTL (seconds) Info   Routing policy Info   Weight

60   ⇕   Weighted   ▼   100   ⇕

1m   1h   1d   The weight can be a number between 0 and 255. If you
                                                                            specify 0, Route 53 stops responding to DNS queries using
Recommended values: 60 to 172800 (two days)   this record.

Health check - *optional* Info   Record ID Info

Choose health check   ▼   East2 test instance

                                                              Cancel   Create records

Figure 2.10 – Creating a weighted record in Route 53

Route 53 will be covered in a bit more detail once we get to *Chapter 13*, *Blue Green Deployments*, and discuss blue/green deployment strategies.

Cloud databases

As you look at the following model, you may be wondering why there are so many databases. This comes from the evolution of the application architecture over the past few decades, where specialization, speed, and scale have all become keys to success in this industry:

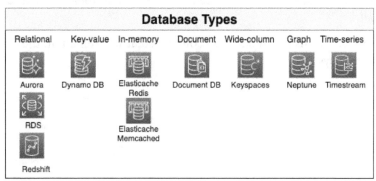

Figure 2.11 – Database types and services in AWS

There is not enough space in this chapter to visit each type of database that AWS offers. However, as part of this foundational overview, we do want to cover some of the relational databases and their basic features.

We will be covering the key-value database, Dynamo DB, in more detail in an upcoming chapter.

Relational databases

The word databases usually brings relational databases to mind, which have rows, columns, and schemas.

Relational databases in AWS come in three main *flavors*, as well as engines that can be classified as part of a community, commercial, or cloud native. Cloud-native engines are used in the Aurora service since they are built based on community engines with cloud-native storage, backup, and replication capabilities.

> **Note**
>
> When we talk about cloud native, we are talking about building and running applications that take advantage of the cloud computing model: loosely coupled services that are dynamically scalable, can run on a multitenant platform, and have flexible networking and storage.

Relational databases tend to follow the ACID set of database properties:

- **Atomic**: All of your transaction succeeds or none of it does.

- **Consistent**: Any data written to the database must be valid according to all defined rules.

- **Isolated**: Ensures that concurrent execution of transactions leaves the database in the same state that it would have been in if the transactions were executed sequentially.

- **Durable**: Once a transaction has been committed, it will remain in the system, even in the case of a system crash.

Relational Database Service

Relational Database Service (RDS) aims to take away the tasks that were previously performed by a database administrator, who had to be on staff but finds that many of their day-to-day duties are starting to be handled by automation and scripting. This includes provisioning a new database, creating backups, scaling out to read replicas, patching and upgrading instances, as well as switching over to a high-availability instance when an incident occurs. RDS streamlines the setup and installation of software and database provisioning. Using the managed RDS service also allows an organization to achieve a standard set of consistency when creating and deploying a database. No custom scripts are needed to deploy, set up, and patch the database since this is all handled in the background.

After deployment, developers and applications have a database endpoint that can be accessed readily, where connections can be made with a client or authentication application, and then queries can be performed on the database.

RDS comes in several different engine formats, including two commercial engines and three popular open source engines. The two commercial engines that are supported are Oracle and Microsoft SQL Server. Both of these commercial engines are used in enterprises. RDS supports access to Oracle schemas via Oracle SQL, along with native tools such as SQL Server Management Studio for Microsoft SQL Server. In the community editions, RDS offers the ability to spin up databases using MySQL, PostgreSQL, and MariaDB. MySQL is one of the most popular community relational databases and AWS runs the community edition of this, defaulting InnoDB tables with the documentation, stating that MyISAM storage does not support reliable crash recovery.

PostgreSQL

PostgreSQL is another extremely popular database, with developers using it for its rich feature set. RDS supports common tools such as pgAdmin or connections via JDBC/ODBC drivers to connect to the database.

After deciding on your engine of choice, you have the opportunity to choose an instance type that will give you varying levels of compute and memory power. There are burstable instances (T family) that are often only used for testing environments. A general (M family) and memory-optimized (R family) instance is preferred when taking your database workloads to a more productionized environment.

One of the differences between RDS and the same type of engines that you would run yourself on EC2 instances is how replicas work. Read replicas are extremely easy to provision by merely clicking a button, either in the same or a different Availability Zone; however, these nodes are read-only and cannot take a write. In the same context, you can make your instances instantly highly available, which replicates your data asynchronously to the copy of the master node in another Availability Zone or region. If an incident occurs, then it will be seamless to your application and end users as the DNS that both use to connect to the database stays the same. This secondary master, however, cannot take any load off your main primary server as it cannot serve any other function (such as being a read slave) except being the failover node. Read replicas can be promoted to a standalone database and at that point, they will have the ability to take writes. However, they will no longer stay in sync with the previous master that they were replicating with.

Aurora

Amazon Aurora was built in response to customers wanting the performance of commercial-grade database engines such as Oracle or Microsoft SQL Server, without having to deal with all the hassle of the licensing handcuffs that came with those products.

The other keynote about Amazon Aurora is that, unlike the other RDS engines that are backed by EBS storage, Aurora built its own storage solution from the ground up after listening to multiple customer requests over the years.

Amazon Aurora comes in either MySQL-compatible or PostgreSQL versions. There are opportunities to run Aurora as a cluster or even run a serverless version of the Aurora database. The main item to know about the serverless version of Aurora is that it provides capacity on-demand as your application or users need for it. This is the distinct difference between provisioned Aurora clusters and the serverless version.

Key-value databases

As applications are conceived, the number of users who will be served is not always known initially. There are usually hopes that at some point in the near future, the popularity of the application will grow from the initial set of users and testers to an exponential scale. To sustain that growth, the underlying database must be able to handle that scale seamlessly. This is one of the features of key-value databases.

We will look at DynamoDB in more detail in *Chapter 5, Amazon DynamoDB*.

In-memory databases

When you are accessing stored items frequently from your primary data store, such as a relational database, there are times when just requesting the data from the database won't provide the user experience that your customers are expecting. This is when an in-memory database such as **Amazon's Elasticache** becomes a viable option.

Elasticache comes in two different engine versions: Redis and Memcached.

Document databases

A document database is one type of non-relational database that allows you to store documents and data in a JSON type format and query that data to find the data. One of the truly unique features of document databases is that there is no fixed schema and that they can have documents nested inside of each other.

AWS offers **Amazon DocumentDB** (with MongoDB compatibility) as a managed database service for those that have either used MongoDB in the past or are looking for the capabilities of a document database. If you have ever been on the operating side of MongoDB before, then you know that while it is powerful and has several advanced features, including self-electing a new master if the current master becomes unavailable, there are complex setup requirements to get it up and running.

It requires (in a production environment) at least three nodes – either two nodes and an arbiter or three master nodes. All of this complex setup goes away with DocumentDB. Amazon takes care of the setup and security and allows you to configure automatic backups that you can then store on S3.

There are a few small things that you are giving up, such as the ability to use admin or local tables, which means you can't use the `mongodump` or `mongorestore` utilities, but there are functions that take the place of these utilities.

Document databases are particularly good for teams who don't want to deal with the administrative aspects of a database, want to simplify the way initial schema values are used using JSON, and just want to start pushing data to the database, which will allow for both simple and advanced querying later.

Message and queueing systems

As you start to build out cloud-scale applications, you need ways to decouple the different tiers of the application so that it can scale independently. This is for several reasons, including making your application more resilient and allowing each tier to scale independently of other tiers. Having the ability to use a managed service to perform these tasks, where you or another member of your team doesn't have to worry about the setup and maintenance of the queues or messaging systems, allows expedited usage of these techniques when moving or developing in the cloud.

We will look at the messaging and queueing systems provided by AWS next and how they can be of benefit to you.

Simple Notification Service (SNS)

Sometimes, you need the ability to simply send messages either from your applications or from other services in several formats, such as email or text message, that can be used in a variety of ways. These can be based on programmatic calls or events that happen in your AWS account. This is where **Simple Notification Service** (**SNS**) comes in handy.

SNS is a publisher and consumer system, where publishers can push a message out to a **topic**. Then, any consumer who is subscribed to that topic can consume that message:

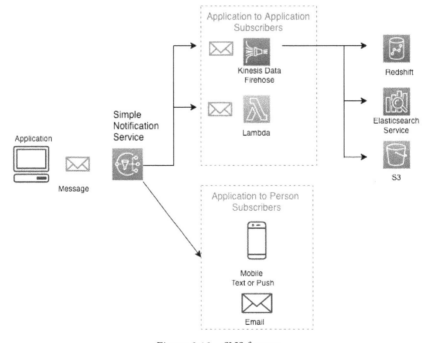

Figure 2.12 – SNS fanout

The SNS topic acts as the channel where a single message can be broadcast to one or more subscribers. The publisher, which is the application in the preceding diagram, only has to send the message once to the topic; each consumer (or subscriber) can receive the message to be processed in a way that works for them. This can be stored for archival purposes in the case of S3, consumed on-demand in the case of mobile SMS or email, or be parsed and acted upon in the case of Lambda.

Simple Queue Service (SQS)

If you are looking for a fully managed queueing service that will allow you to break apart your application in a cloud-native manner, then this is where **Simple Queue Service (SQS)** is a real service fit. It comes in two flavors: **standard queues** and **First In First Out (FIFO) queues**. Standard queues allow for at least once delivery, whereas FIFO queues are capable of one-time delivery regarding messages placed in the queue. The other major difference between these two types of queues is that with FIFO queues, the messages are processed in the order in which they are received. A standard queue tries to preserve the ordering of the messages that are received; however, if a message is received out of order, it will not do any shuffling or re-ordering to keep the original message order.

SQS is a distributed queue system, which means it is spread out across different nodes in the region. This is one of the design features that brings the benefits of SQS, such as being highly scalable, available, reliable, and durable. SQS also allows for **Server-Side Encryption (SSE)**, either through the service key provided by AWS via KMS or a custom key. You can also control who has access to the messages that are either produced or consumed via access policies. There are three main parts of the SQS distributed queue system:

- Producers/consumers (components)
- The queue
- The messages in the queue:

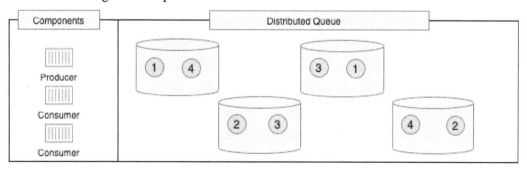

Figure 2.13 – SQS distributed queue

Messages are placed in the SQS queue by producers; they are then distributed across the queue components for redundancy. When the consumer is ready to process a message, it uses either a long or short polling technique to see whether there are any messages in the queue for processing. If it detects messages for processing, then it consumes the message for processing in the queue; however, the message is only flagged as being processed and the visibility timeout begins. Once the consumer has successfully processed the message and deletes the message, the message is removed from all the nodes of the distributed queue.

Messages must be deleted by the consumer once they have been processed successfully. Because of the nature of the distributed system that SQS runs on, once a consumer has pulled down a message for processing, it is then marked so that no other consumers can pull down the same message. There is a timeout period, however, before which the consumer needs to acknowledge the message was processed. If this acknowledgment is never received by SQS, then the message is unflagged and available to another consumer for processing.

Where would you use SNS versus SQS?

When trying to decide which messaging service to use, there is a rule that you can use to help decide which one is going to serve you best; ask the following set of questions:

- Do my messages need guaranteed delivery?
- (If this answer is yes, then SQS is going to be the best choice as there is no option for guaranteed delivery with SNS.)
- Are my messages going to be consumed only by my application?

  (If this answer is yes, then SQS is going to be the best choice.)

- Are my messages going to be consumed by multiple sources?

  (If this answer is yes, then SNS is going to be the best choice.)

By using these three questions, you can determine whether you are dealing with a closed loop of gathering the messages and then processing them for use in the application (SQS) or creating a message to be consumed outside the application (SNS).

Amazon MQ

If you are migrating an application whose previous messaging system was either Apache MQ or Rabbit MQ and you no longer want to manage the instances, then **Amazon MQ** is a great option. Amazon MQ is more of a message broker than just a simple queue.

When you go to allocate your Amazon MQ, you must select either the Rabbit MQ or Apache MQ engine type. Once you have selected your engine type, when using the Management Console to provision Amazon MQ, it will prompt you, through a series of choices, to use a single broker or have an active and standby broker for high availability. It can even create a mesh network of single or active and standby brokers for you using predefined blueprints.

Differences between Amazon MQ and SQS

While SQS is also a managed service from AWS, it is a scalable queue that doesn't require a message broker to be established. Amazon MQ's strength is that it allows previously built applications that have dependencies on APIs such as MQTT, OpenWire, or STOMP to be quickly and easily migrated to the cloud in a highly available fashion with minimal overhead.

Simple Email Service (SES)

Simple Email Service (**SES**) allows you to set up and send emails without most of the complexities of running an SMTP server.

SES only runs out of three regions: us-east-1 (North Virginia), us-west-2 (Oregon), and eu-west-1 (Ireland).

Amazon helps you verify where your mail is coming from. It provides trust to the email that you sent your customers by adding a series of DKIM records to the DNS records we send our emails from.

Trusted Advisor

As the number of resources grows in your AWS account, it can sometimes be challenging to keep track of them all. Challenges start to arise, such as volumes that are sitting around not being used and **Elastic IPs (EIPs)** that are sitting there, not connected to any instance or interface, which is burning through your current budget. Once you enter the dashboard of Trusted Advisor, you will be presented with the four categories that the tool generates for automated checks:

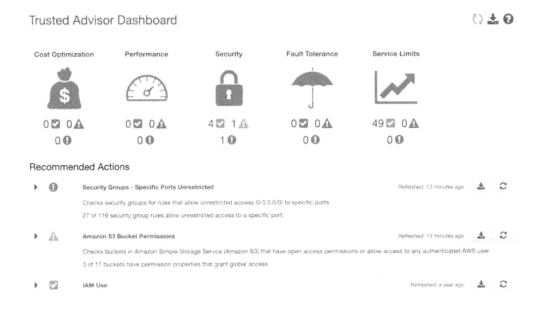

Figure 2.14 – Trusted Advisor dashboard

When it comes to **Fault Tolerance** checks, it can tell you if you have too many instances in a single Availability Zone or region.

The number of Trusted Advisor checks available increases with the support level associated with the account. With an account that has a basic or developer support level, a basic set of checks are displayed. Once the support level is raised to business or enterprise, then the account has access to all 115 Trusted Advisor checks.

Even at a basic support level, Trusted Advisor provides several checks in four areas:

- Cost optimization
- Performance
- Security
- Fault tolerance

> **Note**
> Did you notice that the four basic checks that Trusted Advisor provides can be mapped to four of the AWS service pillars?

On top of this, Trusted Advisor can be set up to send a weekly notification to either the billing contact, the operational contact, the security contact on the account, or any combination of the three contacts listed.

Accessing Trusted Advisor

Trusted Advisor has a graphical dashboard interface that can be viewed, and it even allows you to view potential monthly savings under the cost optimization icon when you view the service from the AWS console for the first time. AWS has worked to refine the console interface to make it easier to use and to be as intuitive as possible, displaying three icons under each major heading icon regarding the number of checks that have no warnings associated (green), investigation recommended (yellow), and then action recommended (red).

You can also access Trusted Advisor from the AWS CLI. This can be useful from a DevOps perspective, especially from a service limit check perspective. You can use your automated pipeline or a periodic Lambda job to check service levels for particular services, and then go out and automatically request a service-level increase or send out a notification to the correct email or distribution list so that you don't face disruption in your build process.

Summary

AWS continues to grow the number of services that it offers all the time. We have gone over quite a few of the services in this chapter, although briefly, but have only covered many of the essential services that are used to host and serve applications running on the AWS platform.

In this chapter, we reviewed the foundational services offered by AWS. It will be these services, along with the others we will look at in the rest of part one, that we will use as the pieces of deployment in our DevOps journey. Having foundational knowledge of these services can let us do more from a development and deployment objective, and it also helps us when digesting the questions being asked on the DevOps professional exam, along with the scenarios being presented to us in our day-to-day professional lives.

In the next chapter, we are going to examine identity and access management, the basis for securing our AWS cloud account and assets.

Review questions

1. If you wanted to bid on Amazon EC2 spare capacity, which of the following would allow you to do this?

    a. Reserved Instances

    b. Auto Scaling

    c. Spot Instances

    d. Savings plans

2. The company you are working with wants to tighten up its network security. However, they don't currently want to have to declare egress for the return traffic. Which VPC construct can you use to help tune their security settings?

    a. Network Access Control Lists (NACLs)

    b. Network Address Translator (NAT)

    c. VPC endpoint

    d. Security groups

3. True/False: Trusted Advisor can notify you of the service limits your account is about to reach.

4. If your company was running a SQL Server instance in their current data center and wanted to migrate it to the AWS cloud, what options are available?

    a. Running the database on EC2

    b. Setting up Direct Connect so that cloud resources can connect to the SQL Server instance

    c. SQL Server on RDS

    d. SQL Server on Amazon Aurora

5. Which of the following statements about FIFO SQS queues is true?

    a. You are not guaranteed to get messages in the order that you sent them.

    b. Messages will be delivered exactly in a FIFO order.

    c. Messages will be delivered exactly once.

    d. Messages can be delivered one or more times.

Review answers

1. c
2. d
3. True
4. a and c
5. b and c

3
Identity and Access Management and Working with Secrets in AWS

With a firm understanding of a myriad of fundamental services under our belt, we now move on to **Identity and Access Management (IAM)**.

Defining the controls and policies that are used to manage access to different **Amazon Web Services (AWS)** services is what the premise of IAM is all about. This can come in the form of users, groups, policies, and even access, with federated access or short-term credentials. There is also the ability to use outside **identity providers (IdPs)** to allow users to access your applications. Understanding how to secure secrets using native AWS tools, especially in the context of a **continuous development (CD)** process, is an enterprise-level skill that is not only present in the DevOps professional exam, but also on the job.

In this chapter, we're going to cover the following main topics:

- Understanding the Shared Responsibility Model in AWS

- IAM roles, groups, users, and policies

- Using **AWS Organizations** as part of your guidance

- Integrating federation with an AWS account

- Storing secrets securely in AWS

- Using **Amazon Cognito** with application authentication

Technical requirements

As we go through this chapter, you should already have an AWS account created, along with the AWS **Command-Line Interface (CLI) version 2 (v2)** installed so that you can follow along with the hands-on activities presented in this chapter. You will also need more than one AWS account if you are going to practice along with the AWS assume-role exercise.

Understanding the Shared Responsibility Model in AWS

Although we touched on it briefly in *Chapter 1*, *Amazon Web Service Pillars*, understanding the Shared Responsibility Model is imperative in order to work with the security of your account, and especially the IAM service.

The following diagram provides an overview of the model:

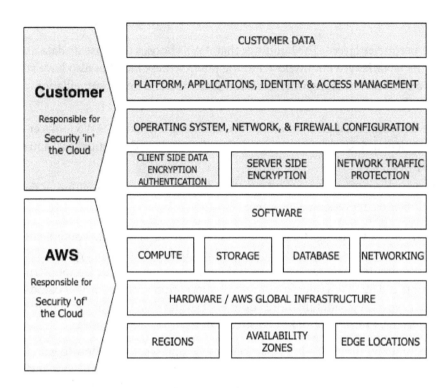

Figure 3.1 – AWS Shared Responsibility Model: Infrastructure as a Service (IaaS)

The essence of this Shared Responsibility Model is about providing flexibility and customer control. With this model there are two major concepts, outlined as follows:

- Who is responsible for *security OF the cloud*:

  This is where AWS holds the responsibility.

- Who is responsible for *security IN the cloud*:

  This is where you as the customer hold the responsibility.

AWS controls the global infrastructure, which includes the data centers that host the servers running all of the AWS services. These data centers are run according to security best practices and a range of security compliance conventions. Although no customer is allowed to visit an actual data center, independent outside bodies certify and audit the centers and practices to ensure that the controls, policies, and procedures are being followed.

There are multiple layers in which AWS secures its data centers, outlined as follows:

- At the perimeter layer—The buildings that AWS chooses to house its data and compute facilities are not disclosed to the public. These facilities also have strict entry requirements, along with intrusion detection and heavy surveillance to protect them.

- At the environmental layer—Before even choosing a location, AWS looks at environmental risks such as severe weather, flooding, and potential earthquake activity.

- At the infrastructure layer—Inside the building, there is a combination of fire detection and suppression equipment.

As a customer, one of the things that you are responsible for is the data you upload, store, and create in your account. You could both store and transmit it unencrypted, depending on what the requirements are for both you and your customers. If this is more stringent than just plaintext data, AWS offers a number of encryption options, or you could manage your own encryption. We will talk more about encrypting data both at rest and in transit later, in *Chapter 19, Protecting Data in Flight and at Rest*.

Your responsibility on the AWS cloud also lies in who and what you allow to gain access to the various services in your accounts. This is where the IAM service plays a key role. By default, AWS will not allow access from any application or user to any other services.

Keeping the operating system up to date with any patches and updates on your **Elastic Compute Cloud** (**EC2**) instances is also part of your responsibility in the Shared Responsibility Model. There are services such as **Systems Manager** (**SSM**) and Amazon Inspector that can help you with this task, identifying crucial security patches and adhering to a maintenance schedule. If you run the actual infrastructure services, then the ultimate responsibility falls to you.

How you allow traffic into your instance via the network layer is also your responsibility. In *Chapter 2, Fundamental AWS Services*, we discussed the **virtual private cloud** (**VPC**) and the tools you have to secure your network using security groups and **network access control lists** (**network ACLs**). Keeping your security rules to only the traffic that you need to allow—and nothing more—prevents outside entities gaining unwanted information from acts such as port scanning and other malicious activities.

This main shared security and responsibility model that was originally published by AWS reflects the responsibilities around the IaaS model. However, with different cloud computing models such as **Platform as a Service** (**PaaS**), in the case of container service offerings and managed service offerings that respond more like **Software as a Service** (**SaaS**), then Amazon takes even more of the responsibility.

With the use of containers (especially in the case of Fargate), the security model shifts more to a PaaS model, so more of the responsibilities are pushed to AWS. No longer are you—the customer—responsible for the operating system, since AWS is managing this aspect. You are still, however, responsible for your customer data and the network traffic in and out of the system, as illustrated in the following diagram:

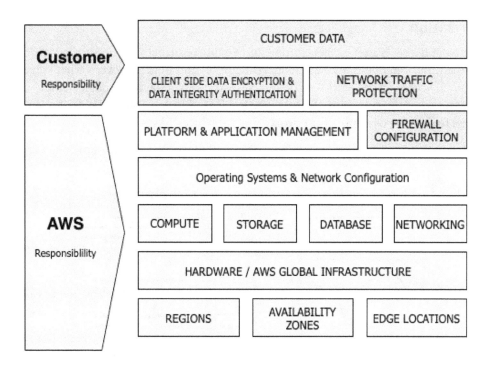

Figure 3.2 – AWS Shared Responsibility Model: container services

One of the main things to understand is that in all of the Shared Responsibility Models, you are responsible for the protection of your data. This includes how many ports it will be exposed to, along with how it will be encrypted or whether it will be left unencrypted.

Authorization versus authentication

As we start to look at IAM, one of the first concepts we need to have a full understanding of is that of authorization versus authentication. Although the two terms may sound similar and are used in conjunction with each other quite frequently, understanding the differences between them is important as we move further.

Authentication

Authentication is the process of verifying who you claim to be. The system is asking who you are and you are going to respond, many times, with a username and password, but other times with a secure session token. Authentication is answering the questions *Who are you?* and *Can you verify who you say you are?*

Authorization

Authorization happens after authentication and establishes what you are allowed to do. Most of the time, it happens after authentication. Rules and policies govern what you are authorized to access. In the world of computing, this can be relayed through a token, such as a bearer token or a **JavaScript Object Notation (JSON) Web Token (JWT)** that grants you access to services or **application processing interfaces (APIs)**.

The processes of authentication and authorization are illustrated in the following diagram:

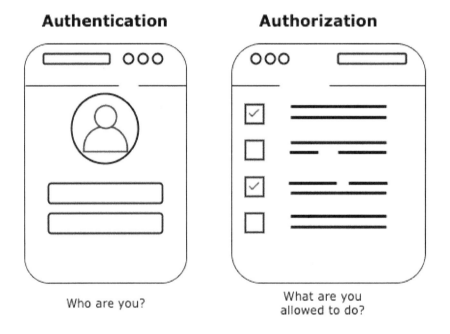

Figure 3.3 – Authentication versus authorization

It can be easy to confuse authentication and authorization since they seem similar, but think of authentication as your picture identification and authorization as a **radio-frequency identification (RFID)** badge that allows you access. Let's take a look at another example for further clarity.

A practical example of authorization versus authentication

At many large office buildings, you need an ID badge with a photo to enter the building. In most cases, you are required to wear the badge at all times to show you have **authorization** to be in the building without being stopped by security. As you move to the elevator, you might need to scan your badge to access the floor you work on. Pressing your ID badge, which contains an RFID chip that is tied back to the policy system and your user profile, lets the central system know which floors you are allowed to access. If you work on the third floor, then you are **authenticated** to press button number 3 and then travel up to the third floor. There might be other doors in the building to which you may or may not have access at all, based on the level of your authentication.

Terms to understand for IAM

As we begin to talk about the IAM service and its individual components, it's important to initially lay out the terms we will be using, along with their definitions. There is no need to memorize these terms for the professional exam, but they are the components of the whole service that you are expected to know from the lower exams. These terms are listed out here:

- **Principal**: An application or person that uses either the AWS root account user, an IAM user, or an IAM role to authenticate to the account and make requests. This is someone or something that can take action on an AWS resource.

- **Resources**: A resource is an item that you can work with inside of an AWS account. An example of a resource could include a Lambda function, an EC2 instance, or a **Simple Storage Service (S3)** bucket.

- **Entities**: An entity can be an IAM user, a federated user, or a user who is coming in from an IdP, or it can be an assumed IAM role in the context of AWS. It is simply the IAM resource object that AWS uses for authentication.

- **Identities**: The resources that are used to identify who is using the services are known as identities in IAM. These are your users, groups, and roles.

Ways IAM can authenticate with a principal

IAM has several ways in which it can authenticate with a principal, outlined as follows:

- **Username and password**—A username and password are the initial way to get into the AWS Management Console.

- **Access key and secret access key**—These are long-term security credentials that can be associated with users or the root user. They can be rotated, and a maximum of two access keys can be associated with any user at any single time.

- **Session token**—You can use an assumed role to take advantage of the **Security Token Service** (**STS**) to pass a token that will allow you or your application to gain the allocated access.

Note

AWS Management Console users can't use their username and password credentials to run any particular program and access AWS services or underlying infrastructure by default. This all comes with the authorization provided by IAM policies.

Authorization concepts with IAM

As you look to add authorization to your users and applications, there are two models to be aware of: **role-based access control** (**RBAC**) and **attribute-based access control** (**ABAC**). We will go over the concepts of each next.

RBAC

Traditional access control is based on grouping users and applications into groups (or roles) and then assigning permissions to the group. Many times, these are job functions, and the permissions that are associated with the group deal with the responsibilities of that job function.

Mapping specific sets of permissions to a role, defining that role as a group in an authentication service such as **Active Directory** (**AD**) or AWS IAM, and then placing users in that group so that they may assume those permissions is defined as RBAC.

ABAC

There are three different types of attributes on which ABAC can base access control, listed here:

- User attributes
- Attributes associated with the system to be accessed
- Current environmental conditions

In AWS, most of these attributes are managed with tags, account information (that is, account number, alias, or **organizational unit** (**OU**)), and federated data associated with the user.

IAM service concepts

One of the most important concepts when using IAM and as you navigate through AWS access is to give only the minimal permissions necessary to do the job, and no more.

IAM roles, groups, users, and policies

Controlling access to your resources in IAM comes down to how you craft your policies that you attach to your users, groups, and roles. Services themselves assume roles, and users are created in IAM and are much more easily managed if placed into groups.

> **Note**
> There is a service limit of 500 IAM users per account.

IAM policies

An IAM policy is a set of permissions expressed as a JSON statement that says which access permissions an entity has.

As you start to dole out the permissions in your account, you will deal with these IAM policies. There are a number of policies that have been pre-crafted by AWS to help you get started, either to attach to users and groups or for services. Many AWS-based policies will have the resource listed as *; however, this allows for any resource to be accessed. There are multiple ways in which we can constrain the amount of resources that can be accessed, even in the case of a simple policy.

Let's look here at a very simple example of an IAM policy that shows how we can give access to all object actions on all S3 buckets in the account, using wildcards:

```json
{
    "Version": "2012-10-17",
    "Statement": [
        {
            "Sid": "AllS3ObjectActions",
            "Effect": "Allow",
            "Action": "s3:*Object",
            "Resource": ["*"]
        }
    ]
}
```

We can, however, restrict which bucket our user can access by changing the resource line of the policy, as follows:

```
{
    "Version": "2012-10-17",
    "Statement": [
        {
            "Sid": "AllS3ObjectActions",
            "Effect": "Allow",
            "Action": "s3:*Object",
            "Resource": ["arn:aws:s3:::my-bucket/*"]
        }
    ]
}
```

There are a number of different policy types that are available for use in AWS, as outlined in the following list. Each has its own role when it comes to permissions:

- Identity-based policies

- Resource-based policies

- Permissions boundaries

- Organizations' **service control policies (SCPs)**

- ACLs

- Session policies

Now we know how IAM policies work, along with the different types of policies available, let's look at some of fundamental concepts around IAM policies and how the different types of policies work together to form effective permissions.

IAM policy concepts

As you start to use IAM, you need to have an understanding of how the service evaluates policy requests based on its internal logic. Here is an overview of the rules that you need to remember:

- By default, all requests are denied (unless you are the root user).
- If any service has been explicitly denied, this rule takes precedence, even if another rule has the service allowed.
- An explicit `allow` in a policy will override the default implicit `deny`.
- If a SCP, session policy, or permissions boundary grants access to a service then it too will override the implicit `deny`.

For a deeper understanding of how the rules are processed, I suggest you read the documentation available here: `https://docs.aws.amazon.com/IAM/latest/UserGuide/reference_policies_evaluation-logic.html`.

In the diagram pictured in *Figure 3.4* are three circles. The bottommost circle is the identity-based policy—that is, where the user or group is granting the permissions for that effective user. However, there are other identity policies at play that can limit—or in some cases expand—the effective permissions for a user. The center of where all the policies intersect—and in this case where the arrow is pointing—is the effective permission set for the user, as illustrated here:

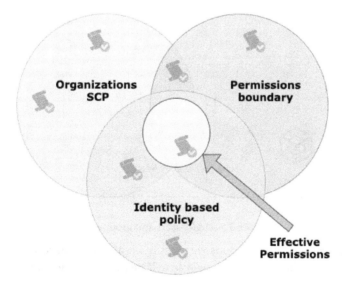

Figure 3.4 – Effective policy permissions

With knowledge of how these different types of polices interact together to form effective permissions, we can now go about creating our identities in the forms of roles, groups, and users.

Identities

When you first create your AWS account, the email account you use to sign up with becomes your AWS root user. Although you have the ability to access the Management Console with this root user, and this user has a full unrestricted set of permissions to all services and resources in your AWS account, it is not recommended that you start creating resources and workloads with this user.

In the following diagram, we can see an example of how the root user is at the top of the AWS account structure, and then our groups are inherited by both users and roles:

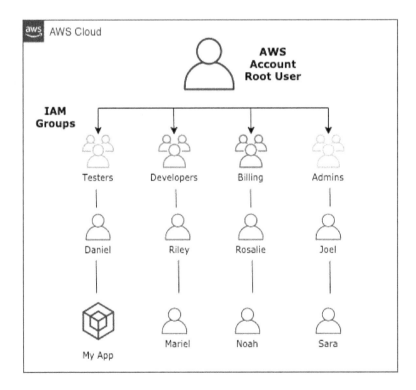

Figure 3.5 – IAM users and groups

Once we have mapped out which groups we want to create in our account, along with the basis of the permissions that each group or role will be inheriting, we can then go about creating these groups.

IAM groups

Here are some important things to know about groups:

- A user can belong to multiple groups, and groups can have multiple users.

- Groups can only contain users, not other groups, hence groups can't be nested inside of each other and inherit privileges.

- Users must be manually added to a group. Upon creation of a user or role, there is no default group where permissions are automatically assigned.

Creating IAM groups

Before we actually add our users, let's create our groups and associate the Amazon-managed IAM policies with those groups, as shown in *Figure 3.5*. We will make the following groups and then associate the following managed policies with them:

Group Name	Managed Policy Name
Admins	`AdministratorAccess`
Billing	`Billing`
Developers	`PowerUserAccess`
Testers	`AmazonEC2FullAccess`, `CloudWatchFullAccess`

Figure 3.6 – Groups and their associated managed policies

Now that we know which groups we are going to set up, let's use our CLI to create those groups, as follows:

1. Open up your terminal and type the following commands so that we can create our first group (`Admins`):

```
$aws iam create-group --group-name Admins
```

After executing the command, there should be a return somewhat like this:

```
{
    "Group": {
        "Path": "/",
        "GroupName": "Admins",
        "GroupId": "AGPAW24Q7QQF4DBRNWTM4",
        "Arn": "arn:aws:iam::470066103307:group/Admins",
        "CreateDate": "2021-03-11T22:36:46+00:00"
    }
}
```

We can create the rest of the groups the same way by substituting the rest of the group names where the current command has the group name `Admins` at the end of it.

2. Now, let's check the current groups in our account, as follows:

```
$aws iam list-groups --output text
```

Notice how I changed the output here—this method for the four groups I'm about to show is not going to take multiple pages with the spacing that JSON formatting takes up. The output should be similar to what is shown here:

```
GROUPS  arn:aws:iam::470066103307:group/Admins   2021-03-
11T22:36:46+00:00       AGPAW24Q7QQF4DBRNWTM4   Admins  /
GROUPS  arn:aws:iam::470066103307:group/Billing 2021-03-
11T22:41:23+00:00       AGPAW24Q7QQF2FJB3ZW27   Billing /
GROUPS  arn:aws:iam::470066103307:group/Developers
2021-03-11T22:54:10+00:00       AGPAW24Q7QQF7NUSSDSWF
Developers      /
GROUPS  arn:aws:iam::470066103307:group/Testers 2021-03-
11T22:51:26+00:00       AGPAW24Q7QQFZ6FIU6WDA   Testers /
```

3. Once we see all the groups listed, we can now start to attach the managed policies we identified before. However, in using the CLI, we need to know the **Amazon Resource Name** (**ARN**) of the policy we want to attach to the group. The good news is that we have already previously identified the exact name of the AWS managed policies we want to attach to our group, and with the CLI we can also combine shell commands such as `grep` to help our search, as illustrated in the following code snippet:

```
$aws iam list-policies --scope AWS|grep
'AdministratorAccess'
```

Notice how we first did the AWS CLI command and then piped the output to the `grep` command to quickly search through the output and find the exact name we were looking for. We were able to use some of the command-line flags to help narrow our initial search; however, there isn't a great option to search for a particular policy name. If you would like to read more about the options for the command, you can find all flags available on the documentation page listed here: https://docs.aws.amazon.com/cli/latest/reference/iam/list-policies.html.

The output should look like this:

```
        "PolicyName": "AdministratorAccess",
        "Arn": "arn:aws:iam::aws:policy/
AdministratorAccess",
        "PolicyName": "AdministratorAccess-Amplify",
        "Arn": "arn:aws:iam::aws:policy/
AdministratorAccess-Amplify",
        "PolicyName": "AdministratorAccess-
AWSElasticBeanstalk",
        "Arn": "arn:aws:iam::aws:policy/
AdministratorAccess-AWSElasticBeanstalk",
        "PolicyName":
"AWSAuditManagerAdministratorAccess",
        "Arn": "arn:aws:iam::aws:policy/
AWSAuditManagerAdministratorAccess",
```

You'll notice that not only one policy ARN came back—a total of four were returned. We are really looking for a policy named `AdministratorAccess`, which was the first returned result. The others were returned because they had `AdministratorAccess` as part of their name.

4. Now that we have found the ARN for our managed policy, we can attach it to our group, as illustrated in the following code snippet. Once the IAM policy is attached to the group, any users we add later will automatically assume those permissions:

```
$aws iam attach-group-policy --policy-arn
arn:aws:iam::aws:policy/AdministratorAccess --group-name
Admins
```

Once you execute this command, it will not give you any return or feedback.

5. We could check from the command line to see if our policy was attached to our group by using the following command:

```
$aws iam list-attached-group-policies --group-name Admins
```

This will then bring back the policy we just affixed to the group, as illustrated in the following code snippet:

```
{
    "AttachedPolicies": [
        {
            "PolicyName": "AdministratorAccess",
```

```
        "PolicyArn": "arn:aws:iam::aws:policy/
AdministratorAccess"
        }
    ]
}
```

At this point, we're done for the moment. We will need to create our users in order to add them to the group.

> **Note**
>
> We could have also done this using a CloudFormation script, which would have made it a lot easier to roll back or clean up any groups we just made. We are going to take a deep dive into CloudFormation and **Infrastructure as Code (IaC)** in *Chapter 7, Using CloudFormation Templates to Deploy Workloads*.

IAM users

An IAM user is a person or application who is credentialed to access and interact with a specified set of AWS services.

When you create your account, you are automatically provisioned as a root account user.

After creating a user, they start with no permissions, nor are they authorized to perform any actions on AWS resources until you or another administrator (or person with IAM permissions) grant them access via an inline policy, attach a managed policy to their user, or add them to a group.

IAM users cannot be associated with more than one AWS account.

The user's section of IAM is also where you store each individual user's **Secure Shell (SSH)** keys to access the AWS Git service CodeCommit. We don't dive deep into CodeCommit until *Chapter 8, Creating Workloads with CodeCommit and CodeBuild* ; however, understanding how the keys, users, and IAM permissions interact is important, especially if you want your users to have the ability to self-manage, add, remove, and rotate their own SSH keys.

If you would like to create more users with specified roles for your account then you can do so with the new roles that you have just created, using the steps outlined in *Chapter 2, Fundamental AWS Services,* to create a user.

Roles

IAM roles allow you to empower services, applications, and other AWS services to have access without hardcoding credentials such as access key and secret access key directly into the application—or in the case of a user, even create an access key and secret access key pair.

Roles are also extremely useful when switching between accounts. If you have multiple accounts—let's say, development, testing, and production accounts—you can use your user in the development account and then create a role in the other two accounts, which are allowed to be assumed by your main user.

Another advantage to roles is that they can be assumed by multiple instances of the same service (for example, EC2/Lambda). Just as in IAM groups, if you need to add more permissions or remove permissions from a role's policy, those changes take effect almost immediately.

In the case of EC2, you would use an **instance profile** to pass a role to an EC2 instance. An instance profile can only contain one role, and this is a hard service limit that cannot be increased. If the permission needs of the instance change, you can remove the current role attached to the instance profile and then attach a different role that contains a distinct set of permissions.

Roles and role assumption are also a way to provide access to third parties who may also have their own AWS account but may need limited access to your account. This could be a third-party partner or an auditor who may just need read-only access.

> **Note**
>
> For this exercise we will need two accounts, as follows:
>
> **Account A**: This will be the account where we will create a role that we will allow to be assumed.
>
> **Account B**: This will be the account we have been working on previously and will use one of our previously created users to assume the role and go to the other account.

Before you start, it would be a good idea to know the account numbers and have them in a text document labeled *Account A* and *Account B*. This will make things easier when we start to go through the process of setting up the role. If you don't know your AWS account number, then the easiest way to find it is to click on your username on the top of the main menu. When the drop-down menu appears, then one of the items in that menu is **Account**, with your account number.

We will first go into *Account A* and create a role to be assumed, as follows:

1. Sign in to the console with a user on *Account A* who has either Admin user privileges or the ability to create IAM users and roles.

2. Navigate to the IAM service at `https://console.aws.amazon.com/iam/home`.

3. In the left-hand menu, click **Roles**.

4. Once the **Roles** screen appears in the main window, click the blue button labeled **Create Role**.

5. The type of trusted identity you want to choose on the next screen is **Another AWS account**, as illustrated in the following screenshot:

Figure 3.7 – Creating an IAM role

6. Enter the account number of *Account B* into the **Account** field and then press the **Next: Permissions** button at the bottom of the page.

7. On the **Attach Permissions Policies** page, in the search box type `S3FullAccess`. This should bring up the `AmazonS3FullAccess` policy. Click the box next to this policy name. After you have checked the box, as illustrated in the following screenshot, then click the blue button at the bottom of the page, labeled **Next: Tags**:

Figure 3.8 – IAM S3 full access policy

8. We are not adding any tags in this exercise, so just click the blue button at the bottom of the screen that says **Next: Review**.

9. Once we get to the **Review** page, we can then name our role `AssumeS3`; if you would like to type a description, you can do so (see the example description in the following screenshot), but this is not necessary. Make sure that everything looks correct and then click the **Create role** button at the bottom:

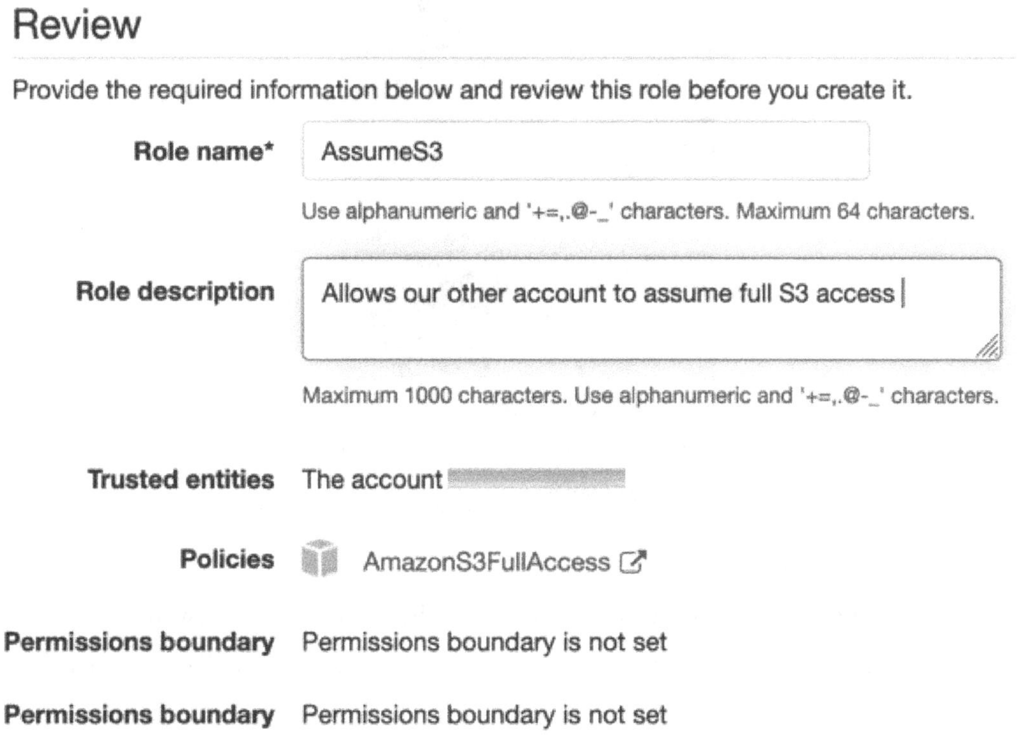

Review

Provide the required information below and review this role before you create it.

Role name*   AssumeS3

Use alphanumeric and '+=,.@-_' characters. Maximum 64 characters.

Role description   Allows our other account to assume full S3 access |

Maximum 1000 characters. Use alphanumeric and '+=,.@-_' characters.

Trusted entities   The account ▬▬▬▬▬▬▬

Policies   🟦 AmazonS3FullAccess ↗

Permissions boundary   Permissions boundary is not set

Permissions boundary   Permissions boundary is not set

Figure 3.9 – IAM assume S3 role review

Now that we have our role set up in *Account A*, we can log out of this account and then log back in to *Account B* so that we can use this new role we created to switch from our main role to the `AssumeS3` role in *Account A*, as follows:

1. Once we have logged in, then click on your name in the top right-hand menu to expose the drop-down menu (we did this previously to find our account numbers).

2. With the drop-down menu exposed, click on the menu item labeled **Switch Roles**, as illustrated in the following screenshot:

Figure 3.10 – Switching roles from AWS Management Console

3. Now, on the **Switch Roles** screen, click the blue button labeled **Switch Role**.

4. On the next screen, you will be presented with three textboxes to fill out, as follows:

    a. In the box labeled **Account**, cut and paste the account number from *Account A*.

    b. In the box labeled **Role**, type in the name of the role we created previously (`AssumeS3`).

    c. In the box labeled **Display Name**, type in `AccountA_S3`, as illustrated in the following screenshot:

Figure 3.11 – Assuming role in Account A

With this information entered, you are now ready to switch roles.

5.   After entering all this information, click the blue **Switch Role** button.

6.   You should now be in *Account A* with the `AssumeS3` permissions signified by the color at the top of the menu bar, along with the display name of the role, as illustrated in the following screenshot:

Figure 3.12 – AWS Management Console showing successful role switch

We've just seen how roles can help us manage the permissions in our own account, as well as giving us the ability to take on a defined permission set in another account. Next, we will look at both ways to put limits in place from an account level with permissions boundaries, along with some IAM best practices.

Permissions boundaries

Permissions boundaries are an advanced feature of IAM that allows you to use AWS managed policies but set the maximum resource limits for those policies. In practical usage, permissions boundaries can be extremely helpful to allow you to take an AWS managed policy that has the permissions you need for a service and attach it to a user, group, or role and then keep that entity from going outside of the resources they need to access using the permissions boundary policy.

If we refer back to *Figure 3.4*, look where the identity-based policy intersects with the permissions boundary policy. In the middle of those two policies is the effective policy for the entity. Most managed policies provided by AWS allow for a full scope in the **Resource** field, denoted with an asterisk (*). For most organizations, both large and small, this is too wide open a permission set. Not only does this present a security concern across different sectors of an organization, but also limiting the access helps mitigate any blast radius that a user could cause in an accidental situation.

IAM best practices

Let's take a look at some best practices when using IAM, as follows:

- Don't use the root account for day-to-day access. Only use the root account for the initial setup of administrative users and in emergency/break-glass scenarios.

- Always set up **multi-factor authentication** (**MFA**) for your root user. Also, it's a best practice to put MFA in place for any account administrators. Try to put MFA in place for all users.

- Set up a password policy that requires users to set a strong password that needs to be rotated on a regular basis.

- Don't set permissions directly to users; instead, use groups or allow users to assume roles—set permission sets to the roles and allow users to assume the roles.

- Don't allow secrets to be left in plaintext in code. Use a secret store such as Secrets Manager or Parameter Store to securely store secrets.

In the next section, we will learn about using **AWS Organizations** as part of your guidance.

Using AWS Organizations as part of your guidance

AWS Organizations is a service that helps you consolidate multiple AWS accounts for ease of management and billing. Not only does AWS Organizations help you create new accounts under the organizational umbrella quickly and easily, but it also provides features for governance that are not available in a standalone AWS account.

The two features of relevance to IAM are OUs and SCPs.

Separation with OUs

To understand OUs, we need to initially look at two basic concepts, as follows:

- Organization

- Root

An **organization** is an entity that you create to unify the different AWS accounts under your control so that you can oversee them as a single unit. There is one master account assigned to an organization that can then branch off into zero or more units. Many times, organizations are organized in a tree-like configuration, with the root account at the top and the OUs and child accounts branching out underneath.

The **root** is the origin container for all other accounts in your organization. If you apply an SCP to the root account, it trickles down and hence gets applied to all organizational accounts and child accounts that are part of the organization.

An OU is simply a container for grouping accounts under the root account. Organizational accounts can be nested inside of other OU units creating the hierarchy of your organization, with the root account sitting at the top.

Features of organizations

Organizations allow you to centrally manage policies across AWS accounts. This helps you improve control over your AWS environment by using **AWS Organizations** to create a group of accounts and then attaching policies to a specified group, establishing that the correct policies are applied across those accounts without a need for any custom scripting.

You can integrate AWS **single sign-on** (**SSO**) to simply provision access for your users to different accounts in your organizations, in a single place.

SCPs allow organizations to administer permission guardrails to AWS services, resources, and regions.

AWS Organizations also allows you to easily set up consolidated billing, using a single payment method for all of the accounts in an organization. This allows you to benefit from pricing advantages such as aggregated usage and volume discounts on different services that would not be available if all accounts were separated.

SCPs

SCPs are a tool that you can use to manage policies throughout a specific OU or throughout your entire organization, through inheritance. They can either enhance or limit the scope of the permissions that your users currently have, and they are at their most effective when combined with advanced IAM constructs such as `Condition`, `ArnNotLike`, `StringNotLike`, and specific regions to provide the guardrails that your organization agrees upon, to ensure that users don't intentionally or unintentionally perform inappropriate actions.

Example of a SCP

The following code snippet provides an example of a SCP that you could attach at the root level of your organization to make sure that if anyone were going to delete an EC2 or **Relational Database Service** (**RDS**) instance, then a multi-factor device would first need to be present for verification:

```
{
    "Version": "2012-10-17",
    "Statement": [
        {
            "Sid": "DenyDeleteandTerminateWhenMFAIsNotPresent",
            "Effect": "Deny",
            "Action": [
                "rds:Delete*",
```

```
                  "ec2:TerminateInstances"
          ],
          "Resource": "*",
          "Condition": {"BoolIfExists":
 {"aws:MultiFactorAuthPresent": false}}
          }
      ]
  }
```

You'll notice how the SCP syntax looks almost exactly like the normal IAM policy syntax, containing an `Effect`, `Action`, `Resource`, and `SID`. This SCP actually has an optional value of a condition, which is where we are checking to see whether the MFA exists and if not, then the action is denied.

Integrating federation with an AWS account

If the users who will be accessing your account already have an authentication method, then there are a few different methods to incorporate federation into your AWS account.

"Identity federation is a system of trust between two parties for the purpose of authenticating users and conveying information needed to authorize access to resources. In this system, an Identity Provider (IdP) is responsible for user authentication, and a service provider (SP), such as a service or application, controls access to resources." (AWS, 2021)

When you incorporate federation, then your users are going to use STS to request a session for authorization into your account.

A session consists of the following:

- An access key
- A secret access key
- A security token
- An expiration date (for when the security token is no longer valid)

A federated token by default will expire in 12 hours, but can be set to time out in as little as 15 minutes or as long as 36 hours.

Federated tokens are handed out from the STS service via a few different request types, a couple of which are listed here:

- `GetFederatedToken`
- `AssumeRoleWithSAML`

When would you use federation?

There are a few scenarios where federation makes perfect sense to use in your environment or accounts. The first would be if you have a **Security Assertion Markup Language** (**SAML**) 2.0-compatible corporate directory. Examples of this would be **AD Federation Service** (**AD FS**), Okta, Shibboleth, and many others. When your directory is SAML 2.0-compliant, then you can configure it to provide SSO access to the AWS Management Console.

Even if the directory isn't SAML 2.0-compliant, you still have the ability to furnish SSO to your users. In this case, though, you will need to produce an identity broker application to provide SSO access.

The second scenario in which you would likely use federation is if your users have identity management through a third party such as a web identity. In this case, the web IdP provides the authentication step, and then you can use federation to grant your users AWS access to the services they need. We will discuss this more when we talk about the use of the AWS Cognito service later in the chapter.

Using AD federation with IAM

When you use AD FS to provide access to your account, the flow happens in the following order:

1. The user (or application) requests a session from the federation proxy.
2. The user is then authenticated via AD to ensure that they are an active user and have permissions to access the AWS cloud.
3. Next, AD determines the quantity of permissions the user is entitled to receive.
4. Then, the federation proxy initiates a `GetFederationToken` request from the STS service in the AWS account.
5. The STS service then responds with the federated token (including the access key, secret key, and session token).
6. The federation proxy then passes the token to the user.
7. The user is then able to make API calls to the AWS account.

Using federation to authenticate to your AWS account is especially useful if you have a large organization and don't want to manage users in both AD and AWS IAM. It can also help you get around the service limit of 500 IAM users per account.

AWS offers a number of different *flavors* of AD depending on your needs, which we will look at next.

AWS Directory Service

AWS provides a multitude of different options for using AD in the cloud according to your needs. There are three levels of the AWS Directory Service offered, and each serves a different purpose.

AWS Directory Service for Microsoft AD (MS AD)

If you are looking to run a managed service Enterprise version of AD in the cloud, then the AWS AD service is an excellent option. This service can handle up to 50,000 users and up to 200,000 AD objects. It also has the ability to be implemented in a highly available fashion across multiple regions. Multi-Region replication is only supported in the Enterprise edition of AWS Managed MS AD.

If you have more than 5,000 users and don't currently have an AD server or the need to manage users and/or AD objects in the AWS cloud, then AWS Directory Service is your best option.

One disadvantage of using this service is if you run into performance issues, you don't have a lot of knobs and levers to pull. This is, after all, a managed service and is supposed to take administrative tasks off your plate. This is also not fitting for web federation.

AD connector

If you already have an AD server on-premises and you want a way to quickly extend this to the cloud, then an AD connector is a good option. An AD connector doesn't cache information in the cloud, but rather serves as a proxy for your on-premises AD server. This also means that you can't make any changes to the AD connector in the cloud. All changes and updates have to take place on the on-premises server.

Simple AD

Simple AD comes in two different sizes: small and large. The small version can support up to 500 users and 2,000 objects and the large version can support up to 5,000 users and 20,000 objects. If these numbers don't support the growth rate of your organization, then you need to consider an alternative solution such as MS AD.

Simple AD does support Kerberos-based SSO authentication and can be spun up using a CloudFormation script.

> **Important note**
>
> Simple AD is based on Samba 4 and isn't true MS AD. It won't support TRUST with other AD systems. It also doesn't support items such as MFA, PowerShell, AWS SSO, or Amazon Chime. While it is a good option if you only have a small set of users or if you are trying to provide a **fully qualified domain name** (**FQDN**) for a small number of Windows machines in the cloud, be mindful of its limitations—especially when reading through exam questions.

AWS SSO

Even with the vast offerings of AWS, there are still other services you or your customers probably want to incorporate as SaaS offerings that require authentication. Rather than making users verify themselves to each and every **service provider** (**SP**), AWS makes this simple with its own SSO service.

This is even the case if you are using AD, be it on-premises or Azure-based AD. You might be using third-party authentication such as Okta or Ping Identity, and SSO can be incorporated with these services as well.

Where AWS SSO really eases the burden from administrators is when you are managing multiple AWS accounts, as you can map your users and their roles quickly and easily to accounts with a single point of entry. There is no longer a need to set up a SAML token or to go through the process of showing users which account to use to switch roles.

Choosing a user identity strategy

With all of these different methods to manage your users and different ways to manage their authorization, you might need some guidelines on when to use each method.

If you have a small group, then that size of team is usually best served by using IAM groups.

If you want to use federation to manage your user logins and you already have an AD in place, then SAML federation via AD is going to be the correct route.

If you want federation and you don't yet have an AD and you want the extra features that AD provides, such as providing **Domain Name System** (**DNS**) management to Windows instances, then you can go to AWS Directory Services.

However, if you want federation and don't want the hassle of setting up and managing AD or the licensing costs, then AWS SSO is going to be your best option.

Storing secrets securely in AWS

Being a DevOps engineer and codifying our infrastructure and applications for deployment through deployment pipelines doesn't take out the necessity of credentialing items such as usernames and passwords or third-party API authentication from our application. It does, however, present a need to authenticate on a much more secure and repeatable basis so that development and testing aren't hindered, along with only sharing access only with those who absolutely need to know. Not every developer who is working on a system needs a username and password to the database if they are all using the same application. You can also limit the management overhead of having to change the credentials every time there is a personal change if you, instead of passing the credentials out to every developer, have the developer request the credentials from a secrets manager. Although there are third-party solutions that perform this task, AWS does provide two different native services that allow for secrets management, and even secrets rotation.

AWS Secrets Manager

You can use Secrets Manager to store, rotate, monitor, and control access to secrets such as database credentials, API keys, and **Open Authorization** (**OAuth**) tokens. One of the main features of Secrets Manager is that it allows for automatic rotation of secrets using AWS Lambda functions.

There is no longer a need to place hardcoded secrets in your IaC or application code, or even to download to your instances at startup and have them stored in an unencrypted vulnerable state. You can now retrieve secrets when you need them simply and securely, with a call to Secrets Manager.

Access to Secrets Manager is also governed by IAM policies. Users or roles must have been granted access to either the Secrets Manager service as a whole or the specific instance of Secrets Manager in order to perform operations such as storing and retrieving secrets.

Another feature of Secrets Manager is that every time a secret is accessed, an audit trail is left. This is a valuable feature in highly compliant industries.

You might be wondering: *Just what is a secret?* Well, in the context of Secrets Manager, a secret is typically a set of credentials (username and password) and the connection details that you use to access a secured device.

The basic structure of a secret in Secrets Manager

In Secrets Manager, a secret contains encrypted secret text and several metadata elements that express the secret and define how Secrets Manager should deal with the secret, as illustrated in the following diagram:

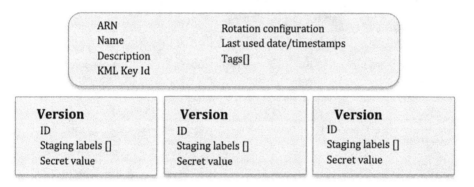

Figure 3.13 – Anatomy of a secret in Secrets Manager

Now that we have seen how Secrets Manager stores and versions the secrets that we place into it for safekeeping, let's go through an exercise of creating and storing a secret using AWS Secrets Manager.

Creating and storing a secret in AWS Secrets Manager

In the following exercise, we are going to store a secret securely using the AWS Management Console and AWS Secrets Manager. Proceed as follows:

1. Open up your browser and go back to the Amazon web console. You may have to log in if your session has expired. In the search bar at the top, type `Secrets Manager` to have the **Secrets Manager** icon appear. Click on the icon to go to the main **AWS Secrets Manager** page, as illustrated in the following screenshot:

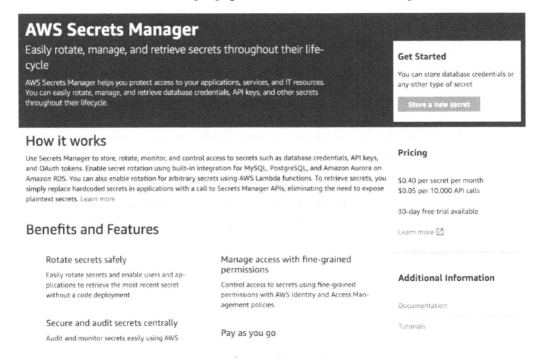

Figure 3.14 – The main Secrets Manager page

2. On either the main `Secrets Manager` page or the **secrets list** page, once you have entered into the **Secrets Manager** section of the console, click **Store a new secret**.

3. On the **Store a new secret** page, choose **Other type of secrets**, as illustrated in the following screenshot:

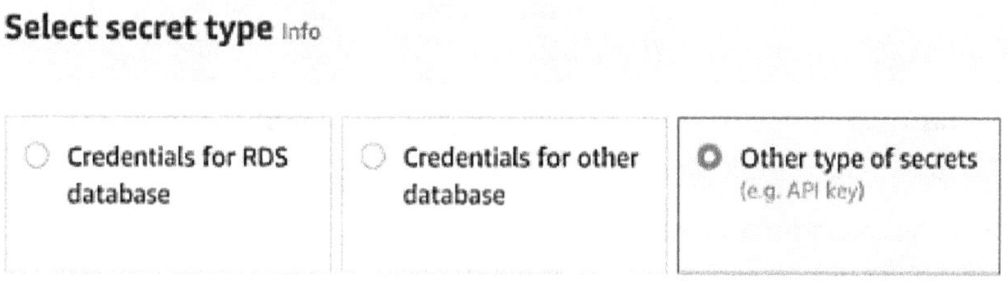

Select secret type Info

○ Credentials for RDS database	○ Credentials for other database	◉ Other type of secrets (e.g. API key)

Figure 3.15 – Choosing the secret type on the Store a new secret page

4. Under **Specify the key/value pairs to be stored in this secret**, choose **Secret key/value** so that you can type in the secret as key-value pairs, as illustrated in the following screenshot:

Specify the key/value pairs to be stored in this secret Info

Secret key/value | Plaintext

username		Remove
password		Remove

+ Add row

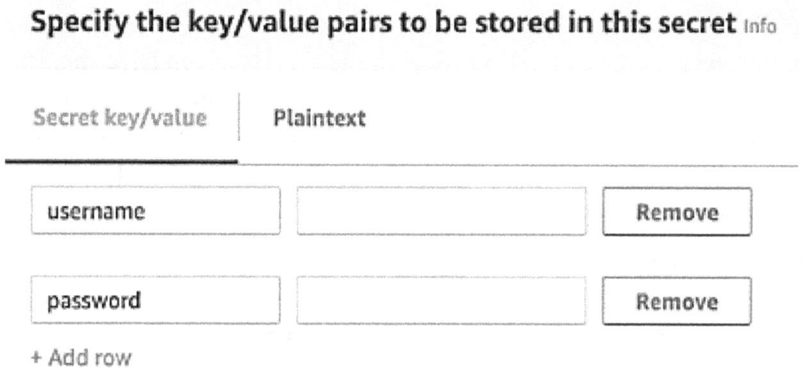

Figure 3.16 – Adding the key pair and values to your secret in Secrets Manager

5. For **Select the encryption key**, choose `DefaultEncryptionKey`, as illustrated in the following screenshot:

Select the encryption key Info

Select the AWS KMS key to use to encrypt your secret information. You can encrypt using the default service encryption key that AWS Secrets Manager creates on your behalf or a customer master key (CMK) that you have stored in AWS KMS.

| DefaultEncryptionKey ▼ | ↻ |

Add new key ↗

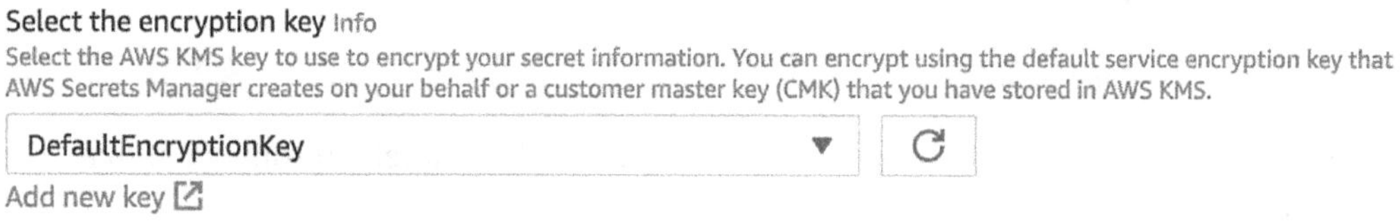

Figure 3.17 – Secrets Manager encryption

6. In the first box, type `username`. In the second box, type `devopspro`.

7. Choose + **Add row** to add a second key-value pair.

8. In the first box, type `password`. In the second box, type `MyTopSecretP@ssword#`.

Once we have typed in the values for our secret, we can view it in a JSON context before moving on to ensure it is correct.

9. Choose **Plaintext** above the boxes to see the JSON version of the secret text that will be stored in the `SecretString` field of the secret.

10. For **Select the encryption key**, leave it set at the default value `DefaultEncryptionKey` and click **Next**.

11. Under **Secret name and description**, for **Secret name**, type `chapter3/ SecretManager`. This stores your secret in a virtual `chapter3` folder.

12. For **Description**, type something such as `Secret Manager Tutorial` and click **Next**.

13. We don't need to rotate this secret, so choose **Disable automatic rotation**, and then click **Next**.

14. On the **Review** page, you can check all of the settings that you have chosen. If you are using the secret within an application, be sure to review the **Sample code** section that has cut-and-paste-enabled code that you can put into your own applications to use this secret to retrieve the credentials. Each tab has the same code in different programming languages.

15. To save your changes, choose **Store**.

Retrieving your secret from Secrets Manager

Now that we know how to use the AWS Management Console to save a secret, we will use the AWS CLI to retrieve the secret and see that it has been stored successfully. Proceed as follows:

1. Open up your terminal if you closed it previously, and type the following command so that we can retrieve the secret:

```
$aws secretsmanager list-secrets
```

After executing the preceding command, there should be a return somewhat like this:

```
{
    "SecretList": [
        {
            "ARN": "arn:aws:secretsmanager:us-east-
2:470066103307:secret:chapter3/SecretManager-qBEqSh",
            "Name": "chapter3/SecretManager",
```

```
            "Description": "Secret Manager Tutorial",
            "LastChangedDate": "2021-03-
15T21:55:53.828000-04:00",
            "Tags": [],
            "SecretVersionsToStages": {
                "83e08e0f-00d6-46d2-b05a-223041403f19": [
                    "AWSCURRENT"
                ]
            },
            "CreatedDate": "2021-03-15T21:55:53.783000-
04:00"
        }
    ]
}
```

2. Now that we can see our secret with the name `chapter3/SecretManager` stored in Secrets Manager, we can go about retrieving the secret by running the following command:

```
$aws secretsmanager get-secret-value --secret-id
chapter3/SecretManager
```

Since we are authorized to get the value of the secret with our user, we should see the secret being returned as follows:

```
{
    "ARN": "arn:aws:secretsmanager:us-east-
2:470066103307:secret:chapter3/SecretManager-qBEqSh",
    "Name": "chapter3/SecretManager",
    "VersionId": "83e08e0f-00d6-46d2-b05a-223041403f19",
    "SecretString":
"{\"username\":\"devopspro\",\"password\":\"MyTopSecretP@
ssword!\"}",
    "VersionStages": [
        "AWSCURRENT"
    ],
    "CreatedDate": "2021-03-15T21:55:53.810000-04:00"
}
```

You will see both the username and password returned unencrypted in the `SecretString` field for you to use. You should also notice that it has returned the current value by default. We could have also asked Secrets Manager to return us back to the previous version of the secret stored if we needed that version of the secret.

We just used AWS Secrets Manager to successfully create, store, and retrieve a secret. One of the reasons why we could retrieve the secret is because our IAM role gives us the permissions to do so. Next, we will look at the other service that AWS offers for safekeeping of secrets: SSM Parameter Store.

SSM Parameter Store

Parameter Store is a managed service from AWS to store strings. These can be plaintext strings or, in the case of using it as a secrets manager, it can store encrypted data. Access to the parameters is governed by IAM policy access.

Parameter Store is more flexible than Secrets Manager and if you store standard parameters there is no charge, but there is a limit of 10,000 parameters.

Secrets and values stored in Parameter Store can be accessed from many different AWS services, including the following:

- EC2
- **Elastic Container Service** (**ECS**)
- Secrets Manager
- AWS Lambda
- CloudFormation
- CodePipeline
- CodeBuild
- CodeDeploy

If you or your team are looking for a central way to separate your secrets and configuration data from your code base, along with the ability to store different values for different environments, then Parameter Store could be a good option for you.

Secrets Manager versus Parameter Store

As we look at the different services that allow us to store secrets securely in AWS, we might wonder why we would choose one over the other. Let's take a quick comparison of the two services and see the main differences, as follows:

- If you are looking for an option that won't add any more to your budget, then SSM Parameter Store is the right choice for you and your organization.

- If you want the ability to store plaintext secrets, then SSM Parameter Store has this ability. AWS Secrets Manager cannot store values in plaintext.

Both services are proportionately secure as both can use **Key Management Service (KMS)** to securely encrypt values at rest.

If you want the ability to automatically rotate secrets, then AWS Secrets Manager needs to be your service choice. (With Parameter Store, you would have to rotate the secrets manually.)

Using Cognito with application authentication

Cognito is a service that allows web and mobile applications to authenticate to your platform using either username and password or third-party identification social IdPs such as Amazon, Facebook, or Google. The three main functions of **Amazon Cognito** are user management, authentication, and synchronization. Cognito can be used to coordinate the authentication of external users to your applications via third-party IdPs. You can also federate your users with any SAML-based IdP.

Some of the advantages of using Cognito include the following:

- Ease of setup, whereby you can have an authentication and authorization system that integrates into your AWS services up and running in a matter of minutes, versus the amount of time it would take you for custom programming to do the same task.

- You can quickly and easily integrate MFA into **Amazon Cognito**.

- Cognito metrics are easily monitored using the CloudWatch service, with no extra effort or coding.

- Any API calls made to the Cognito service are recorded to CloudTrail (assuming you have turned on CloudTrail for the account and region).

Where would you use Cognito?

As you build out an application for external customers (customers that are not part of your company) and you want these customers to have the same experience no matter what type of device they are using, this is a case to use AWS Cognito. Cognito allows users to have a similar login experience whether they are using a mobile device, tablet, laptop, or **virtual machine** (**VM**).

Another case for using Cognito is if you want to quickly incorporate **two-factor authentication** (**2FA**) such as email or **Simple Message Service** (**SMS**) authentication into your application.

User pools

The first main component of Cognito is user pools. Essentially, user pools present sign-up and sign-in services. Understand these facts about user pools:

- They provide sign-up and sign-in services.
- User pools have a built-in **user interface** (**UI**) that is customizable.
- They allow for sign-in via social IdPs such as Amazon, Google, Facebook, and Apple, along with SAML IdPs from your user pool.
- They allow for security features such as MFA authentication, can check for compromised credentials, and have account takeover protection.
- User pools have phone and email verification.
- User pools allow for customized workflows along with user migration via AWS Lambda triggers.

Once you have successfully authenticated a user with a user pool, then Cognito issues a JWT that can be used to authorize access to either your AWS credentials or your APIs, as illustrated in the following diagram:

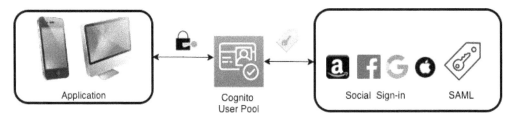

Figure 3.18 – Cognito user pool authentication

With an understanding of user pools and what function they serve, we can look at the other main component of **Amazon Cognito**: identity pools.

Identity pools

The second main component of **Amazon Cognito** is that of identity pools. Identity pools provide the credentials to access AWS services. With an identity pool, you can create unique identities for your users that give them temporary access credentials to AWS services.

If you look at the workflow shown in *Figure 3.17*, you will see that the user initiates the login (usually from an application on their device) with the web IdP. Once authenticated with the web IdP, a Get Id call is passed to Amazon Cognito, which then validates the request. The app will then make a GetCredentialsForIdentity request. Cognito will once again validate this request. At this point, Cognito will call out to the STS service and get a short-lived token for the services the app is authorized for and will then pass that back on to the application, as illustrated in the following diagram:

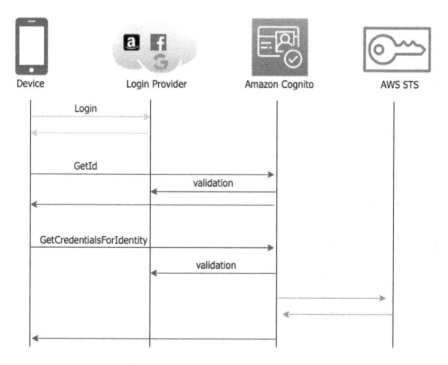

Figure 3.19 – Cognito identity pool authorization flow

Identity pools also allow you to define your own custom developer-authenticated identities in addition to the social IdPs supported for authentication. You can do this in the Cognito console by choosing **Custom**, setting your own name, and then using the sample code provided by AWS as the base for creating your new custom federated identity pool.

> **Tip to remember**
>
> User pools provide the authentication, and identity pools provide the authorization.

Summary

In this chapter, we learned about authentication and authorization methods in AWS. We talked about how to authorize users either with username and password or via federation using the IAM service and then authenticating them using IAM policies. **AWS Organizations** also showed us how we could further restrict the authorization settings using SCPs. We looked at different models of federation with services such as AD, AWS SSO, and Cognito. We also looked at how to securely store secrets as we start to work with our applications using Secrets Manager or SSM Parameter Store.

In the next chapter, we will wrap up our section on AWS fundamentals with a look at the NoSQL service DynamoDB.

Review questions

1. In the Shared Responsibility Model presented by AWS for IaaS, who is responsible for operating system security and patching?

2. What are the major differences between a user and a role from the choices presented here?

    a. A role can be assumed by multiple principals; a user cannot be assumed.

    b. A role uses long-term credentials.

    c. A role can be part of a group.

    d. A role does not use long-term credentials.

3. Which comes first: authorization or authentication?

4. Which native service in AWS that stores secrets offers automatic secret rotation?

5. A company wants to extend their current AD into the AWS cloud but doesn't want to manage more servers. Which service is the best choice for them?

    a. AWS Simple Directory Service

    b. AWS Cognito

    c. AWS User Pools

    d. AWS AD Connector

Review answers

1. The customer is responsible for operating system patching and security.

2. a and d.

3. Authentication.

4. AWS Secrets Manager.

5. d.

4
Amazon S3 Blob Storage

Amazon S3 is one of the core tenants of Amazon Web Services. However, in the context of being a DevOps professional, there are certain nuances about the service that you must not only familiarize yourself with but also become comfortable with implementing.

Amazon's **Simple Storage Service (S3)** is the entry point for many users and companies looking to get into the cloud. Some of the main features it provides are being highly available, exceedingly durable, extremely performant, and easily managed, along with the ability to be thoroughly secure. One of the major features of S3 is its eleven 9's of durability (99.999999999%), which means that it doesn't lose objects or data. Once you upload an object, before it returns the 200 success status, that object must be copied to multiple systems in multiple Availability Zones to prevent data loss in an array of scenarios.

In this chapter, we're going to cover the following main topics:

- S3 concepts
- Using lifecycle policies in S3
- Using S3 events to trigger other AWS services
- S3 access logs
- S3 endpoints

S3 concepts

Before we dive into S3, let's at least briefly talk about the three distinct types of cloud storage:

- **Object storage** – Data is saved as an object and is bundled with the associated metadata of that object.

- **File storage** – Data is stored as a single piece of information in a folder structure.

- **Block storage** – Data and files are separated into blocks. Each of these blocks is then stored as a separate piece of data.

S3 is an object storage service, and although it seems to have a folder structure, this is really just the metadata that is tagged to the object in key/value pairs so that the data can be categorized more efficiently.

Once an S3 bucket has been created, then not only is it ready for data, but it is also at that point almost infinitely scalable. There are also a number of helper services that AWS has created to assist you in moving data into S3. These range from streaming solutions such as Amazon Kinesis to **SSH File Transfer Protocol** (**SFTP**) alternatives such as AWS SFTP, and even bulk data load services such as AWS Snowball and Snowball Edge:

Figure 4.1 – S3 options for data transfer

By default, AWS accounts are allowed to provision up to one hundred S3 buckets. This is a soft limit, and if you need more buckets this can be raised by placing a service limit increase ticket.

Interacting with S3

Users have a few ways to interact with S3, starting with the AWS console. This will show you a graphical listing of your buckets and display your objects in a folder-like format based on the object tags that you have given the different objects.

As you become more comfortable with S3 and its capabilities, you might find that the console requires more than one click to perform simple tasks such as deleting a file, and this is where the CLI and knowing its commands can become your ally.

In the AWS CLI, there are a few base commands that you can use to interact with S3:

- There is the base `aws s3` command – this command gives you the ability to perform nine basic operations, such as creating a bucket, listing buckets or their contents, and even creating a pre-signed URL for access to a bucket.

- There is the `aws s3api` command – this command provides a different set of secondary commands for the items in the base `s3` command.

- The `aws s3control` command allows you granular access to the S3 control plane.

- And finally, there is the `aws s3outposts` command – this command provides access to S3 on AWS Outposts.

S3 naming guidelines

Every S3 bucket name across all of AWS must be unique, not just in your account.

When you create your bucket, there are a few rules that you must follow in order for the name to be compliant with S3:

- A bucket name must be at least 3 but no longer than 63 characters long.

- Bucket names can only be made up of numbers, lowercase letters, dots (.), and hyphens (-).

- A bucket name must start and end with either a number or a letter.

- Bucket names cannot start with the name `xn-`.

- A bucket name may not be formatted like an IP address (such as `192.168.2.1`).

It's also Amazon's recommendation that you do not use dots (.) in your bucket names unless you are using the S3 bucket for static web hosting.

Bucket names for enterprises

The preceding rules are the minimum guidelines for naming S3 buckets. As you move into the real world, most large organizations that are already in the AWS cloud have a bucket naming scheme. Some small and medium firms still do not have organizational standards relating to how to name their S3 buckets, but this can be a mistake.

S3 buckets are very easy to create, not only for developers but for almost anyone that has access to the service. The issue arises when you start to find buckets that are named like `mytest-bucket123`. Once again, this comes back to those initial principles of trying to figure out things like who the owner of this data is and whether it needs to be replicated for safekeeping or can be safely deleted.

As you move to an enterprise naming scheme, you and your organization need to come to a consensus regarding a uniform naming standard for the buckets in the accounts for which you are responsible:

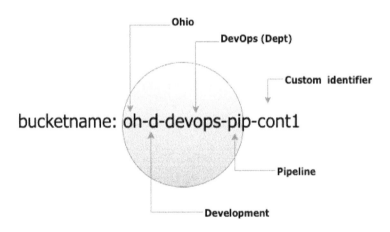

Figure 4.2 – Enterprise bucket naming example

Using an abbreviated schema such as `<region>-<environment>-<department>-<product>-<identifier>` creates a unique name for each bucket that still stays within AWS naming standards.

This helps to quickly identify who owns each bucket and allows teams and account managers to sort and search through resources quickly and easily, not only identifying who created the buckets by the product or project they belong to. This is shown in *Figure 4.2* and with the bucket named `oh-d-devops-pip-cont1`. This is bucket name is shorthand for `ohio-development-devops-pipeline-containers`.

Creating an S3 bucket

If you want to back up your instance either for point-in-time recovery purposes or to use in a launch configuration with autoscaling, then you need to create an AMI image:

1. Launch an EC2 instance.

2. We need to have an instance to back up and create the AMI:

    ```
    $ aws s3 MB s3://devopspro-beyond -region us-east-2
    ```

    > **Note**
    > You will need to use a different bucket name than the one shown in this example. You can name your bucket anything you like as long as you stick to the S3 naming guidelines.

3. If it was successful, then you should see the following output:

    ```
    make_bucket: devopspro-beyond
    ```

    You now have a bucket we can work with. If you wanted to see the bucket in the command line, we could list it with the following command:

    ```
    $ aws s3 ls
    ```

    This bucket that we just created is now ready to hold files, media, logs, or whatever you'd like to store in it at this point.

Moving data to S3

At this point, we have now created at least one bucket. As we go through later exercises, you'll notice that some of the AWS services create buckets in your account, such as **CloudFormation**.

If we were simply trying to move items one at a time or a folder at a time, then we could use the AWS Management Console or the CLI using the aws s3 copy command.

If we had a server that was generating logs, and we wanted to put those logs into an S3 bucket for either storage, backup, analysis, or a combination of those options, then we could use the aws s3 sync command. The s3 sync command will sync all objects in the stated folder with the specified bucked in S3. This works extremely well in concert with a cron or cron-like job for enacting the command on a schedule.

When trying to move a whole server or data center, it can be time-consuming to try and push all the files and objects across the wire. This is where the **Snowball** family of services comes into play. A **Snowball Edge Storage Optimized** (**SESO**) allows secure transport of your data with up to 80 TB of usable hard disk drive storage, which, after being uploaded to the device and shipped to AWS, is then offloaded to an S3 bucket that you designate.

S3 inventory

Using the list command to see the different objects in your bucket is helpful until you start to have tens of thousands to millions of objects in your buckets. At that point, it can be helpful to have a more powerful tool at your disposal. **S3 inventory** was created for just this purpose. The S3 inventory tool creates a report, which is then delivered to another bucket and provides information about your objects on items such as the following:

- Creation date

- Storage class

- Encryption status

- Replication status

- Object size

You have the option to encrypt the reports either with SSE-S3 or with a **Key Management Service** (**KMS**) key of your choosing.

Once the reports have been delivered, then you can use tools such as Amazon Athena to query the reports using Standard SQL, looking for trends or anomalies.

S3 inventory does come with a small cost to generate the reports (less than .003 cents per million objects); however, it should not be assumed that the tool comes with the S3 service for free.

S3 storage tiers

The Amazon S3 service has a collection of different storage tiers that can serve different needs as well as different cost structures. The default storage tier is the standard storage tier, although this is not always the correct choice for storing your data, especially if your organization is looking for long-term storage and/or cost savings.

You can create lifecycle policies to move your objects from one storage tier to another or even be deleted if the object is unneeded after a set period of time.

S3 Standard

Once you initially set up a bucket, by default, without specifying any other storage tier, it will be in the S3 Standard tier. This is a highly available general access storage policy that provides millisecond access to objects when requesting their retrieval. Although this is the costliest of all the storage tiers, S3 Standard storage is extremely inexpensive when compared to other types of storage services like file and block storage.

Key points to remember about S3 Standard:

- The Standard tier provides high throughput and low latency and performance for object uploads and downloads.

- If no other storage tier is indicated, then Standard is the default storage class.

- Crafted for 99.99% availability during a given year.

- Ideal for objects that need frequent access.

- Ideal for use cases such as data lakes, cloud-native applications, websites, and content distribution.

S3 Intelligent-Tiering

There are times when you might think that the algorithm that Amazon has derived to move your objects to the different tiers of storage may be more efficient than any you may be able to come up with. This is a perfect case of when to choose S3 Intelligent-Tiering. With Intelligent-Tiering, AWS will move your objects automatically between the frequently accessed tiers and the infrequently accessed tiers based on your usage and then charge you accordingly.

Key points to remember about S3 Intelligent-Tiering are as follows:

- Designed to optimize storage costs by automatically moving objects to the most cost-effective storage tier.

- Designed for longer storage of at least 30 days (minimum 30-day charge) and Intelligent-Tiering takes 30 days to start to figure out access patterns.

- It stores objects in two access tiers and optimizes that storage based on frequently and infrequently accessed objects.

- There is no performance impact, and there are no additional fees when Intelligent-Tiering moves objects between tiers.

- Crafted for 99.99% availability during a given year.

- Optimized for data lakes and other datasets where the access patterns are unknown.

S3 Standard Infrequent Access (S3 Standard-IA)

If you have data that you don't access frequently but still need to be able to retrieve it in real time, Standard-IA is an option to consider. There are some points that need to be considered when thinking about this storage option, such as the files need to be stored for a minimum of 30 days before deletion (or be charged for the 30 days), along with having a minimum file size of 128 KB.

Key points to remember about S3 Standard-IA are as follows:

- Designed for files over 128 KB (smaller files will be charged as if they were 128 KB).

- Designed for longer storage of at least 30 days (minimum 30-day charge).

- There is a higher GET, PUT, COPY, POST, LIST, and SELECT charge than Standard but a lower storage cost, so it is designed for infrequent access, as the name states.

- Objects are available to access in real time with no delays.

- Crafted for 99.99% availability during a given year.

- Copies of data are stored in multiple Availability Zones.

S3 One Zone Infrequent Access (S3 One Zone-IA)

S3 One Zone-IA has many of the features of Standard-IA but at a lower price because the data is being stored in only one availability zone instead of a minimum of three. This is not a good option for critical data but can present large cost savings for files that are infrequently accessed and can be re-created if necessary.

Key points to remember about S3 One Zone-IA are as follows:

- Ideal for data that can be re-created or object replicas when setting *cross-region replication*.

- Designed for longer storage of at least 30 days (minimum 30-day charge).

- Objects are available for real-time access.

- Crafted for 99.95% availability during a given year.

- Data is subject to loss stemming from data center outages caused by disasters such as floods or earthquakes.

S3 Glacier

The S3 Glacier storage tier provides you an option for a low-cost, durable storage archive with low fees for data retrieval. Unlike the Glacier service from AWS, there is no need to wait for days for your objects to appear back in your S3 bucket. S3 Glacier has 3 tiers of retrieval speeds. The first is an expedited tier that can bring your objects back in just 1-5 minutes. The second is the standard retrieval tier, which restores objects in 3-5 hours, and the third is the bulk tier, which takes around 12 hours to restore objects to your bucket.

Key points to remember about S3 Glacier are as follows:

- Designed for longer storage of at least 90 days (minimum charge of 90 days).
- Crafted for 99.9% availability during a given year.
- Objects can be locked via the **VAULT LOCK** feature.
- Glacier retrieval times can be configured from minutes to hours.
- Appropriate for low-cost data archival on infrequently accessed objects, especially for compliance purposes.

S3 Glacier Deep Archive

Like the Glacier service, if you have items that are rarely retrieved but need to be retained, then Glacier Deep Archive can be a practical solution to your storage problems. Often, there are cases such as moving from a tape backup system to a digital tape backup system where you would only be retrieving the data once or twice per year and could withstand waiting 12 hours for data retrieval. These controls come with deep savings because storage in Glacier Deep Archive costs only $1 per TB per month.

Key points to remember about S3 Glacier Deep Archive are as follows:

- Designed for long-term digital storage that may be accessed once or twice during a given year
- Crafted for 99.9% availability during a given year
- Designed for longer storage of at least 180 days (minimum 180-day charge)
- An alternative to on-premises tape libraries

> **Note**
> S3 Glacier and S3 Glacier Deep Archive are storage classes within S3 and, as such, the object stays within the S3 service.

Using lifecycle policies in S3

As we just talked about the different storage classes available with S3, not all data that you store in S3 needs to be in the standard tier at all times, as shown here:

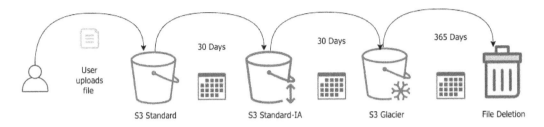

Figure 4.3 – An example S3 lifecycle policy

Depending on how long you need to access your data on a regular basis, you can move your objects to different tiers using object lifecycles.

Creating a lifecycle policy

The following exercise will use the AWS console and the bucket we created previously in this chapter:

1. Log in to the AWS console (in the account in which you created the S3 bucket earlier in this chapter). Make sure you have logged into the account with a user that has S3 full access rights.

2. Navigate to the S3 page (`https://s3.console.aws.amazon.com/`).

3. If you have more than one S3 bucket, click on the bucket that you made in the previous exercise.

    This bucket should currently have no objects and no current lifecycle policy associated with it.

4. Once you're on the bucket's main page, click on the **Management** tab on the main screen.

    This will bring the **Lifecycle rules** section to the middle of the main screen. It should have a **(0)** right after the **Lifecycle rules**, meaning that there are no **Lifecycle rules** currently associated with this bucket:

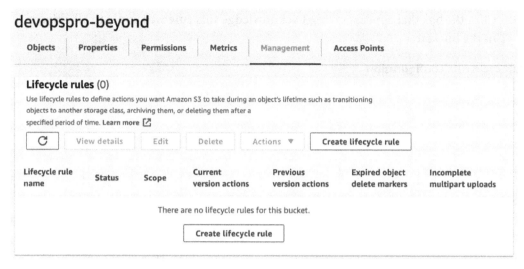

Figure 4.4 – S3 lifecycle rules via the Management tab

5. Now we can click on the button in the **Lifecycle Rules** section labeled **Create lifecycle rule**. This will bring us to the area where we can start to create our own lifecycle rules.

    We are going to create a rule that deletes objects in our bucket after one day. We could create multiple rules that follow the path in *Figure 4.2*, but if you remember the key points of the Infrequent Access storage tier, any object that isn't kept for at least 30 days will be charged for 30 days of storage. Rather than have those charges when testing out how to create lifecycle rules, we will just create a rule that will delete the objects after a certain period of time instead.

6. Call the rule `devopspro-1day-delete`.

7. Under the rule scope, click the option that says **This rule applies to all objects in the bucket**.

8. Click the box that appears saying **I acknowledge this rule will apply to all objects in the bucket**:

Lifecycle rule configuration

Lifecycle rule name

devopspro-1daydelete

Up to 255 characters.

Choose a rule scope
○   Limit the scope of this rule using one or more filters

◉   This rule applies to *all* objects in the bucket

⚠   **This rule applies to *all* objects in the bucket**
      If you want the rule to apply to specific objects, you must use a filter to identify those objects. Choose
      "Limit the scope of this rule using one or more filters". Learn more ↗

      ☑ I acknowledge that this rule will apply to all objects in the bucket.

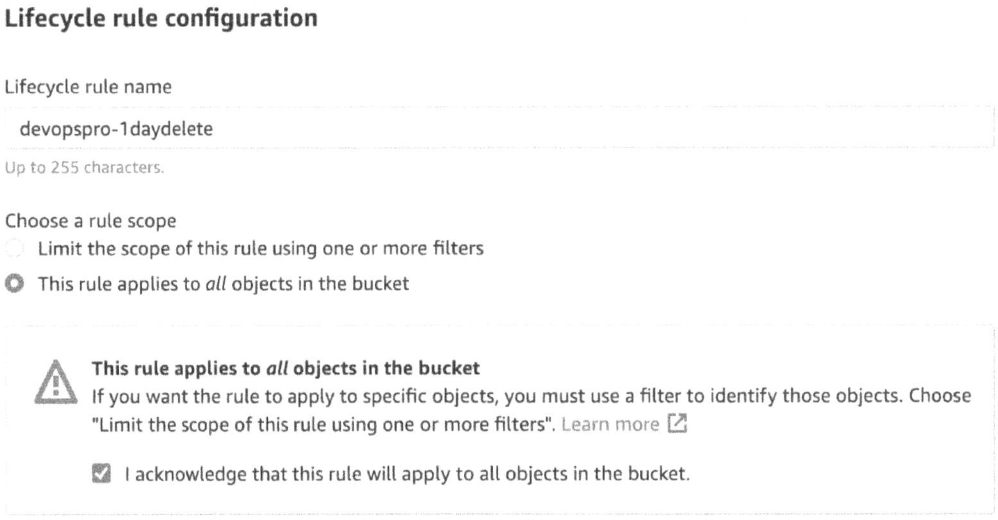

Figure 4.5 – Configuring the lifecycle rule

9. Under the **Lifecycle rule actions**, check the box that is labeled **Expire current versions of objects**.

10. When the **Expire current version of objects** box appears set the number of days after object creation to 1.

11. Click on **Create rule** at the bottom of the page:

Figure 4.6 – Create rule button

We have now created our lifecycle rule and, in order to test it out, we need to upload a file into the bucket and then wait for a day so that we can see it be automatically deleted.

S3 endpoints

Prior to the creation of S3 endpoints, all data being accessed from S3 traversed the public internet. If you had private information that you were passing from a private S3 bucket to a resource in a private subnet in your **Virtual Private Cloud** (**VPC**), then not only did this pose some security risks, but it also required some extra networking to allow the resources in the private subnet to talk to the internet so that the S3 buckets that you wanted to access could be uploaded to and downloaded from.

If we have resources in a private subnet of a VPC that do not have a public route to the internet via a NAT instance or a NAT gateway, then we would not be able to access items in our S3 buckets without setting up that NAT instance, or we can make a more secure connection by using an S3 endpoint.

An S3 endpoint, which is a gateway endpoint, allows us to add an entry to the route table of our **VPC**. By adding this endpoint, we can now bypass the public internet with both our public and private instances and protect the privacy of our data being passed along the route. This is a much more secure solution for transporting your data from EC2 instances and other services residing within your VPC than using the public route.

S3 access control

Once you have data and objects uploaded into your bucket, unless you have a bucket with public access, then you might want to restrict who can access the objects within the bucket. Starting small, you may just allow the default access controls to whoever has authorization into the account. They may access the data in any non-public bucket. As you move to most corporate environments, there will be segments of data that will need to be cordoned off from one business unit to another. One product team would most likely have no need to access the data stored by another data team. Along those same lines, data being stored by the financial and business departments will probably need to restrict any technology members from accessing and possibly deleting the data.

This is where the access controls of S3 come into play. There are two main methods for implementing access controls: using S3 bucket policies on who and what can access the individual objects themselves and then by using IAM controls to limit users, groups, and resources that can access the buckets individually, or by using controls such as tags in an **attribute-based control model**.

It's generally a good idea to pick one method of access or the other and not try to mix the two since this can lead to some very frustrating sessions of trying to troubleshoot permission issues:

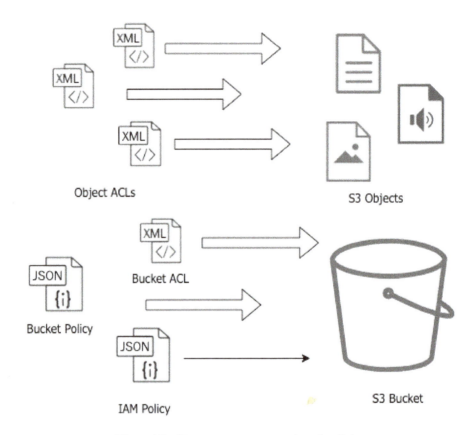

Figure 4.7 – S3 resource versus user-based policies

You should also know that you must explicitly make an S3 bucket public as all S3 buckets are private by default and block public access.

Resource-based policies

If you and your organization prefer to restrict at the object level, then you can use **Access Control Lists (ACLs)** either at the bucket access level or at the object access level.

User-based policies

Many, on the other hand, would rather control access to their S3 buckets with IAM policies. This allows you to control entitlement from the bucket and folder level, along with the ability to construct more complex conditions in the IAM policy based on a tag, `VPC-id`, `source-IP` address, and other factors.

Cross-account access

If you need a user or resource in one account to be able to access the objects in another account, then you can set up a cross-account access role as we did in the IAM exercise in *Chapter 3*, *Identity and Access Management and Working with Secrets in AWS*.

S3 access logs

When storing different objects in S3, especially those that are to be downloaded by various users and groups, you might want to know who is accessing the different files, when, and from what location.

Users can capture all the access logs and records of who is accessing various objects in a bucket via a simple setting in S3. You cannot store the logs in the same bucket as the items that you are tracking, so you need to either create an entirely new bucket expressly for the purpose of capturing the logs or designate a previously created bucket in your current account to hold the logs.

Logs are not pushed to that bucket in real time as items are accessed, as would be the case for logs on a web server. Amazon pushes the logs in batches in a best-effort approach.

If you don't want to set up an entirely different bucket to capture these logs, and if you have **CloudTrail logs** turned on for the account, then you can gather IAM user information on S3 API calls.

Encryption options with S3

S3 allows encryption at rest for the objects it stores. The default option when you store an object is to store it unencrypted. If you are working in any kind of environment that requires compliance, then you will most likely need to encrypt the objects you are storing.

If you have decided that your objects stored in S3 need to be encrypted, then you do have options. You can choose between *server-side encryption* and *client-side encryption*. There are some key questions to ask before making this decision:

- Do you need to manage the encryption key?

- Where is the encryption key going to be stored?

- Who is going to do the encryption and decryption of the data?

Server-side encryption

AWS has made the process of encrypting your objects and data in S3 storage easy with their server-side encryption options:

- `SSE-S3`: Using the `SSE-S3` option allows you to use the AWS S3 master key to encrypt your objects and data. This allows your data to be stored encrypted at rest without a lot of management or extra configuration on your or your team's part in the setup. You simply upload your objects to the S3 bucket of your choice, and then once they are received successfully, the S3 service handles the encryption of those objects. By the same token, when an object is requested by a service or user, and as long as that service or user has the proper authorization to access the object, then the S3 service decrypts the requested object.

- `SSE-K:MS`: Integrating the **Key Management Service** (**KMS**) into server-side encryption adds a small cost, but a few more features and benefits over just using the default encryption key provided with the S3 service. You now have another layer of granular control of the customer keys and of which IAM entities are allowed to access that key. KMS also provides an audit trail of who has accessed the key. And one of the main features is that you have the control to rotate the key if needed.

Client-side encryption

When choosing client-side encryption, the full responsibility for encrypting and decrypting falls on you the client. This method involves encrypting the objects before they reach S3. You are also responsible for any master/child key management along with key rotation. Client-side encryption is a good choice if your organization needs total control of both the master keys and the encryption algorithm used.

We are going to take a deep dive into the protection of data in flight and at rest in *Chapter 19, Protecting Data in Flight and at Rest*.

Using S3 events to trigger other AWS services

The S3 service can notify other services when certain things happen to objects in a bucket. Two of the most common scenarios are if an object was uploaded or if an object was deleted from a particular bucket. The S3 bucket can then notify one of three other AWS services of what has happened and the bucket where the event occurred. The three services that allow S3 event notifications are as follows:

- AWS **Lambda**
- Amazon **Simple Queue Service (SQS)**
- Amazon **Simple Notification Service (SNS)**

You can arrange for notifications to be issued to SQS or SNS when a new object is added to the bucket or overridden. Notifications can also be delivered to AWS Lambda for processing by a Lambda function:

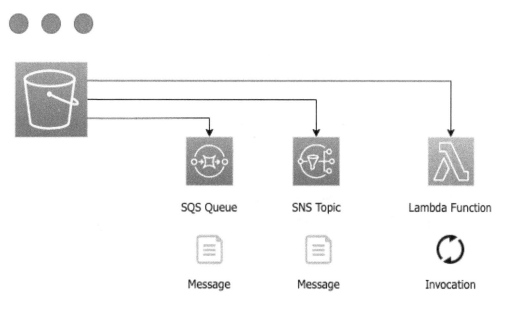

Figure 4.8 – S3 Event Flow

Let's think about this and how we would use this from a DevOps perspective. AWS Lambda is an extremely powerful tool that we will explore in detail in *Chapter 12*, *Lambda Deployments and Versioning*, and it can be used to invoke almost any other AWS service. In our current scenario, we could have a customer who is using the AWS SFTP service to upload a file to an Amazon S3 bucket. That bucket could trigger a bucket event to AWS Lambda. The Lambda function could then kick off an AWS Pipeline build that would process the file, which, on passing or failing, sends a notification to the development team of a new build available for deployment.

> **Note**
>
> In order to use S3 events, you must grant the S3 principle the necessary permissions in order to use the requested services. This includes the permission to publish to SNS queues or SQS topics, as well as the ability to invoke Lambda.

Triggering an S3 event

We will go through the exercise of using our previously created bucket to add an event trigger whenever an object is uploaded to the bucket. This event will be something simple to start with: an email notification to ourselves. In order to send that email, we will need to create an SNS topic and then subscribe to that topic with our email. Then we can go back to our bucket and add the bucket event configuration so that whenever an object is uploaded, it will send us a notification.

Now that we know that we have to set up an SNS topic, let's use our CLI to create that topic and subscribe so we can get the emails once something has been uploaded to our bucket:

1. Open up your terminal and type the following commands so we can create the topic:

```
$aws sns create-topic --name s3-event
```

If the topic is created successfully, then it should return something like this:

```
{
    "TopicArn": "arn:aws:sns:us-east-2:470066103307:s3-event"
}
```

2. Now that we have our topic, we need to subscribe using our email address:

```
$ aws sns subscribe \
    --topic-arn arn:aws:sns:us-east-2:470066103307:s3-
event \
    --protocol email \
    --notification-endpoint devopsproandbeyond@gmail.com
```

This should return a JSON statement telling you that the subscription is pending:

```
{
    "SubscriptionArn": "pending confirmation"
}
```

3. Now we need to go to our email account and find the email that the SNS service has just sent, and then click on the link that says **Confirm Subscription**.

    Now let's log into our account and go to our S3 bucket so that we can configure the **Event Notifications**.

4. The final step for our SNS topic to be ready is to add an IAM role that allows it to receive the notification events from the S3 bucket.

    We will use the following policy and then need to fill in the following values before saving them to a file:

    • **Account Number**

    • **Region**

    • **Bucket name**

    We will create the following policy and then add this to the topic in the **Access** section once we log into the AWS console:

```
{
    "Version": "2012-10-17",
    "Id": "s3-event-sns-ID",
    "Statement": [
      {
        "Sid": "s3-publish-ID",
        "Effect": "Allow",
        "Principal": {
          "Service": "s3.amazonaws.com"
        },
```

```
        "Action": "SNS:Publish",
        "Resource": "arn:aws:sns:region:account-num:sns-
   topic",
        "Condition": {
          "StringEquals": {
            "aws:SourceAccount": "account-num"
          },
          "ArnLike": {
            "aws:SourceArn": "arn:aws:s3:::bucket-name"
          }
        }
      }
    ]
  }
```

5. Let's now log into our AWS account and go to the SNS service so we can update our access policy. This way, the SNS topic has the correct permissions to interact with the S3 event notification.

6. Once on the SNS service, choose **Topics**.

7. In the **Topics** menu, you should see the topic we created via **CLI – s3events**. Click on the topic name so we can get to the configuration.

8. Once inside the topic, we need to press the **Edit** button near the top-right of the main page:

> **Edit**

Figure 4.9 – Edit button near the top of the page

9. Now find the **Access Policy** box and click on the arrow to expand. We will cut and paste the JSON policy we made earlier into this section. Once you have replaced the previous value with the new access policy, click the orange **Save changes** button at the bottom of the page.

10. Now we can go to the S3 console (https://s3.console.aws.amazon.com/).

11. Find the bucket that you made previously in this chapter (ours was named devopspro-beyond). If you haven't made a bucket already, you could choose any bucket you have in your account or create a new bucket quickly. Click on that bucket name so that you are brought to the main bucket page.

12. Once you are on the main bucket page, then click the **Properties** menu item in the horizontal menu in the main window frame:

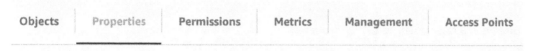

| Objects | Properties | Permissions | Metrics | Management | Access Points |

Figure 4.10 – S3 horizontal menu with Properties highlighted

13. Now scroll down the screen until you find the panel named **Event Notifications**:

Event notifications (0)   Edit   Delete   Create event notification
Send a notification when specific events occur in your bucket. **Learn more** [↗]

| Name | Event types | Filters | Destination type | Destination |

No event notifications

Choose **Create event notification** to be notified when a specific event occurs.

Create event notification

Figure 4.11 – S3 Event notifications with no created notifications

14. Now click on the **Create event notification** button.

15. Use the following configurations for our S3 event test:

- **Event Name** – S3 Test
- **Prefix** – *[leave blank]*
- **Suffix** - .txt

- **Event Types** – Check the box labeled **Put**:

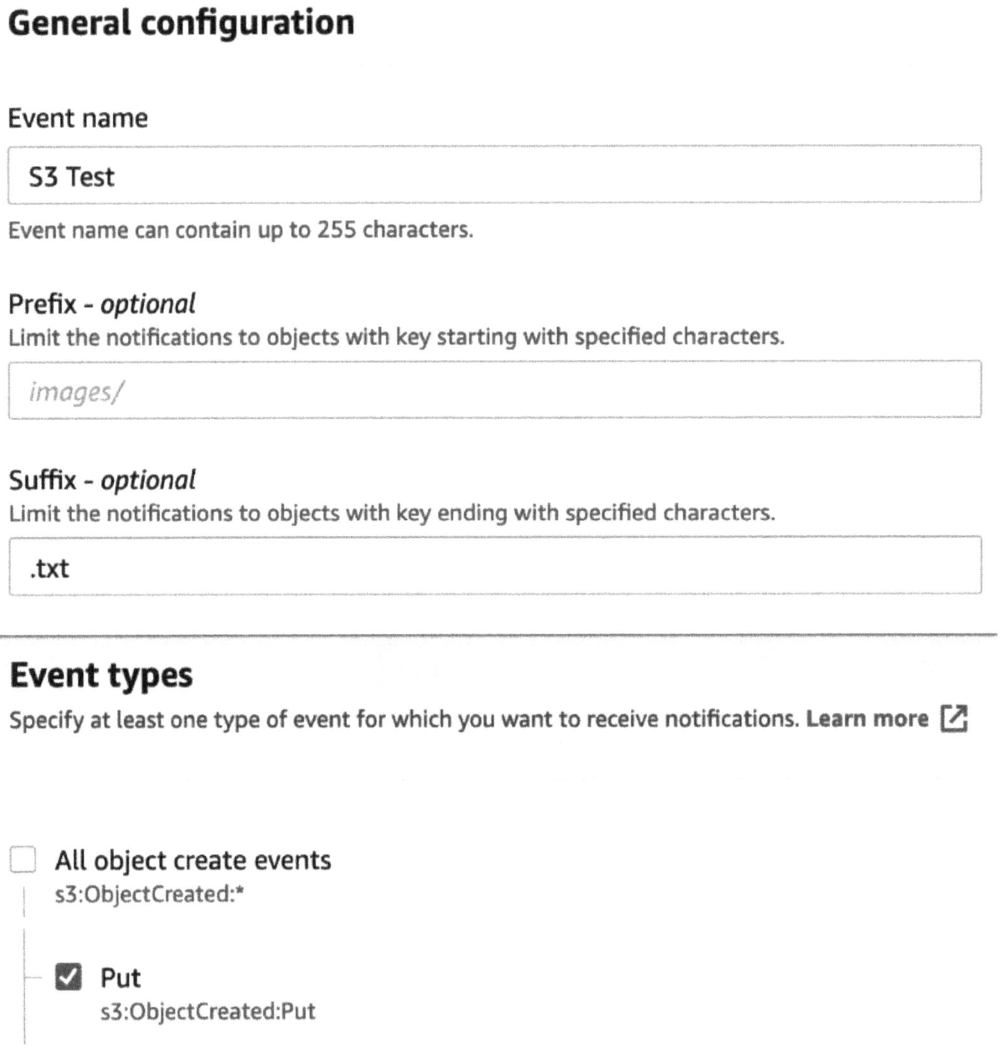

General configuration

Event name

S3 Test

Event name can contain up to 255 characters.

Prefix - *optional*
Limit the notifications to objects with key starting with specified characters.

images/

Suffix - *optional*
Limit the notifications to objects with key ending with specified characters.

.txt

Event types

Specify at least one type of event for which you want to receive notifications. **Learn more** ↗

☐ **All object create events**
s3:ObjectCreated:*

☑ **Put**
s3:ObjectCreated:Put

Figure 4.12 – Configuring an S3 event notification

16. Scroll down to the bottom of the page where you see **Destination**.

17. Choose the radio button that is next to **SNS Topic**, then from the drop-down menu, select the **SNS Topic** that we just created (s3-event):

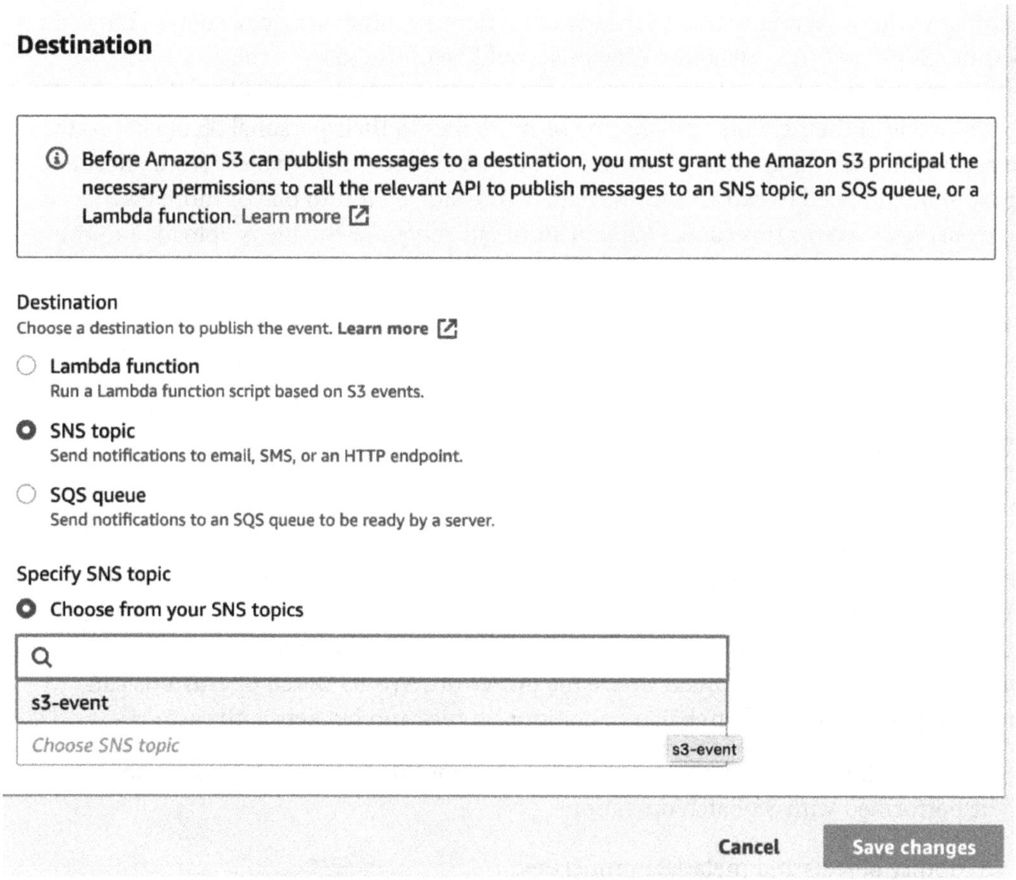

Destination

ⓘ Before Amazon S3 can publish messages to a destination, you must grant the Amazon S3 principal the necessary permissions to call the relevant API to publish messages to an SNS topic, an SQS queue, or a Lambda function. Learn more ↗

Destination
Choose a destination to publish the event. **Learn more** ↗

○ **Lambda function**
Run a Lambda function script based on S3 events.

● **SNS topic**
Send notifications to email, SMS, or an HTTP endpoint.

○ **SQS queue**
Send notifications to an SQS queue to be ready by a server.

Specify SNS topic

● Choose from your SNS topics

🔍

s3-event

Choose SNS topic   s3-event

Cancel   **Save changes**

Figure 4.13 – Choosing the destination for our S3 event notification

18. Click the orange **Save changes** button.

19. It's now time to upload a text file and test that we receive an email notification.

20. All we need to test is any text file (remember we configured the event to only be configured on .txt files and no other types of files). We will use the CLI to upload our file:

```
$aws s3 cp test.txt s3://devopspro-beyond/test.txt
```

If you uploaded your file successfully, then you should see this output:

```
upload: ./test.txt to s3://devopspro-beyond/test.txt
```

21. Once the file is uploaded, you should receive an email notification at the email address you used to subscribe to the SNS topic.

Now that we have seen how we can trigger operations on other services, such as **Lambda**, **SNS**, and **SQS**, we can think about how this would be of use to us in the real world. In the case of SNS, you may have a client who has an account and would like to be notified whenever one of their clients uploads one or more files to their personal S3 bucket so they can review the files. In the case of Lambda, you may be receiving invoices from another department and need to extract out the data before storing it into one or more data stores, and by using S3 events this can all happen automatically once the file is uploaded to the bucket.

In the next section, we will look at S3 Batch operations and see how, with the help of a manifest file, we can process a few or a few thousand files at once using just the S3 service.

S3 Batch operations

Having a good tagging strategy is part of recommended AWS account hygiene. Just like other initiatives, many of these strategies evolve over time. There may come a time when you or your organization feels the need to change some of the mandatory tags on the objects in your current set of S3 buckets. If you have been running in AWS for any period of time then there are most likely too many objects to re-tag by hand and so you are left trying to devise a solution. This is where the power of AWS **S3 Batch operations** can come into play. It can perform batch operations on files and buckets with ease.

S3 Batch operations allow you to do more than just modify tags. The following operations can be performed with S3 Batch operations:

- Modify objects and metadata properties.
- Copy objects between S3 buckets.
- Replace object tag sets.
- Modify access controls to sensitive data.
- Restore archive objects from Glacier.
- Invoke AWS Lambda functions.

Once the job has been created, then it goes through a series of statutes before it either reaches the completed or failed state. The following table describes the different statuses available to an Amazon S3 Batch job:

Status	Description
New	A job begins in a new state when you create it.
Preparing	Amazon S3 is processing the manifest object and other job parameters.
Suspended	The job requires confirmation, but you have not yet confirmed that you want to run it. Only jobs created using the console require confirmation.
Ready	Amazon S3 is ready to begin running the requested object operations.
Active	Amazon S3 is executing the requested operations listed in the manifest.
Pausing	The job is transitioning to the Paused state.
Paused	A job can become paused if you submit another job with higher priority.
Canceling	The job is transitioning to the Cancelled state.
Canceled	The request was canceled and was successful.
Failing	The job is transitioning to the Failed state.
Failed	The job has failed and is no longer running.

S3 Batch hands on-example

To test out the power of S3 Batch, we are going to take 75 files, upload them to our bucket, and then use AWS Batch to add a tag to each of the files almost instantly.

> **Note**
>
> If you don't want to re-create all the files for this exercise, then simply go to the GitHub repository for this book; there are 75 small files available in *Chapter 4, Amazon S3 Blob Storage*, in the batch subfolder.
>
> Also, since all the files have the `.txt` extension, you may want to turn off the S3 event notification or unsubscribe from the topic before uploading all of the exercise files to the S3 bucket.

We will now use a hands-on example with S3 Batch to update the tags on a number of files at once. If you have a mandatory tagging strategy in place and files are missing some of those tags, then this can be an efficient way of managing those changes rather than trying to either write a custom script to perform the task or changing the tags on the files manually:

1. Before we start, in order for the job to be able to execute, we need to ensure that we have an IAM role. Let's first log in to the AWS Management Console and navigate to the IAM service.

2. Create a role for an AWS service, choose **S3**, and then at the very bottom of the page, choose **S3 Batch Operations**:

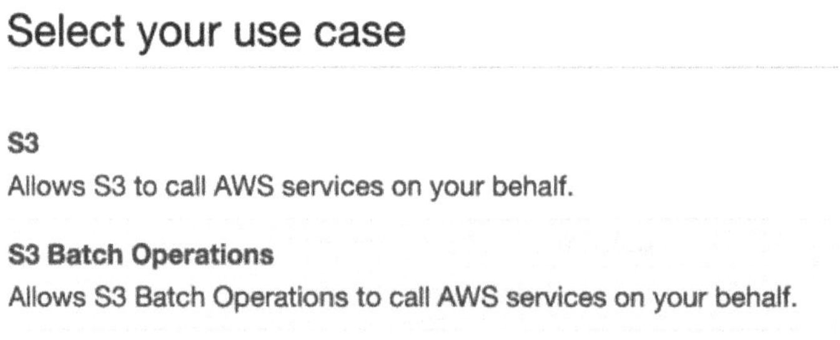

Select your use case

S3
Allows S3 to call AWS services on your behalf.

S3 Batch Operations
Allows S3 Batch Operations to call AWS services on your behalf.

Figure 4.14 – Selecting the use case for S3 permissions in IAM

3. Click on the blue button labeled **Next: Permissions**.

4. When you get to policy, click on the **Create Policy** button. This will open up a new tab in your browser. Instead of a prebuild policy, click on the **JSON** button and then cut and paste the `JSON` code from GitHub named `S3_batch_IAM.json`. You will need to replace the name of your S3 bucket in all the places where it asks for variables. (The variable names are notated as `<<TargetResource>>`, `<<ManifestBucket>>`, and `<<ResourceBucket>>`.) Unless you are storing the manifests and reports in different buckets, then just replace the same bucket name in each value. When you are done, you can click the blue button labeled **Next: Tags**. There is no need for any tags at this moment, so just click the blue button labeled **Next: Review**.

5. Now we can name and save our role; a good descriptive name is something like `S3-Batch-Tagging-Role`. Add the description if you desire and then click the blue **Create Policy** button.

6. Go back to the other tab where we were creating the role and search for the policy we just created, named **S3-Batch-Tagging-Role**. If it doesn't appear on the first search, you may need to click the refresh icon at the top right of the search bar. Click the box next to the policy and then click the blue button at the bottom of the page labeled **Next: Tags**. Once again, there is no need for any tags for clicking through to the **Next: Review**. We can name this role S3-Batch-Tagging. Once this role has been created, we need to take a note of the ARN so that we can use it in our batch command later.

7. Download the 75.txt files from the GitHub directory (or create your own set of files) into a single directory so that they can be uploaded into the S3 bucket you created earlier.

8. Next, we will use the s3 sync command to quickly move the files from our local directory:

```
$aws s3 sync . s3://devopspro-beyond
```

9. We will also need to download the manifest (.csv file) from the GitHub repository in order to start our batch job. In the manifest, you will need to replace the current bucket name, devopspro-beyond, with the bucket name where you have uploaded your objects. Once you have changed those values, make sure that you upload the manifest to the S3 bucket, as S3 Batch needs to read the manifest from an S3 location when using a CSV file and not from a locally sourced file.

10. The final report also needs a *folder* in our bucket to reside in. We will use the s3 cp command to move a file into the new folder and have it ready to receive the final report:

```
$aws s3 cp manifest.csv s3://devopspro-beyond/final-
  reports/manifest.csv
```

11. Now that our objects and manifest have been uploaded, we can go back to the AWS management console and start the batch job. Going back to the browser window where you had previously made your IAM role, navigate to the **S3 service**.

12. On the left-hand menu, click on **Batch Operations**:

Batch Operations

Figure 4.15 – Batch Operations menu item

13. Click on the orange button on the right-hand side that is labeled **Create Job**.

14. On the Manifest page, select the radio button that says **CSV**. For the **Manifest object** field, click **Browse S3** and find the `manifest.csv` file in your S3 bucket. The manifest ETag will populate automatically:

Manifest

Manifests must only reference objects in a single S3 bucket. To generate a new manifest, **configure an S3 inventory list for a bucket or prefix**

Manifest format

○ S3 inventory report (manifest.json)

◉ CSV

    CSV format must be either 2 or 3 columns in the following order: bucket name, object key, and optionally version ID.

☐ Manifest includes version IDs

Manifest object

| s3://devopspro-beyond/manifest.csv | View 🗗 | Browse S3 |

Format s3://bucket/prefix/object. **Learn more** 🗗

Manifest object version ID - optional
For objects in a bucket with bucket versioning enabled, you can enter a version ID to use a previous version of the object. If you don't specify a version ID, Batch Operations uses the most recent version of the object. **Learn more** 🗗.

| Enter version ID |

Manifest object ETag
The ETag is used to verify that you have selected the correct manifest object.

cb9dca274fbe770a53d63e75062016a1

Figure 4.16 – The manifest information for S3 Batch

15. Click the orange **Next** button at the bottom of the page.

16. Under **Operation**, choose the radio button next to **Replace All tags**.

17. This will make another set of options appear. For the **Key**, enter TAG, and for the **Value**, enter Chapter4. Once you have done this, click on the orange **Next** button:

| Key | Value | |
| TAG | Chapter4 | Remove |

Figure 4.17 – Adding the key and value for what to change

18. For the completion report, browse to the `final-reports` folder that we created earlier by uploading a copy of the `manifest` file via the CLI:

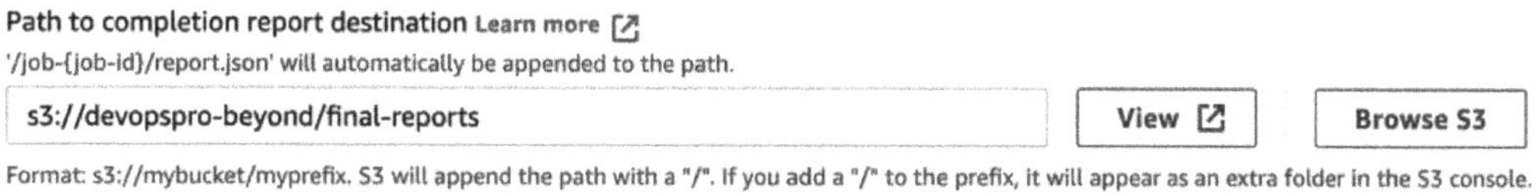

Path to completion report destination Learn more [↗]
'/job-{job-id}/report.json' will automatically be appended to the path.

| s3://devopspro-beyond/final-reports | View [↗] | Browse S3 |

Format: s3://mybucket/myprefix. S3 will append the path with a "/". If you add a "/" to the prefix, it will appear as an extra folder in the S3 console.

Figure 4.18 – Choosing the destination for the S3 Batch reports

19. Under permission, choose from the existing IAM roles and then, in the drop-down box, select the **S3-Batch-Tagging-Role** that you created at the beginning of this exercise. Click on the orange **Next** button at the bottom of the page.

20. On the review page, scroll all the way to the bottom and click on the orange **Create job** button.

21. Once you have created the job you will now be brought back to the S3 batch main screen. It may take a minute or two for the job to finish being created, but once it has, you can select the radio button to the left and then click on the button labeled **Run job**. This will start processing all the tags on the files in your manifest.

22. We can now go back to the AWS Management Console and navigate to our S3 bucket so we can look at our files to see if the **Delete** tag has been added.

> Note
> The manifest that has been uploaded in the GitHub repository has the S3 example bucket name. You will need to change the bucket name on the manifest for the S3 bucket that you have created before uploading and running your batch job.

S3 replication

Even with its high durability guarantees, there are numerous cases where you need to devise a plan to protect your data in case of regional outage or loss. You might even want to copy your original data, which is restricted by access policies, somewhere else so that another team can access that data. This is where **S3 replication** comes into play. It gives you the ability to asynchronously copy your data to another bucket. There are two versions of S3 replication available:

- **Cross-Region Replication (CRR)**: This is where the objects in the bucket are replicated into a separate bucket that has been created in a different region than the primary region in which the original bucket was created.

- **Single Region Replication** (**SRR**): In SRR, objects are still replicated to a new separate bucket from the originating bucket, but both buckets are in the same geographical region.

S3 versioning

In the S3 service, you can keep track of how files change over time using the versioning feature. While this feature does add additional cost, it is especially useful as a way to help restore deleted objects.

Once you enable versioning, each object in that S3 bucket gets a value for the version ID. If you haven't enabled versioning on the bucket, then the version id for the objects in the bucket is set to null:

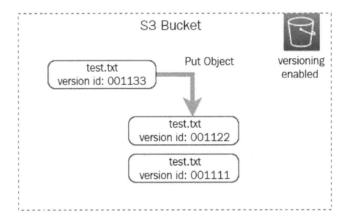

Figure 4.19 – An S3 bucket with versioning enabled showing version ids

Once you upload a subsequent version of the object with versioning turned on, Amazon produces a new version id for the new version of the object and then places that newer version of the object in the bucket.

Summary

In this chapter, we covered the AWS S3 service and many of its features. We examined not only the basics of creating buckets and how buckets are secured with the different types of access policies but also how you can encrypt your data at rest using different encryption methods from AWS. We also saw how to trigger workflows using bucket events to do things such as kick off our DevOps pipeline. Now that we have a firm understanding of object storage, we will move on to the serverless NoSQL database DynamoDB.

Review questions

1. You have a department in your company that needs an S3 bucket configured where the objects are accessed on a weekly basis and need to be both durable and reliable. Which S3 storage class should you use to configure the bucket?

2. You have five departments in your organization. Three of the departments are product teams, one is the accounting department, and one is the HR department. The accounting department has decided to migrate their files to S3 from the data center in order to save costs. How can you be sure that only members of the accounting department have access to the accounting files and no one else?

3. A healthcare company is preparing for an internal audit. They need to make sure that all of their files stored in S3 have been encrypted and that the keys are rotated no less than once per year. The company has been in business for over 15 years and has recently (in the last 5 years) made a digital push to move the majority of its files onto the cloud. This has resulted in over 1.5 million documents, including billing records, patient information, business information, and other documents being stored. What is the most effective way to check for this continually?

Review answers

1. S3 Standard.

2. Make sure that an IAM group has been created just for members of the accounting department. Create an IAM (user-based) policy that allows members of the accounting group to have full permissions on the accounting bucket. You could go a step further and create a policy boundary that explicitly denies access to the accounting bucket and place that on all other groups.

3. Create an S3 inventory report. Use AWS Athena to query for files that are not encrypted.

5
Amazon DynamoDB

As application architectures demand scalability and shift focus to serverless design patterns, developers look to data stores that are flexible, scalable, and have low management overhead. DynamoDB has become a proven, trusted solution for these types of features. However, it has continued to evolve, and many of the features that sprouted from this service have relevance in the DevOps professional exam.

While knowing how to calculate the read to write ratio when provisioning a DynamoDB database isn't a topic of focus in the professional-level exam, understanding how this core AWS service fits into deployments and scenarios is. Having a firm understanding of the capabilities of DynamoDB, along with its features, will help you answer the exam questions, as well as allow you to implement solutions in your career engagements.

In this chapter, we're going to cover the following main topics:

- Understanding the basis and background of DynamoDB
- Understanding DynamoDB data modeling
- Inserting and accessing data in DynamoDB
- Understanding DynamoDB Streams
- Using the **DynamoDB accelerator** (**DAX**)
- Authenticating and authorizing in DynamoDB
- Monitoring DynamoDB

Understanding the basis and background of DynamoDB

DynamoDB is a NoSQL database. This means that it is not only SQL and, more importantly, that DynamoDB doesn't need a fully structured schema to enter data. Its flexibility and performance are what drives many to DynamoDB, along with its pay-per-use pricing model and high availability and scaling.

DynamoDB origins

In 2007, Amazon published a whitepaper authored by the future AWS CTO, Werner Volgels, and others called *Dynamo: Amazon's Highly Available Key-value Store.*

You can still find this paper today at `https://www.allthingsdistributed.com/files/amazon-dynamo-sosp2007.pdf`.

As Amazon built out its e-commerce platform, it was trying to solve issues such as the following:

- Partitioning
- High availability for writes
- Handling temporary failures
- Recovering from permeant failures
- Membership and failure detection

The databases at the time were just not performant enough, and the e-commerce website was starting to see bottlenecks, especially at the database layer.

NoSQL versus relational databases

Relational databases have been around since the 1970s. Relational databases are good when you need to enforce data integrity and can structure data using the **Structured Query Language** (**SQL**). Relational databases are optimized on the premise that storage is one of your most limiting factors. Once you run out of storage or disk drive space, you will need to procure more storage space. Knowing this fact is one of the reasons for the use of primary keys and joins when using relational databases. Using the ID of the data, and then retrieving the columns of the actual data needed using table joins, allows the data to only be stored a single time, hence saving space on the systems, as shown here:

Employees

EMP_ID	FIRST_NAME	LAST_NAME	DEPT
5001	David	Burr	Accounting
5002	Wanda	Harris	Public Works
5003	Davis	Johnson	Accounting

Query: Accountants per City

CITY ID	EMP_ID	FIRST_NAME	LAST_NAME	NAME
3001	5001	David	Burr	Springfield
3002	5003	Davis	Johnson	Arlington

Cities

CITY_ID	NAME	BOOKEEPER_ID	POPULATION
3001	Springfield	5001	15,251
3002	Arlington	5003	26,144

Figure 5.1 – Relational database tables

With storage in the cloud, storage and its associated costs are no longer a limiting factor. This brings us to NoSQL, or Non-SQL (sometimes referred to as not only SQL). Rather than storing data in tables, NoSQL systems store their data alternately, frequently, as in the case of DynamoDB, in JSON documents.

NoSQL databases bring about the flexibility that relational databases just can't provide. Today's modern applications, such as web apps and gaming systems, along with mobile applications, need this type of flexibility, along with the ability to scale to meet their users' needs, as well as provide high performance when retrieving and inserting the data when requested.

Unlike traditional relational databases, DynamoDB is more like a key store that is extremely efficient in both retrieving and storing data.

Core components of Dynamo

Let's take a look at the main components that DynamoDB is comprised of:

- The table and its attributes
- The primary key
- Secondary indexes

Let's get started.

Tables and their attributes

A **table** in DynamoDB is comparative to a database in other database systems, rather than just a table in relational database systems. A table is a collection of data on a specific topic. Each table in Dynamo is a separate store of items, and this is where the data is stored.

Each table contains zero or more **items**. Items have different fields and attributes.

The following is a DynamoDB table that contains items:

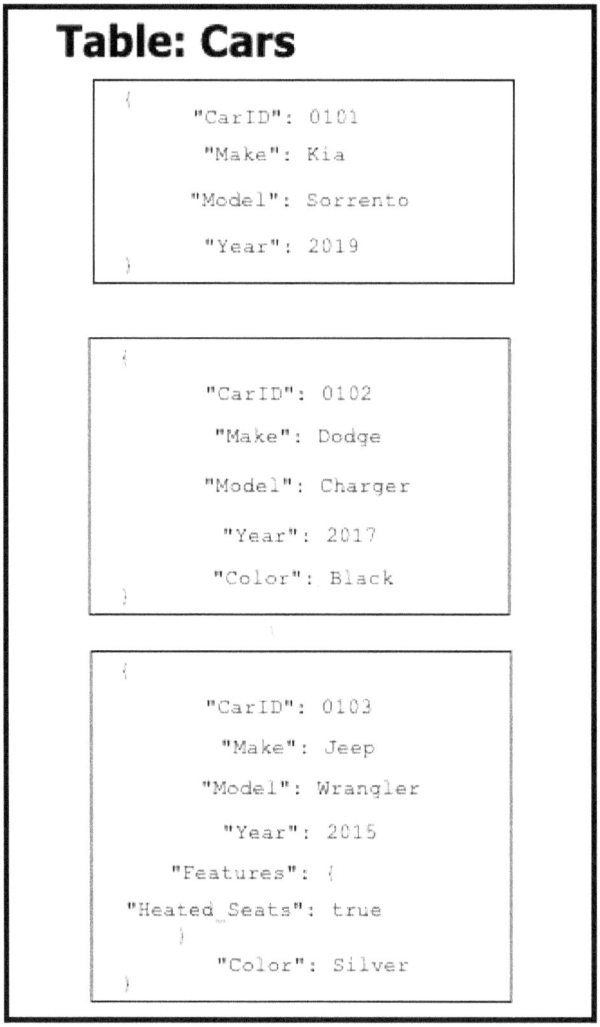

```
Table: Cars

{
        "CarID": 0101
        "Make": Kia
        "Model": Sorrento
        "Year": 2019
}

{

        "CarID": 0102
        "Make": Dodge
        "Model": Charger
        "Year": 2017
        "Color": Black
}

{
        "CarID": 0103
        "Make": Jeep
        "Model": Wrangler
        "Year": 2015
    "Features": {
    "Heated_Seats": true
        }
        "Color": Silver
}
```

Figure 5.2 – Dynamo DB table and items

If you look at the preceding table, you will see a few of the following things:

- Each item has a primary key. This is the unique identifier for the item (`CarID`).

- There is no set schema for the items. As long as the primary key is present, any of the other attributes can or cannot be present.

- The majority of the attributes only have one value. However, the last item has a features field. This is a nested value that can hold multiple attributes.

The primary key

As you are creating a table in DynamoDB, you must specify the **primary key**. This primary key is a unique identifier for each item in the table, which means no two items can have the same primary key.

With the primary key, you can reference distinct items in the table. DynamoDB uses the primary key as the data for its own internal hashing algorithm. This hash is used to determine which partition should be used to store the attribute.

Secondary indexes

Secondary indexes are optional keys that can be used to query against. There are two types of secondary indexes supported by DynamoDB:

- Global secondary index
- Local secondary index

We will take a deeper look at secondary indexes later in this chapter.

Other pertinent Dynamo information

As you start to think about your table, it's also important to note that there isn't a single server instance or machine hosting Dynamo. Once added, the data is distributed across multiple instances, which allows the key scaling and performance features of DynamoDB to be used. Write calls do not return as successful until the data has been redundantly stored.

Understanding DynamoDB data modeling

If you have ever designed a relational database, then you are familiar with schemas such as star schemas. Each table needs to have a specified attribute and if that attribute has no value, then a null is kept in its place.

DynamoDB uses partitions. These partitions can be either hot partitions or cold partitions.

Every item in DynamoDB requires at least one attribute, and that is the partition key. This partition key is used by Dynamo to hash your data and place it in memory. To achieve optimal performance in DynamoDB, we need to choose a partition key that allows DynamoDB to spread its searches across the disk and not let a single partition get too *hot*.

This is best demonstrated with a bad example of a partition key, such as date. If you are trying to gather lots of data all from the same date, then the hash value of the single date will be stored in the same partition. Different dates may be stored across different partitions since their hash will be different; however, when querying all the events that happened on a single date, that singular partition will become hot, and this can cause performance problems.

Some examples of high-quality partition keys are as follows:

- Location ID
- Department ID
- Customer ID
- First letter of the last name

In the preceding examples, the data would spread across different partitions when it comes to both reading and writing, as shown here:

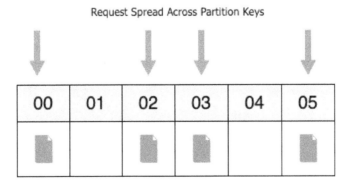

Figure 5.3 – Keys evenly spread across partitions

Read and write capacity

When you create your table, you must specify both your read and write capacity values. Once specified, DynamoDB reserves the resources needed to handle that capacity and divides it evenly across the partitions.

There are two different types of capacity units that need to be specified:

- **Read capacity units (RCU)**: This is the number of strongly consistent reads per second that your table can handle. It can contain items up to 4 KB in size.

- **Write capacity units (WCU)**: This is the number of consistent writes per second in 1 KB units that your table can handle.

Adaptive capacity

If you have a workload that is imbalanced in nature, DynamoDB has a feature called **adaptive capacity** that helps minimize throttling. The best part about this feature is that it is enabled for every DynamoDB table automatically at no additional cost. There is no need to go into the settings and turn an adaptive capacity setting on or off.

Although you may provision 10 **Write Capacity Units (WCUs)** per partition, there may be one partition that is receiving more writes than the others. If the total capacity of the table has not been exceeded, then DynamoDB can use the adaptive capacity feature and allow the *hot* partition to continue to receive the writes before throttling:

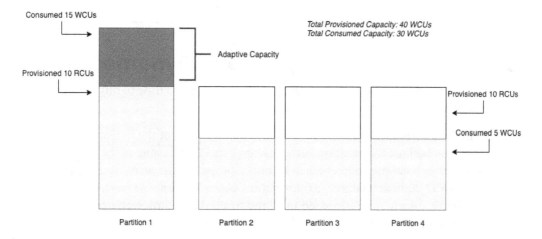

Figure 5.4 – Adaptive capacity example

The preceding diagram shows an example DynamoDB table consisting of 4 partitions. When created, we allocated a total of 40 WCSs across the table. **Partition 2**, **Partition 3**, and **Partition 4** are only consuming 5 WCUs each, for a total of 15 WCUs, thus leaving another 25 WCUs free, as allocated from our initial provisioning. **Partition 1**, however, has the most activity and is consuming 15 WCUs, or 5 over the allocated 10 per partition. The adaptive capacity feature takes into account that there is extra WCU capacity and adjusts without throttling the table.

Data types available in DynamoDB tables

DynamoDB allows various types of data to be inserted into attributes.

Number, string, Boolean, binary (which would need to be `base64-encoded`), and null values are all supported data types for attributes. These are all examples of single values that can be inserted into an attribute field.

DynamoDB also allows sets of items to be inserted into an attribute. These sets can contain numbers, binaries, or strings. Sets must be all of the same type, so you can't mix a set of numbers and strings, and sets do not preserve order.

Somewhat like a document database, DynamoDB allows `JSON` documents to be added as attributes, and they can nest up to 32 layers deep.

> **Note**
> Primary keys must be either string, number, or binary.

Inserting and accessing data in DynamoDB

Now that we've covered the history and theory of DynamoDB, it's time to put our hands on our keyboards and actually get into the data.

For our example, we will create a fictional database to keep track of the projects at our company. This can include information such as `ProjectID`, the name of the project, who the project owner is, what the contact email is for the project or the team, and even other information such as build and language information. Since DynamoDB has a flexible schema, not all this information is needed in all the rows. We do, however, need to declare our primary key and then, depending on what we query, our secondary key.

Our schema will look like the following `JSON`:

```
{
Project_ID,
```

```
Department,
Owner,
< optional information (like language or build id) >,
Contact_Email
}
```

With our schema defined, we can start creating our table.

Creating tables in Dynamo DB

We can now open up our terminal and create our table using the following command:

```
$aws dynamodb create-table --table-name projects \
--attribute-definitions AttributeName=Project_
Name,AttributeType=S \
--key-schema AttributeName=Project_Name,KeyType=HASH \
--provisioned-throughput
ReadCapacityUnits=5,WriteCapacityUnits=5
```

If you compare the previous command with our schema, you will notice that we have only defined one column in the table: `Project_Name`. This is because this column is our hash key (the primary index). The other fields can be defined later and are not necessary. Those fields will be populated once the data has been inserted, either in bulk or row by row. It's important to know that now, all the strings in the `Project_Name` field must be unique; otherwise, they will be rejected from being inserted as duplicate values.

You may have also noticed that, at the end of the statement, we allocated five read capacity units, along with five write capacity units, to our table initially.

Inserting data into DynamoDB

In the GitHub directory for this book, under `chapter five`, we have several files to upload to the table that we have created. We have three different JSON files to download for this exercise:

- `project_item.json`
- `projects.json`
- `projects_bulk.json`

We are going to do all the inserts via the CLI. Although you can do single-line and bulk inserts in the **Amazon Management Console** (**AMC**), we want to concentrate on the ability to script our commands so that we have can automate them later, if needed.

The first type of insert we are going to do is a single item insert into the table. For this, we are going to need the `project_item.json` file, so if you haven't already downloaded the file, take a look at its structure, as shown here, to see what is going on with this JSON file:

```
{
    "Project_ID": {"N": "0100"},
    "Dept": {"S": "Test Team"},
    "Dept_ID": {"N": "0001"},
    "Project_Name": {"S": "Serverless Forms"},
    "Owner": {"S": "Jerry Imoto"},
    "Builds": {"NS":   ["2212121"] },
    "Language": {"S": "python" },
    "Contact": {"S": "test_team@testcompany.com" }
}
```

Before we upload the file, we want to look at some of the notations of the file. You'll note that the data types are notated before each field. Here, we have used string fields, notated with S, numerical fields, notated with N, and finally, for our builds, we have a numerical list denoted with NS.

We can now open up our terminal and run the following command to add the item to the DynamoDB table that we created earlier. Make sure that you navigate to the same directory where you have downloaded the files before you run the following command:

```
aws dynamodb put-item \
    --table-name projects \
    --item file://project_item.json \
    --return-consumed-capacity TOTAL
```

Once you have executed the previous command, you should get a return statement similar to the following:

```
{
    "ConsumedCapacity": {
        "TableName": "projects",
        "CapacityUnits": 1.0
    }
}
```

```
}
```

Congratulations! You now have an item in your DynamoDB table. Just a single item is not great for querying, however, so we need to add some more data. Adding more than a single item to a table via the CLI requires both a different command and a different format for the file that is going to be used.

In the `projects.json` file, we have provided `10` items that you can quickly add to your table via the `batch-write-item` command. You'll also notice that, in the `batch-write-item` command, unlike the `put-item` command, you don't need to specify the table. This information is specified in the table itself.

Scanning data

Now that we have loaded the test data into the table, we can use a table scan to see those data entries. A scan operation will return either all the items in a table or in the index specified.

If you haven't already done so, then go back to this book's GitHub repository and download the file named `scan-values.json`, since we will be using this file in the next exercise.

Go back to your terminal (or reopen it if you closed it previously) and type in the following command:

```
aws dynamodb scan \
    --table-name projects \
    --filter-expression "Dept = :d" \
    --expression-attribute-values file://scan-values.json
```

We're not going to show all the output from this command, but one of the things to look at is at the bottom of the command, where it will show you the **Count** of the number of rows returned and the **Scanned Count** of the total number of rows scanned.

> **Note**
>
> This is an expensive operation in terms of reads as you are going through all the data in the table. If you know the data you need, it is much better to perform a query and only call back the items and records that you need.

What is a scan in DynamoDB?

When you execute the `scan` command in a DynamoDB table, every item in that table or secondary index gets read. If you don't specify any filter conditions, then the scan will return all the items in a single scan, so long as the results are under 1 MB of data.

Querying data

What if we only want to retrieve certain values from our table? Or if we want DynamoDB to give us a count of how many records match a certain criterion? In this case, instead of using a scan, it's much more efficient to use a query.

Before we start to query, make sure that you have downloaded the file from `chapter five` named `query-values.json` from this book's GitHub repository.

Let's open up our terminal once again so that we can perform our query and see what results are returned:

1. First, make sure that you are either in the same directory where you downloaded the `query-values.json` file, or copy the file to your current working directory.

2. Second, make sure that you have created the `projects` DynamoDB table; otherwise, the query will not be successful.

3. In your terminal window, type the following command:

```
aws dynamodb query \
    --table-name projects \
    --projection-expression "Dept" \
    --key-condition-expression "Project_Name = :v1" \
    --expression-attribute-values file://query-values.
json \
    --return-consumed-capacity TOTAL
You should receive a result like the following one:{
    "Items": [
        {
            "Dept": {
                "S": "Training"
            }
        }
    ],
    "Count": 1,
    "ScannedCount": 1,
```

```
    "ConsumedCapacity": {
        "TableName": "projects",
        "CapacityUnits": 0.5
    }
}
```

Notice that, in the value for `projection-expression`, we noted the fields that we wanted to return from the table. We already knew the project name (we had to because it is our primary key), so we were trying to look up what department this project belonged to. This is one of the ways that a query is much more efficient than scanning all the data when searching and returning values in DynamoDB.

Secondary indexes in Dynamo, both global and local

As you may have noticed, in our previous query, we had to use the primary key when performing our command. In our table's case, we are using a primary key (`Project_Name`). A table in DynamoDB can have many different indexes, which allows your applications to have a myriad of different query patterns to use, without having to resort to the scan operation. These indexes can contain all or just a subset of the data that is included in the table.

> **Note**
>
> You must declare the index in your query for DynamoDB to know that you are performing the action against that particular index. If you don't declare a specific index, then the query will go against the table itself.

Local Secondary Index (LSI)

A **Local Secondary Index** (**LSI**) gives you the opportunity to change the sort key that was originally defined in your table. LSIs must always use the same primary key as the table where it was created.

LSIs also share the same throughput as the table it was created on. As a final note about LSIs, they **can only** be created at the time of table creation. If you did not create one when you created your table, you either need to drop and recreate your table, create a new table and then migrate your data, or use a **Global Secondary Index** (**GSI**) instead. Also, you can't delete a local secondary index without deleting the base table.

GSI

If we wanted to use a different primary key to search with, then we would have to create a GSI. Another key difference between GSIs and LSIs is that unlike LSIs, which have to be declared and created at the same time as when the table is being created, a GSI can be created at any point in time. This is especially helpful if you notice that either your queries are not requesting the right information, in which case you only want to bring back a subset of the information that you could store in the GSI, or a certain subset of queries that you are requesting need to have more RCUs or WCUs allocated.

Now, if we wanted to see our GSI after creating it, we could run the following command:

```
aws dynamodb describe-table \
    --table-name projects \
    --query "Table.GlobalSecondaryIndexes"
```

This will show us any GSIs, if any, that have been created on our table.

Understanding how to create indexes for optimal query performance is one of the key concepts to harnessing the power of DynamoDB. Next, we will move on to other features that DynamoDB can provide, such as replicating items using streams.

Understanding DynamoDB Streams

There may be times when you have a table in DynamoDB and you want to either be updated when a change comes in or have an event-driven process happen. This was the exact reason why AWS created **Dynamo Streams**. Streams are a time-ordered sequence of item modifications, such as insert, update, and delete operations.

When a stream in DynamoDB writes data, it does so in a strict ordering format. This means that as you write data to the table, pending the configuration settings you have set for the stream, it will push out the items in the same order in which they were written to the table.

Global tables

There are times when you need to either have a high availability plan in place in case of a regional outage for a service such as DynamoDB, or must have quicker local access to your data from another region besides where you originally created your data.

Global Tables, even though they are replicas of an origin table, are all owned by a single account.

When setting up a global table, the first step is to create an initial table in your primary region.

Then, you need to enable DB Streams on that initial table.

Next, for every region where you want to replicate your global table, you must set up that same table with streaming enabled in a different region.

Then, in the original region, you can define the global table's setup, which is made easy by the AWS Management Console.

If there are requirements in your organization to maintain a certain level of uptime that has been relayed via **service-level agreements** (**SLAs**), then you can take care of this using a combination of Streams and Global Tables. Next, we will examine how to speed up query time in DynamoDB by using the DynamoDB accelerator.

Using the DynamoDB accelerator (DAX)

DynamoDB, by itself, even with thousands of requests, can deliver millisecond latency response times. This satisfies the needs of many companies and customers, but there are some applications that need microsecond performance, and this is where **DynamoDB Accelerator** (**DAX**) comes into focus.

DAX is almost equivalent to a turbo button for your DynamoDB table. It caches the most frequently requested results and then makes them available via an endpoint. There is no need to spin up a third-party cache, no need to manage nodes, and it's as easy to implement. You simply go to the DAX menu under the main DynamoDB page and then spin up your DAX cluster:

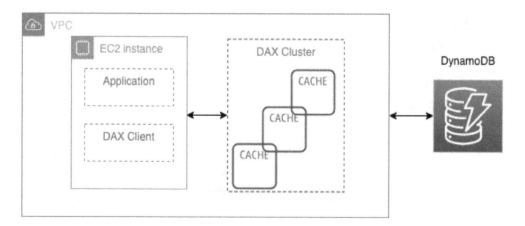

Figure 5.5 – DAX example architecture

Industries such as Ad-Tech can utilize the caching properties and speed of DAX when designing and implementing their systems to go out and place bids for digital advertising real estate. Those millisecond latencies in placing bids can equate to real dollars in that industry.

Knowing how to handle queries in DynamoDB that are taking longer than expected is what we just covered with DAX's caching abilities. Now, we will look at how to authorize access in DynamoDB.

Authenticating and authorizing in DynamoDB

As with other services in AWS, Dynamo DB allows fine-grained access control via the IAM service. You can either allow or disallow users at the service, table, or attribute level, depending on how you have structured your IAM policies.

AWS recommends, as a best practice, that you use the principle of least privilege and only allow users to access the data in the tables that they require versus blanket access.

Web Identity Federation

DynamoDB is an especially popular backend database for mobile and game developers. This can result in thousands of users needing to access even a single table. In this use case scenario, it is impractical to try to create an IAM user for each and every user:

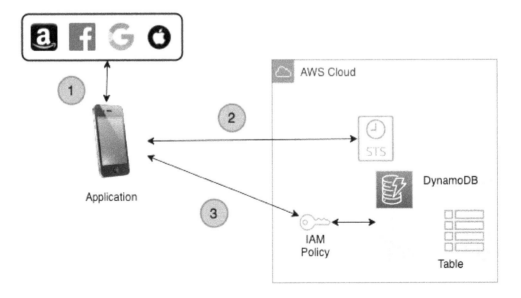

Figure 5.6 – Web Identity Federation to a DynamoDB table

The best way to understand how users who are authenticating via a web identity provider, such as *Facebook* or *Google*, gain access to data in a DynamoDB table is to look at the steps that it takes to grant authorization:

1. The application authenticates to a web identity provider such as Amazon. That identity provider then returns an identity token.

2. The application then calls the security token service to gain a temporary access credential based on a role that has been previously defined.

3. That role should give access to not just the full DynamoDB table, but the user's items via the IAM policy.

As DynamoDB is the preferred data store for mobile applications hosted on AWS, we looked at how to authorize web identity users to access specific data using a combination of IAM and STS tokens. Next, we will move on to monitoring our DynamoDB tables to make sure that we don't get caught off guard by throttles.

Monitoring DynamoDB

When you look at what metrics you need to concentrate on when monitoring DynamoDB, there are a few that come into focus:

*   Our GET and PUT requests are in the range of what we expected

*   We are not being throttled by either read or write capacity

From the DynamoDB console itself, we can get quite a few metrics regarding our table's health.

First, we can see both read and write capacity at a glance.

There is also a section in the metrics of any table that you choose for basic latency showing four areas of interest: GET, PUT, Query, and Scan Latency.

Let's log in to the **Amazon Management Console** and take a look at these basic metrics for ourselves:

1. Log in to the **Amazon Management Console** using the account you previously created for the projects table in DynamoDB.

2. In the search box at the top, type DynamoDB so that the service name appears in the results. Click **DynamoDB** to be taken to the DynamoDB service. On the top-right-hand side, double-check that you are in the correct region where you created the table.

3. Once you are at the main DynamoDB screen, click **Tables** on the left-hand side menu. Once clicked, you should see the table you created earlier, named **projects**:

Figure 5.7 – Showing the DynamoDB Tables menu option on the left-hand menu

4. Click on the table named **Projects** in the main window. This will open another pane, adding more information about your table.

5. From here, we can select the **metrics** tab from the top tab bar in the rightmost pane:

Figure 5.8 – The top menu bar once in the DynamoDB table

6. If, in the **metrics** display, you don't see much activity, then try running the queries from the CLI that we created previously in this chapter. This should prompt the metrics meters to start showing more PUT and GET objects of the data.

Contributor Insights

Enabling **Contributor Insights** for your DynamoDB table allows you to gain much better insight into the items that are being accessed and throttled. This allows you to make adjustments to your table or schema accordingly, if needed.

Once you have enabled **Contributor Insights**, DynamoDB creates two rules on your behalf if you only have a primary key:

- Most accessed items (partition key)
- Most throttled keys (partition key)

If you have a global secondary index or sort key, there will be two additional rules created for you by DynamoDB. These are specific to secondary indexes:

- Most accessed keys (partition and sort keys)
- Most throttled keys (partition and sort keys)

Summary

In this chapter, we looked at the NoSQL service known as DynamoDB. We explored how tables are created and partitioned, along with how to write and query data from DynamoDB. We also examined how to stream data from tables to other sources, along with using the **DynamoDB Accelerator** (**DAX**), a specialized cache, to help speed up our queries.

We have now finished Part 1, *Establishing Fundamentals*, where we reviewed **Amazon Web Services**. Next, we will move on to *Part 2, Developing, Deploying, and Using Infrastructure as Code*. In the upcoming chapters, we will incorporate many of the services that we have just discussed and then put them into action.

Review questions

1. Our application is storing project data in a DynamoDB table. You need to run a query to find all the builds that were run by a specific department last month. Which attributes would you use in your query?

    a. The partition key of `Build_Date` and the sort key of `Department`

    b. A composite primary key made up of `Department` and `Build_Date`

    c. The partition key of `Department` and the sort key of `Build_Date`

    d. The partition key of `Build_Date` and the sort key of `Dept_ID`

2. Which of the following AWS services provides an in-memory write-through cache that has been optimized for DynamoDB?

    a. ElastiCache

    b. CloudFront

    c. DAX

    d. Athena

3. A scan operation in DynamoDB is used for which of the following scenarios?

    a. To return the entire contents of the table, filtered on the primary or sort key

    b. To find items in a table based on a primary key attribute

    c. To find items in a table based on a sort key attribute

    d. To return all the items in a table

4. Your developers have created a DynamoDB table and seem to find that the performance always slows down after 20-25 minutes of their testing process. They can see from the basic monitoring on the AWS console that their requests are being throttled. What can you do to help pinpoint the issue?

    a. Increase the **Read Capacity Units (RCUs)** on the table so that the queries are no longer throttled.

    b. Enable **Contributor Insights** on the table so that the keys that are being throttled the most are shown.

    c. Add enhanced **Cloud Watch** monitoring with alarms whenever throttling occurs.

    d. Add **adaptive capacity** to the table so that the extra **Read Capacity Units (RCUs)** are spread evenly across partitions that are becoming hot.

5. Which of the following items would make a good partition key?

    a. `OrderID`

    b. `Ship_From_Location`

    c. `Total`

    d. `Product_Brand`

Review answers

1. b
2. c
3. d
4. b
5. a

Section 2: Developing, Deploying, and Using Infrastructure as Code

In this part, we are going to apply the concepts required to automate a CI/CD pipeline. You will learn about incorporating testing, managing artifacts, and deployment/delivery strategies.

This part of the book comprises the following chapters:

6
Understanding CI/CD and the SDLC

The **software development life cycle (SDLC)** section of the exam is the heaviest weighted of all the sections. Understanding the concept of SDLC, as well as **continuous integration (CI)** and **continuous deployment (CD)**, is paramount for passing the **Amazon Web Services (AWS) development-operations (DevOps)** exam. There are multiple stages in the SDLC, and there are specific AWS and third-party services that map to these stages.

Knowing the role that the AWS services play—along with key third-party tools—is also essential, not only to pass the exam but also to know for your day-to-day duties as a DevOps engineer.

In this chapter, we're going to cover the following main topics:

- Introduction to the SDLC
- Development teams
- Understanding the different types of deployments

Introduction to the SDLC

The SDLC consists of the following six basic cycles or stages:

- Source

- Build

- Test

- Deploy (for release)

- Monitor

- Plan

These stages are depicted in the following diagram:

Figure 6.1 – Phases of CI/CD

The first four of these stages fall on the development side of DevOps. The fifth falls on the operations side, and the final stage is done as a team exercise. You may notice in the preceding diagram that the planning phase is absent. Once again, this is a phase that, although a vital part of the SDLC, is not part of either the CI or CD process.

One of the initial things we need to understand in the context of this session is the use of the acronym *CI/CD*. As we think about the CI stage, we are talking about the first three steps of the pipeline.

CI

CI is a software development practice through which developers regularly merge their code changes into a central repository. After this, one or more automated builds are initiated, and tests are run against both the code that was committed and the previously existing code base.

AWS native tools for CI

Next, we will look at some of the tools that AWS offers within its ecosystem to help with CI.

AWS CodeCommit

AWS CodeCommit allows you to host private source control repositories in a highly scalable fashion.

These are the main benefits of CodeCommit:

- **Ability to collaborate**: Software teams can work together on a code base using known Git functions such as pull requests, branches, and merging.

- **Encryption**: CodeCommit repositories are automatically encrypted at rest using AWS **Key Management Service (KMS)**. Code transfers to and from the repository are also encrypted via either **HyperText Transfer Protocol Secure (HTTPS)** or **Secure Shell (SSH)**.

- **Access control**: Being fully integrated with the AWS **Identity and Access Management (IAM)** service allows you to specify which users have access to which repositories, without having to navigate through a third-party system.

- **High availability**: AWS CodeCommit is backed by **Amazon Simple Storage Service (Amazon S3)** and **Amazon DynamoDB** for code-and-commit storage. This is a highly redundant and scalable setup to ensure that your repositories are accessible.

- **Notifications**: CodeCommit can be integrated with **Amazon Simple Notification Service (Amazon SNS)** so that events significant to the repository can be broadcast to the correct channels. Notifications sent by CodeCommit include a status message and a link to the corresponding repository.

AWS CodeBuild

As you look to compile the code you have committed in your source control repository and create software packages for deployment, **AWS CodeBuild** allows teams to customize their build-and-test process using the **YAML Ain't Markup Language (YAML)** language.

These are the main benefits of AWS CodeBuild:

- **Fully managed**: With CodeBuild, there is no need to set up a separate build server. This means that there are no more software patches or updates, and nothing to manage. Jobs are set up and submitted, and then run.

- **Secure**: CodeBuild is integrated with the IAM service and, hence, users can be assigned to specific build projects only. Any artifacts that CodeBuild produces are encrypted with AWS KMS.

- **Scalable**: CodeBuild scales automatically according to the number of jobs that have been submitted at any one time. There is no need to think about vertical or horizontal scaling when a big burst of build jobs or tests is about to happen, as CodeBuild handles all this automatically.

- **Enables CI and CD**: As part of the **AWS Developer** services, CodeBuild naturally integrates into other CI/CD tools offered, such as **CodeCommit** and **CodePipeline**. Integration has also been done with other tools in the ecosystem—**Jenkins**, for example, has the ability to use CodeBuild as a scalable worker node.

AWS CodeArtifact

As software builds get more regimented, companies and teams start to look for a way to ensure that everyone is using the same approved packages and versions of packages. This is where the managed artifact repository **CodeArtifact** comes into play.

If you are interested in the security of your build process, then CodeArtifact offers a number of features that help your development teams create a safer environment. Firstly, packages and artifacts needed in the build process can be accessed from a **virtual private cloud** (**VPC**) using AWS PrivateLink endpoints. This means that if you have the necessary libraries and items needed for the build stored on your CodeArtifact service, then these files can be transferred to your functions and instances without a need to traverse the public internet.

Secondly, with the CodeArtifact service, you as an account administrator have the ability to approve packages for use. This approval process can also be automated using a combination of the CodeArtifact **application programming interfaces** (**APIs**) and the Amazon EventBridge service.

Thirdly, many package repositories have recently been placing download limits on their servers. Having a deployment or build fail due to the fact that the repository is not currently taking download requests from the current **Internet Protocol** (**IP**) address is not only frustrating—it can also become a real impediment for deployment. For example, if you are building your instances in real time versus having pre-built **Amazon Machine Images** (**AMIs**) and you have the need for certain node packages from public **Node Package Manager** (**npm**) servers, then if you are in an autoscaling state and are trying to scale to meet the traffic demands of your customers, this can become more than just a nuisance. However, if you have your packages stored on AWS CodeArtifact, then you are not limited by any third-party servers and can build your functions and instances, pulling down the required packages as many times as required, as illustrated in the following screenshot:

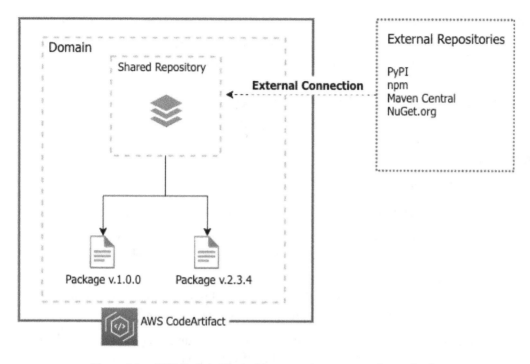

Figure 6.2 – AWS CodeArtifact with connection to external repositories

You can configure CodeArtifact to go out to public repositories and fetch popular packages for storage and versioning for your teams. CodeArtifact can work with package managers many developers are both familiar and comfortable with, including `pip`, `yarn`, `npm`, `Maven`, and `Gradle`.

These are the main benefits of CodeArtifact:

- **Store and share artifacts securely**: If integrated with the KMS service, the artifacts you store in CodeArtifact can be encrypted.

- **Reduce operational overhead**: CodeArtifact quashes the need for the setup and maintenance of an artifact server. It is a highly available service that scales automatically to the number of artifacts stored.

- **Publish and share packages**: CodeArtifact allows you a central place to publish the packages your team creates, eliminating the need to hunt around on the internet.

- **Consume packages from public repositories**: CodeArtifact can be set up to automatically download packages from public repositories such as the `npm` registry, `NuGet.org`, and **Maven Central** with a couple of clicks instead of complex scripting.

Continuous delivery

If the deployment to production occurs with continuous delivery, there will most likely be a manual approval process rather than an automated deployment.

AWS native tools for continuous delivery

Let's take a brief look at some of the AWS tools used in continuous delivery.

AWS CodeDeploy

These are the main benefits of CodeDeploy:

- **Automated deployments**: CodeDeploy totally automates software deployments.

- **Ease of adoption**: CodeDeploy can easily be integrated into your existing deployment tooling, such as Jenkins, GitHub, or AWS CodePipeline.

- **Consolidated control**: In one view, CodeDeploy can both show you the status of your deployment and provide the ability to send push notifications to one or multiple members of your team, letting them know the pass or fail status of builds.

- **Minimize downtime**: In allowing you to introduce your changes incrementally, CodeDeploy helps preserve the availability of your application. It can also help the process of rolling back to a previous version should an issue be discovered.

AWS CodePipeline

AWS CodePipeline allows you to automate the building, testing, and deploying steps and phases that you go through while you produce your software and **Infrastructure as Code (IaC)**. It can be integrated with third-party services such as GitHub.

These are the main benefits of CodePipeline:

- **Rapid delivery**: CodePipeline provides quick feedback to you and your team as you move your code through a structured process to deployment. Flaws can be found and fixed without too much effort.

- **Easy to integrate**: If you have existing pieces already in use for your CI/CD process, then CodePipeline allows you to incorporate those items without any hassle. This can include a Jenkins server that is already set up to run tests running in the cloud or on-premises or even a third-party source code repository such as GitHub.

- **Configurable workflow**: Each software release has a slightly different process depending on the tests configured or the service that it is being deployed. CodePipeline provides you the ability to customize your steps in a variety of ways, including with the **AWS Management Console** interface, using the **command-line interface (CLI)** or one of the available AWS **software development kits (SDKs)**, or even by crafting your pipeline in a **CloudFormation template**.

- **Improved quality**: CodePipeline allows you to automate all your processes in an easy-to-follow flow so that steps are not missed when doing a deployment. Having tests run automatically allows for consistency in your code and provides developers with instant feedback.

CD

With CD, there is no manual approval process as the code revisions are pushed into the production environment. Instead, testing practices and guidelines are relied on in order to ensure the integrity of code meets quality checks before being automatically deployed to the production environment. Any revisions that don't meet these guidelines then get failed as a build process, and feedback is given to either the individual developer or the development team. This initial feedback can be simple in nature, such as a notification that a build has failed, and can be in the form of an email or a **Short Message Service (SMS)** message that could be sent using the **SNS service**. An alternative to this could even be posted to a messaging service such as **Slack** or **Microsoft Teams**, using a combination of SNS and **AWS Lambda** to post the message.

> **Note**
> Continuous delivery is *not* CD. This is a common mistake that many make. You need to understand, however, that with continuous delivery every commit that passes a set of automated tests is then pushed into the production environment.

Let's take a look at how the SDLC translates into the different developer tools provided by AWS. You can see from the following diagram that almost every stage has its own dedicated service. The exceptions are the build and test stages, where AWS CodeBuild can perform both of these tasks:

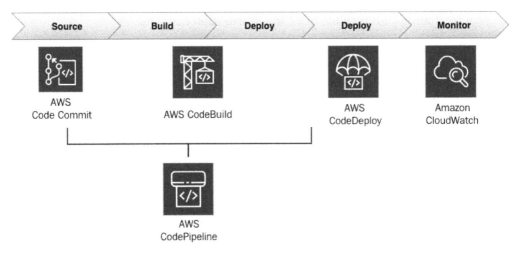

Figure 6.3 – AWS tools used in CI/CD

As we look at CI and continuous delivery from an **Amazon code tools** point of view, there are several tools that can help you achieve these tasks.

Testing

Testing plays a vital role in the SDLC. It can help you improve the reliability, scalability, performance, and security of your application.

> **Note**
>
> Did you notice how many of the things that testing provides are pillars of AWS? Although it may seem faster to leave testing out of your initial pipeline design, it's a vital part of ensuring the stability, safety, and capabilities of your workload.

If you have potential weaknesses in your code, testing is one of the ways that you can discover many of them. This can prevent you from deploying a fatal flaw into your production environment. As you automate this process as a DevOps engineer, the initial outlay to build the pipelines can be a bit of an effort. Once the pipeline and tests have been established, then multiple types of code—including both IaC and application code—can be both tested and deployed in a rapid and repeatable fashion.

Types of affiliated tests

There are a variety of different types of tests available to ensure that your code base is functioning as it should. Each type of test performs a specific task, and some take more time to run than others. Here's a diagram showing the different testing stages:

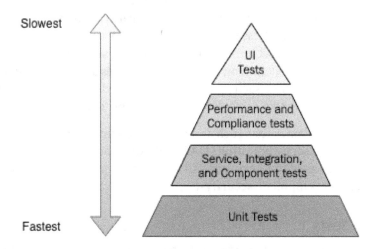

Figure 6.4 – Different stages of testing

Looking at the stages in *Figure 6.4*, we can see the relative speed at which tests can be performed. We can now examine each of the testing types in more detail.

Unit tests

Unit tests are small in the amount of code they span but provide developers instant feedback, especially since they can be run directly from a developer's individual machine. They can be added to a CI/CD pipeline, where they can be used to proactively prevent any known defects from entering into the code base.

Service integration and component tests

Integration tests are used to check that all the different components in a system work together. These tests can also include tests of any third-party services or APIs. Examples of integration tests would be having the web layer insert a test value into a database to ensure connectivity or send a notification after an event.

Performance and compliance tests

Once you have built your application, you need to know that it will handle your estimated traffic effectively. This includes scaling out and in with your resources as need be and checking the system while it is under sizable load. **Performance tests** will help accomplish this task, exposing a lack of performance or bottlenecks that the system might display under normal usage or high usage. Two of the most popular types of performance tests are **stress** tests and **soak** tests. A stress test will simulate many users on the resources for a small duration of time or for a set number of routines, such as playing back web server logs to emulate past users. A soak test, on the other hand, gives a steady stream of users over a much longer period of time such as a week or more, trying to expose items such as memory leaks or bad routines that don't show up under short durations.

Adding security and compliance tests such as **static application security testing (SAST)** and **dynamic application security testing (DAST)** allows known vulnerabilities that have been reported and discovered to be exposed. Adding these types of tests to your CI/CD pipeline is part of the **development-security-operations (DevSecOps)** umbrella.

User interface tests

At the top of the pyramid are **user interface (UI)** tests. These tests are not just testing the **graphical UI (GUI)**, but the system as a whole. Tests can encompass routines such as ensuring users can log in correctly or reset a password. They may contain activities such as uploading and downloading files or visiting specific pages that touch datastores and middleware components.

Although manual testing may be introduced out of necessity during this stage, it's better to keep manual tests to a minimum.

Maturing throughout the process

When initially starting to build your CI system, you may only have a few steps, such as pulling code from a source code repository and deploying to a test server.

As time goes on, the code base will gain a greater amount of test coverage and settings will be tuned so that team members will have much more confidence in the deployment process, so much so that they come to rely on it as just another day-to-day tool.

As you mature, you will find that you are starting to gain the following benefits:

- **Speed**: Teams become much more self-sufficient with automated pipelines and release processes. There is no more waiting for a particular person or team to install developed software packages or updates.

- **Reliability**: Automating your process eliminates dependency on a single person who knows the process. Any team member who starts the process will feel confident that it will run the same each and every time it is invoked.

- **Consistency**: Creating a pipeline with standard steps for your deployments makes for a consistent process each time a process is initiated. This prevents any steps such as testing from being forgotten or skipped.

- **Scalability**: The frequency of updates can—and usually does—increase as your organization begins to scale. Automated pipelines that have steps to build, test, and deploy out to different environments help you scale without adding additional staff.

- **Efficiency**: Moving your testing from a manual to an automated process allows you to not only speed up the testing process but also allows the testing team to concentrate on developing new and improved tests versus spending their time manually testing the system.

Now that we have looked at the actual SDLC process, we will now look at optimizing the setup of teams for the CI/CD process.

Development teams

It's a recommendation from AWS to have three development teams for implementing a CI/CD environment: **an application team**, **an infrastructure team**, and **a tools team**. Amazon preaches the concept of two-pizza teams, meaning that no team should be larger than could be fed by two pizzas. Smaller teams lead to better collaboration, have full ownership of their features or application, have full accountability, and—last but not least—this all aligns with the DevOps model of agility.

The application team

Application team members are the ones responsible for creating an application itself. Members of this team are skilled in one or more programming languages, but also have (or come to have) a deep understanding of the platform and system configuration.

Application team members have responsibility for creating a backlog of items that need to be worked on, as well as creating a task for working on stories. In addition to having the programming skills to create and maintain an application, this team should also become skilled in automation to allow members to create their own pieces of the pipeline once the tooling has been created.

The infrastructure team

In order for the application team members to run their applications, they need some infrastructure to run them on. Even if an application is of a serverless nature, it still needs to have permissions created in IAM. If it's not a serverless setup, then there are servers to provision and manage.

With the **infrastructure team**, this is done via IaC either via CloudFormation, using scripts with the AWS CLI, or with the AWS **Cloud Development Kit** (**CDK**).

Many times, this team will also be responsible for things such as **Active Directory (AD) servers** and **single sign-on (SSO) integration**, especially because they are tightly coupled with the IAM permissions.

The tools team

The **tools team** both builds and manages the CI/CD pipeline. This team must be proficient in building and integrating all of the different pieces and components of the pipeline so that the dependent application teams have a smooth process. Although the tools team is not part of the two-pizza team, it is responsible for creating tooling and systems that enable the other teams to perform their jobs.

The team may choose to implement tools and services such as AWS CodeCommit, CodePipeline, CodeBuild, and CodeDeploy, along with other third-party tools such as Jenkins, GitHub, Bitbucket, Artifactory, and other related tools.

Many organizations will classify the tools team as the DevOps team, but this is a misnomer as DevOps is more of a practice of people and processes versus the use of tools.

Not every tools team is ready to make a complete buy-in to the AWS toolset, and AWS understands this. There is even an entire whitepaper around the automation server Jenkins. Having an understanding of how third-party tools interact and complement the AWS ecosystem is essential knowledge for the DevOps professional exam.

With our team members now having areas they can concentrate on to maximize their effectiveness, we will move on to the types of deployments and see how to choose a deployment strategy that best fits our needs.

Understanding the different types of deployments

When you think about deployments, especially as we have been talking about the SDLC, you may think we are talking about application code. However, as you move to automate more and more of your systems in AWS, deployments can take on multiple meanings. Deployments can mean application code, but they could also mean infrastructure code, configuration code, or other layers.

There are five main types of deployment strategies to consider when dealing with deployments on AWS. Each method has its own advantages and disadvantages.

When choosing a deployment strategy, these are the main things you need to consider:

- How quickly can you deploy?
- Are any **Domain Name System (DNS)** changes needed?
- Would there be any impact with a failed deployment?
- What would the rollback process be?
- Where would the code be linked to? (New or existing instances?)

With this in mind, let's look at the five different deployment strategies in depth.

In-place deployments

When you perform an **in-place deployment**, you are updating instances that have already been deployed to an environment. A load balancer can be used to deregister each instance while doing the deployment, perform a health check, and then place the healthy instance back into service. In-place deployments can be done all at once or they can be done as a rolling deployment.

Let's take a look at the pros and cons of in-place deployments, as follows:

Pros	Cons
• Quicker deployment type • Low cost due to no repeated infrastructure • Available for **Elastic Compute Cloud (EC2)** and Lambda	• Has the possibility for downtime • Rollback is timely as you need to reinstall resources

Table 6.1 – In-place deployment pros and cons

Immutable and blue-green deployments

In the case of **blue-green deployments**, a whole new set of infrastructure is created and often tested before doing a DNS switch to the new environment. This is the safest type of deployment method, but it takes time and is also the costliest since you are bringing up two full environments for a period of time until you do the DNS switchover. You then have an option to take down the secondary environment once it is not being used at that moment to save on costs, or you can keep it running in the background to save on deployment time and use it as a failover environment. In the case of a deployment failure, your customers would never even know, since when using blue-green deployments you only switch the DNS once your second (or green) environment is up and healthy.

Immutable deployments refer to deploying code for a whole new set of resources using new configurations or new application code. This task is made much simpler in the cloud versus on-premises hardware as resources can be provisioned using simple API calls.

The following table shows both the pros and cons of immutable deployments:

Pros	Cons
• Zero downtime deployment • Zero downtime rollback • Available for EC2 and Lambda	• Requires a new set of resources and hence doubles the bill • Requires a DNS change to switch from one environment to another • Is one of the slower deployment types

Table 6.2 – Blue-green deployment pros and cons

See *Chapter 13, Blue Green Deployments*, for a much deeper look at blue-green deployments.

Canary deployments

A **canary deployment** is a type of blue-green deployment strategy where a predetermined number of instances are initially upgraded with either the updated application code or operating system updates. At that point, a portion of the traffic is then shifted—for example, 20% of the traffic or customers with a certain cookie who have been chosen for the canary deployment. This initial traffic is then watched over a period of time to see if any issues arise. If the deployment has no major issues, the rest of the instance or application fleet is then upgraded. At that point, all of the traffic has been shifted to the new application or operating system upgrades. If there do seem to be any issues with the initial deployment, then the traffic can easily be routed back to the instances that do not have the upgrades, either by setting the routing policy in Route 53 to 0% on the new instance **Canonical Names (CNAMEs)** or by taking the new instances out of the load balancer. At that point, the new instances are taken down out of service, and any launch configurations that were previously updated can be updated to use a previous version of a working AMI.

In the following table, we compare the pros and cons of using a canary deployment method:

Pros	Cons
• Better for large updates • If an issue arises then only a small percentage of users are affected • Zero downtime deployments • Easy rollback—just take canary instances out of the load balancer/DNS and all traffic goes back to the original instances • Available for EC2 and Lambda	• Long time for deployment, as canary instance needs to be monitored • Happens in stages, hence takes more time • Extra cost involved with launching the canary resources

Table 6.3 – Canary deployment pros and cons

Rolling deployments

When using a **rolling deployment**, not all of the instances are updated at the same time. This strategy prevents any downtime since if a process fails, only a portion of the group is upgraded at any one particular time. As the initial instances are sent to the deployments, they must come back both healthy and online before further instances will be engaged in the new deployment.

It's important to note that since not all members of the group are deployed at the same time with either the application code or system upgrades, then there could be multiple versions running for a user to experience. Using sticky sessions can help, but not eliminate, trying to provide a seamless experience to a customer during a single session while deployment is happening.

Although it's not a long list of pros and cons, take a look at the following table to see the benefits and drawbacks of using a rolling deployment method:

Pros	Cons
• Zero downtime deployment • No DNS change required	• Rollback takes longer as you are redeploying to existing instances

Table 6.4 – Rolling deployment pros and cons

Linear deployments

In a linear deployment, traffic is shifted equally in increments across resources over a set number of predefined increments. A linear deployment is a subset of the blue-green deployment type. Instead of deploying to the instances or resources that your application is currently running on, you are first standing up a new set of infrastructure and then over time shifting your traffic from the old code base to a new one, using a service such as **Route 53** and weighted routing. This way, you can keep a close eye on your new environment, and if any issues arise then you can quickly shift all traffic back to the original environment, with no downtime.

Linear deployments can also be done with Lambda aliases and Fargate containers in order to shift a percentage of your traffic to a new version of your code.

For example, if your deployment window is 1 hour and you want to spread 100% of your traffic in equal increments to new resources in that hour, then your linear strategy may go something like this:

- Minimum 0-16% of traffic shifted over to the new resources
- Minimum 10-32% of traffic shifted over to the new resources
- Minimum 20-48% of traffic shifted over to the new resources
- Minimum 30-64% of traffic shifted over to the new resources
- Minimum 40-80% of traffic shifted over to the new resources
- Minimum 50-96% of traffic shifted over to the new resources
- Minimum 60-100% of traffic shifted over to the new resources

There are both pros and cons to using a linear deployment method, so let's compare them here:

Pros	Cons
• Allows you to slow down deployments and dictate how many users will be affected at once • Available for EC2 and Lambda • If an issue arises, then only a small percentage of users are affected	• Requires a new set of resources and hence doubles the bill • Requires a DNS change to switch from one environment to another • Happens in stages, hence takes more time

Table 6.5 – Linear deployment pros and cons

We will be covering Lambda in depth in *Chapter 12*, *Lambda Deployments and Versioning*.

All-at-once deployments

In this type of deployment, all the traffic is shifted from the original environment to a new environment at the same time. This is the fastest of all deployment methods. This also means that any rollbacks that need to happen would also take the most amount of time, as the code would need to be redeployed to all instances. You *can have downtime* with this type of deployment if you have an issue as you wait for your rollback to come back online.

The advantages and disadvantages of using an *all-at-once deployment method* are outlined in the following table:

Pros	Cons
• Ability to roll back quickly if there is an issue • Zero downtime • Available for EC2 and Lambda	• Speed, although this is the quickest type of blue-green deployment • Extra cost of standing up a dual set of resources

Table 6.6 – All-at-once deployment pros and cons

> **Note**
> An all-at-once deployment can also be called an in-place deployment. Be familiar with both terms as either can show up on test questions.

Review questions

1. A medium-sized software company has hired you as a DevOps consultant to help set up its deployment pipeline. The staff want to be able to push their tested code into their production environment in a quick manner but do not want the possibility of dealing with downtime for their customers. Their DNS is hosted on a third-party service, and changes to the DNS would require a change ticket. Which deployment method would you recommend implementing?

    a. Blue-green deployment

    b. In-place deployment

    c. All-at-once deployment

    d. Rolling deployment

2. A medical device company is looking to set up its development pipeline using Jenkins to deploy its code base in an automated fashion. Since this is only the development environment, they want to keep costs to a minimum and would be fine if the application team needed to redeploy in case of a failed deployment. Which strategy should they use?

    a. Blue-green deployment

    b. In-place deployment

    c. All-at-once deployment

    d. Rolling deployment

3. A mobile gaming company is trying to speed up its production time with all the new enhancements it has been developing for its most popular game. Staff have noticed on social media that around the dates of the last two releases, users complained of an increased number of glitches. Some of the glitches reported were known issues to teams that were working on the development of the game. The gaming company already has an automated deployment pipeline set up with AWS CodePipeline, and its code is stored in AWS CodeCommit. What is the most cost-effective way to reduce the number of glitches being deployed in each release?

    a. Spin up a new environment and run full UI tests before releasing the code to the production environment.

    b. Add a step in the current CodePipeline to spin up an EC2 instance that runs the Jenkins software and can use **Simple Systems Manager (SSM) Parameter Store** to download the current CodeCommit repository, after which it runs unit tests to pass or fail the build.

c. Add a CodeDeploy step to the current AWS pipeline that runs the current set of unit tests connected to an AWS SNS topic so that on failure, the current build is failed and the development team is notified.

d. Add a CodeBuild step to the current AWS pipeline that runs the current set of unit tests connected to an AWS SNS topic so that on failure, the current build is failed and the development team is notified.

4. A research company is working on a confidential project and the management team wants to be aware of any progress made as soon as it happens. The developers are using AWS CodeCommit for their source code versioning, along with CodeBuild to run unit tests. Which measures can you put in place to allow the management team to get the updates they desire? (Choose all that apply)

    a. Create an SNS topic for the management team and add all their emails.

    b. Have AWS CodeCommit push notifications to an SNS topic any time that either a commit has been made or a feature branch has been merged with the master.

    c. Have CodeCommit create a daily report of commit activity and then push the report to S3 so that the management team can view it from a bucket to which they have access.

    d. Enable notifications on AWS CodeBuild to an SNS topic for when a job passes or fails.

5. A growing company currently has a Jenkins server that runs on EC2. Developers are complaining that they are waiting too long for their builds to get started and complete. You have been asked to help the tools team in coming up with a solution that can scale with the growth and speed of the development team but can also be implemented in the quickest and most cost-effective manner possible. Which solution will least need to be managed by the tools team?

    a. Create an AMI from the Jenkins server and use the AMI to create three additional worker nodes, using the current Jenkins system as the master.

    b. Rebuild the Jenkins server as a containerized system using **Elastic Kubernetes Service (EKS)** Fargate and worker-node plugins to scale as needed.

    c. Integrate AWS CodeBuild into Jenkins to allow for automatic worker-node creation once the queue gets over a level of 1.

    d. Create an AMI from the Jenkins server and use the AMI to create a launch configuration for an autoscaling group that will launch a new Jenkins instance when the queue gets over a level of 1.

Review answers

1. d

2. b

3. d

4. a, b, d

5. c

Summary

In this chapter, we looked at the **SDLC, CI, continuous delivery**, and **CD**. We also started to look at the tools AWS offers that can help us with these different stages of the SDLC. Then, we took a look at the different types of teams and what their job responsibilities consist of. Finally, we reviewed the different types of deployment strategies available in AWS and how they can best be used.

In the next chapter, we will be taking a deeper dive into AWS's CloudFormation IaC service. We will see how to create reusable resources, along with the scripting methods available in CloudFormation templates.

7
Using CloudFormation Templates to Deploy Workloads

CloudFormation templates offer DevOps engineers an easy way to create, manage, and provision related resources automatically. They also allow you to stand up the same infrastructure over and over again quickly, be it for development, testing, production, or disaster recovery. They are not only an essential concept to know of, but also to understand for the DevOps professional exam.

In this chapter, we're going to cover the following main topics:

- Essential CloudFormation topics
- Creating nested stacks with dependencies
- Adding a helper script to a CloudFormation template
- Understanding how to detect drift in previously created stacks
- Using the **Cloud Development Kit** (**CDK**) as an open source framework

Technical requirements

In dealing with the CloudFormation templates, this chapter will be a bit more hands-on than previous chapters, which have focused more on theory. You should be comfortable with the YAML syntax. And at this point, you should be comfortable with both the AWS Management Console as well as the CLI. Most of the templates discussed in this chapter are too large to type out entirely on the following pages, as some CloudFormation templates can become a few thousand lines in length. We have included the templates discussed here in the GitHub repository for this book in the `Chapter-7` section: `https://github.com/PacktPublishing/AWS-Certified-DevOps-Engineer-Professional-Certification-and-Beyond/tree/main/Chapter-7`.

Essential CloudFormation topics

As we look to make repeatable pieces of infrastructure that can be built with automation, having the correct **Infrastructure as Code** (**IaC**) tool allows for the implementation of these repeatable processes. It is these repeatable processes where the CloudFormation service comes into play. CloudFormation is IaC that can be checked into source control systems such as **CodeCommit** or **GitHub**. The fact that it is a piece of code allows it to be versioned and collaborated on with multiple team members. It can also be added to an automated build pipeline as part of a CI/CD process.

CloudFormation templates have the flexibility to be built in either the **JSON** or **YAML** programming languages. Using YAML allows you to put comments and use short codes in your template. However, you do need to stick to the strict YAML formatting guidelines.

Once you have created a template, especially one with the proper mappings and without hardcoded references, then it has the ability to be reused over and over again. This means that you can use it both in multiple regions and across numerous accounts.

The anatomy of a CloudFormation template

A CloudFormation template can contain many different sections. There is only one mandatory section when creating CloudFormation templates, and that is the *resources* section. Sections don't necessarily need to be added in a particular order; however, as you look at different examples, there is a logical order in the way that sections are structured since some sections reference other sections.

Format version

This is the version that the CloudFormation follows. It is an *optional* section. If added, it is usually the first section of the template.

Description

The description is a small section of text that tells users about the template. It must always follow the format version of the template. Descriptions are beneficial as you start to gather a more extensive collection of templates and then have the ability to quickly know what the template's purpose is and what type of resources it will create. It is an *optional* section.

Metadata

This section contains objects that provide additional information about the template. It is an *optional* section.

Parameters

This is a section of values that can be passed into the template at runtime. Default values can also be defined in lieu of needing user input. These values can then be used when creating your resources. You can also refer to parameters from the **Resources** and **Output** sections of the template. This is an *optional* section.

Rules

Rules are used to validate one or more parameters that are transferred to a template at runtime. Rules can help enforce things such as making sure that a large enough EC2 instance is being spun up in a production environment or that particular subnets are used. Rules very often validate the input of parameters . This is an *optional* section.

Mappings

Using the intrinsic `Fn:FindInMap` function allows CloudFormation to find the corresponding key to the matching value. One of the most frequent scenarios for the use of the mappings section is when declaring the correct **Amazon Machine Image** (**AMI**) for particular regions being used, ensuring that the template is reusable in more than a single region. This is an *optional* section.

We will look a bit more into mappings when we look at the intrinsic `Fn:FindInMap` function.

Conditions

Although CloudFormation templates don't have a lot of logical programming available to them, conditions inside of the template allow for certain values to be assigned or certain resources in the template to be created based on values being present. A good example would be if the stack were being created in the test environment, then it would create a database. However, if it were being created in the production environment, it would not. This is an *optional* section.

Resources

This section actually declares the resources and their properties that will be created by the CloudFormation template. Resources can be a whole variety of AWS services from EC2 instances to IAM users to even **chatbots** and **Step Functions**. There are few services that AWS offers that cannot be created via CloudFormation templates. This is a *required* section.

Outputs

Outputs allow for the declaration of a key-value pair along with a corresponding description that can be consumed by an end user once the stack has completed creation or can be consumed by another stack. In the case of an end user, you might want to output the URL or IP of an EC2 instance that has just been spun up so that you don't have to go hunting around the console for it. Alternatively, in the case of creating outputs for other stacks to consume, **Amazon Resource Names** (**ARNs**) are one of the main items that can be used in other stacks as reference points. This is an *optional* section.

Launching CloudFormation templates

Once you have your template ready to go, then you can launch it from either the AWS Management Console or the **Command-Line Interface** (**CLI**). We will go through the process of using the template located in the GitHub repository named `sqs-queues.` `yml`. If you would like to follow along with the following exercise, then go to the repository and download this template. Alternatively, you can use one of your own CloudFormation templates, bearing in mind that the parameters and resources will not be the same as we go along, but the process will still follow the same steps:

1. Take the downloaded template and upload it to your S3 bucket, which we created in *Chapter 4, Amazon S3 Blob Storage*. (In our case, our bucket's name is `devopspro-beyond`.):

    ```
    $aws s3 cp sqs-queues.yml s3://devopspro-beyond/
    sqs-queues.yml
    ```

2. Now that the template has been uploaded, log in to the AWS Management Console and navigate to the CloudFormation service page at `https://us-east-2.console.aws.amazon.com/cloudformation/`.

3. Click the orange button on the right-hand side labeled **Create stack**.

4. Under **Pre-requisite – Prepare template**, make sure that the radio button is already selected for the **Template is ready** option:

Prerequisite - Prepare template

Prepare template
Every stack is based on a template. A template is a JSON or YAML file that contains configuration information about the AWS resources you want to include in the stack.

| ● Template is ready | ○ Use a sample template | ○ Create template in Designer |

Figure 7.1 – CloudFormation prerequisites

5. Under **Specify template**, make sure that that the **Amazon S3 URL** option is already selected and then in the URL field, type the value of the location of your template, which in our case is `https://devopspro-beyond.s3.us-east-2.amazonaws.com/sqs-queues.yml`:

Specify template

A template is a JSON or YAML file that describes your stack's resources and properties.

Template source

Selecting a template generates an Amazon S3 URL where it will be stored.

| ● Amazon S3 URL | ○ Upload a template file |

Amazon S3 URL

https://devopspro-beyond.s3.us-east-2.amazonaws.com/sqs-queues.yml|

Amazon S3 template URL

Figure 7.2 – CloudFormation Specify template screen

6. Click the orange **Next** button at the bottom of the page.

7. We now need to fill out a few details before we can proceed. We'll start with the **Stack name**. Name your stack `SQSqueue` (no spaces are allowed).

8. Next, fill out the rest of the **Parameters** by adding your address for the **AlarmEmail** field and the name of the queue for the **QueueName** field. You can leave the other two values at their defaults. After you have done this, press the orange **Next** button at the bottom of the page:

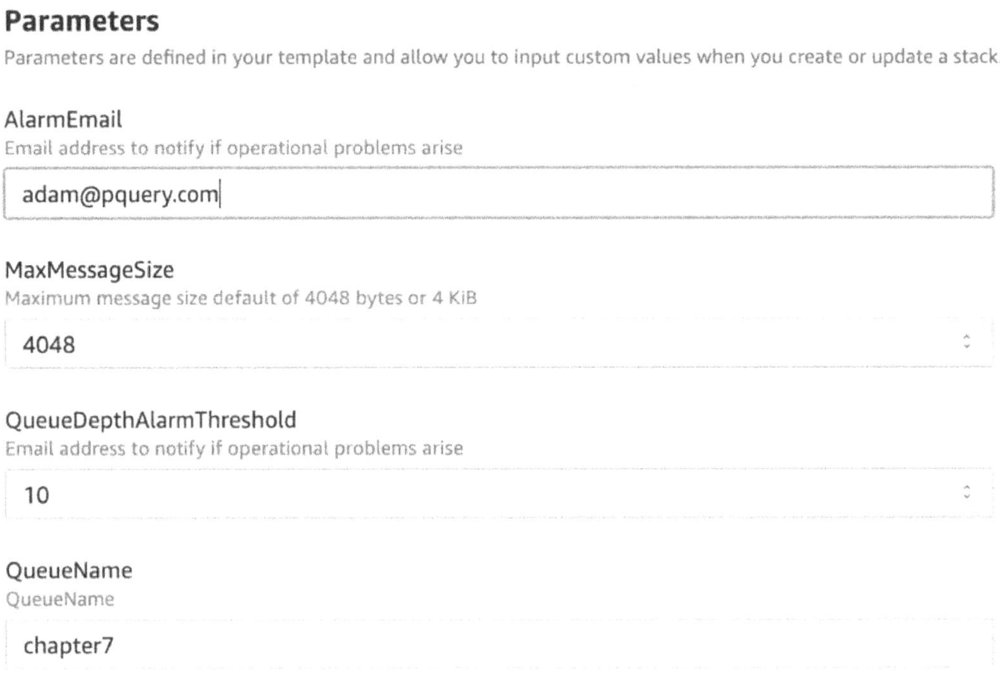

Parameters

Parameters are defined in your template and allow you to input custom values when you create or update a stack.

AlarmEmail
Email address to notify if operational problems arise

adam@pquery.com

MaxMessageSize
Maximum message size default of 4048 bytes or 4 KiB

4048

QueueDepthAlarmThreshold
Email address to notify if operational problems arise

10

QueueName
QueueName

chapter7

Figure 7.3 – Entering the parameters for the template

9. On the next page, labeled **Configure stack options**, scroll down to the bottom of the page and click on the orange **Next** button.

10. We are now at the point where we can review what we have entered, and if everything looks correct, scroll down to the bottom of the page and click the orange button, which is labeled **Create Stack**.

Once we start the creation process, we will be taken to the **Stacks** screen, where we can see the **CREATE_IN_PROGRESS** notification while our stack is being created.

There is a menu at the top of this section that allows you to see the specific resources that each stack creates, specific stack info, any outputs that you have declared for the stack to show after its creation, the parameters you entered when creating the stack, the actual template used to create the stack, and change sets used on the stack.

When you launch a CloudFormation template, that template is saved in an S3 bucket, which AWS creates in the region in which you are launching the template if you have not previously saved the template into an S3 bucket. Each time that you either update the template or launch a new version, a new copy of the template is added to this bucket.

Using the CLI to launch templates

After going through all the steps from the Management Console to launch a template, we can see that if we are going to automate our deployments as part of a CI/CD pipeline, then doing this each and every time would not be practical. The AWS CLI does have a `deploy` command that allows us to launch a stack with one line. Before we run the command, be sure that you have downloaded the YAML template from the GitHub repository in the `Chapter-7` folder. Once downloaded, open up your terminal window and either copy or move that template to your working directory or change your working directory to where you have downloaded the template:

An example of the CLI command to launch a template is shown as follows:

```
$aws cloudformation deploy --template my-template.json --stack-
name CLI-stack --parameter-overrides Key1=Value1 Key2=Value2
```

With this single command, we have replicated all of the steps that we went through before. As noted previously, our template should be in an S3 bucket with a random name since we uploaded it as part of the creation process, and the CloudFormation service stored it on our behalf.

Using change sets

If your CloudFormation stack needs to be updated, sometimes you may like to know and understand how the existing resources currently running will be affected. Change sets allow you to preview how changes will impact the currently running resources before any changes take place, and then allow you to cancel the update if the changes would be detrimental to current resources, or proceed if the change would execute as expected.

> **Note**
>
> Change sets don't tell you if your CloudFormation will run successfully upon execution. They can't foresee account limit updates that you might run into, or determine whether you possess the correct IAM permissions to update a resource.

Steps to execute a change set

Let's look at both the mandatory and optional steps required for executing a change set on a stack:

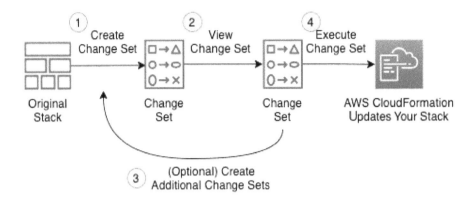

Figure 7.4 – CloudFormation change sets

With the basis for how to execute a change set under our belts, we will now update our original CloudFormation stack – SQSqueue – with a change set.

Using a change set to update our stack

If you haven't already done so, then download the template named sqs-queues_ change_set.yml from the GitHub repository in the Chapter-7 folder. This is the file that we will use to create our change set. Also, make sure that you have your browser open to the AWS Management Console and have navigated to the CloudFormation service, then proceed with the following steps:

1. From the CloudFormation Stacks page, you should see your SQSqueue stack in a CREATE_COMPLETE state. Click on the stack name to be taken to the page with details about the stack. This is where we can perform the change set:

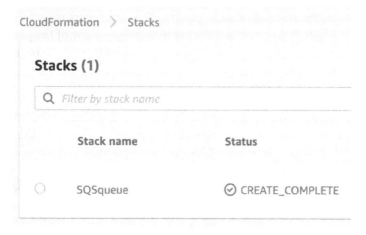

Figure 7.5 – Our previously created CloudFormation stack

2. On the horizontal menu under the stack name, you will see a number of choices, including **Change sets** on the very right-hand side. Click on the **Change sets** option:

Figure 7.6 – CloudFormation stack menu with the Change sets option highlighted on the right

3. Once in the **Change Sets** section, click on the button labeled **Create change set**.

4. At this point, a new screen will appear with the heading **Prerequisite – Prepare template**. You will be presented with three choices. Select the choice in the middle that reads **Replace current template**:

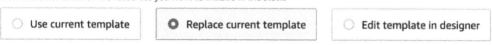

Figure 7.7 – CloudFormation change sets Prepare template screen

5. This will make another set of selections appear underneath the initial set of radio buttons. This time instead of uploading our template to S3 prior to creating the change set, we will upload our template (`sqs-queues_change_set.yml`) here using the **Choose file** button and selecting the YAML file that we downloaded previously from the GitHub repository. After you have uploaded the file, click the orange button labeled **Next**.

6. The next screen, **Parameters**, should confirm the parameters we entered previously when we first created our template. We are not changing any of the parameters at this time so just click the orange **Next** button at the bottom of the page.

7. On the next screen, **Configure stack options**, we do not make any changes. You can scroll down to the bottom of the page and click the orange **Next** button.

8. Now, finally, on the **Review** page, we will have to check one more option at the bottom of the page acknowledging that our new stack is going to create an IAM role needed for the Lambda function to be added. Under the **Capabilities** heading, click the check box acknowledging that this new template is creating IAM permissions. Once you have done this, you can click the orange button labeled **Create change set**.

9. A window should now pop up, giving you the option to name this change set before its actual creation. You may name the change set if you like, or just click the orange **Create change set** button at the bottom right.

10. Once the change set has been created, then you should see a status of CREATE_ PENDING as CloudFormation does the calculations of what is actually about to be changed in the current stack based on what has been requested by the new template that you have created with the change set. Once completed, then the orange **Execute** button will appear on the top right-hand menu. Scroll down and look under the **Changes** header to see the two changes that will be executed on your stack. Once you are satisfied with the changes, scroll back up and click the orange **Execute** button:

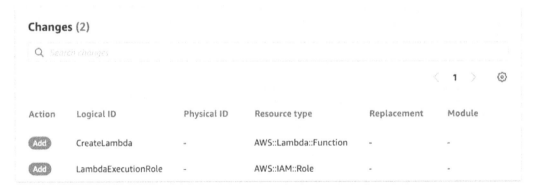

Figure 7.8 – Changes to be executed on the stack by the change set

11. You will now be taken back to the **Events** section of your SQSqueue stack and see an UPDATE_IN_PROGRESS status until the update is complete.

Just because you have created a change set doesn't mean that you have to execute it. You could have multiple change sets sitting in your **Change set** tab waiting for execution as you and your team think about which changes should be implemented. It's important to note that unless you delete a change set, then anyone with the permissions to your stack could execute it.

At this point, we have updated our original stack to add a Lambda function along with the IAM role, which the Lambda function needs in order to operate. Everything went smoothly during our operation, but what happens if we have conflicts, missing information, or errors in our template? We will find out next as we talk about the rollback features of CloudFormation.

Rollback features of CloudFormation

If the operations fail when creating or updating a stack in CloudFormation, then the stack will roll back to its previous state. There is also a feature available known as a **rollback trigger**. These triggers allow you set user-defined alarms that CloudWatch can monitor and, in the case of failure, roll back the stack.

The monitoring period can be set to last from 0 to 180 minutes while your stack is being created or your change set is updating it. We will look at an example of creating a change set update next that includes a rollback trigger.

We could create a CloudWatch alarm to watch over our stack by using the CLI command like the following:

```
aws cloudwatch put-metric-alarm --alarm-name "SQS_stack_errors" \
    --alarm-description "example alarm" --namespace "SQS_log_
errors" \
    --metric-name Errors --statistic Maximum --period 10 \
    --evaluation-periods 1 --threshold 0  \
    --treat-missing-data notBreaching  \
    --comparison-operator GreaterThanThreshold
```

Once we've created that alarm, we can use the ARN returned back to us for use in our rollback trigger. If you need to find the ARN, use the CloudWatch `describe-alarms` command:

```
aws cloudwatch describe-alarms --alarm-names "SQS_stack_errors"
```

We need to create a piece of JSON code and push it to a variable in order to pass that ARN to our command-line option:

```
RB_TRIGGER=$(cat <<EOF
{
  "RollbackTriggers": [
    {
       "Arn": "arn:aws:cloudwatch:us-east-
2:470066103307:alarm:SQS_stack_errors",
       "Type": "AWS::CloudWatch::Alarm"
    }
  ],
  "MonitoringTimeInMinutes": 5
}
EOF
)
```

This will now give us the information that we need in order to create the rollback trigger:

```
aws cloudformation create-change-set \
    --change-set-name "SQS-UPDATE" \
    --stack-name "SQSqueue" \
    --template-url "https://devopspro-beyond.s3.us-east-2.
amazonaws.com/sqs-queues_change_set.yml" \
    --change-set-type "UPDATE" \
    --parameters ParameterKey=QueueName,ParameterValue=chapter7
\
    --capabilities CAPABILITY_IAM \
    --rollback-configuration "$RB_TRIGGER"
```

Currently, only a CloudWatch alarm is supported as a monitor for a rollback trigger.

We will cover CloudWatch alarms in greater detail in *Chapter 15, CloudWatch Metrics and Amazon EventBridge.*

Intrinsic functions in CloudFormation

CloudFormation has several functions built into it that work with both JSON and YAML templates to expand the power and capabilities of the templates. We have the ability to stack or combine intrinsic functions.

> **Note**
>
> Intrinsic functions can only be used on certain sections of a CloudFormation template. As of the publishing of this book, intrinsic functions can be used in resource properties, outputs, metadata attributes, and update policy attributes.

We will now look at some of the more common intrinsic functions, along with some examples of their usage. There are additional functions available for use in the YAML templates. In the function headings, the short form, will be shown if available after the original code, separated by a pipe:

- `Fn::FindInMap | !FindInMap`

  The `FindInMap` function returns a value based on the key in the `Mappings` section:

  ```
  Mappings:
    RegionMap:
      us-east-1:
        HVM64: "ami-032930428bf1abbff"
      us-east-2:
        HVM64: "ami-027cab9a7bf0155df"
      us-west-1:
        HVM64: "ami-088c153f74339f34c"
      eu-west-1:
        HVM64: "ami-015232c01a82b847b"
      ap-southeast-1:
        HVM64: "ami-0ba35dc9caf73d1c7"
      ap-northeast-1:
        HVM64: "ami-0b2c2a754d5b4da22"
  Resources:
    EC2Instance:
      Type: "AWS::EC2::Instance"
      Properties:
  ```

```
        ImageId: !FindInMap
          - RegionMap
          - !Ref 'AWS::Region'
          - HVM64
        InstanceType: t2.small
```

- `Fn::GetAZs | !GetAZs`

  The `GetAZs` function will return the list of Availability Zones in a given region. This can be especially helpful for creating dynamic templates that move from region to region:

```
    PublicSubnet1:
        Type: AWS::EC2::Subnet
        Properties:
          AvailabilityZone:
            Fn::Select:
            - 0
            - Fn::GetAZs: ""
```

- `Fn::GetAtt`

  The `GetAtt` function is helpful for getting an attribute, especially an ARN, from a resource that was created previously in the template. This function in particular is one that makes CloudFormation templates especially dynamic:

```
    SourceSecurityGroupName: !GetAtt EC2.SourceSecurityGroup.
    GroupName
```

Condition functions

The optional conditions section contains statements that define the circumstances where resources may or may not be created or configured based on dynamic options.

Scenarios for the use of conditions in CloudFormation templates include the following:

- You are trying to use the same template for both the DEV and PROD environments without having to change anything except for possibly parameter values.

- You want to specify the number of EBS volumes to mount via a drop-down list.

- You want to create a CloudWatch dashboard or not based on user selection.

> **Note**
>
> During a stack update, you cannot update conditions by themselves. You can update conditions only when you include changes that add, modify, or delete resources.

You can use condition functions to help evaluate items such as AWS-provided variables or input gathered from parameters and then create additional resources if certain conditions apply:

- `Fn::And`

  This function returns `true` if all the specified conditions passed to it return `true`.

- `Fn::Equals`

  This function compares two different values and then returns `true` if the values are equal.

  In the following example, the template will or won't create `Public Subnets` based on a value passed in the `Parameters` section of the template:

```
Conditions:
    PublicSubnetsCondition:
        Fn::Equals: [ !Ref CreatePublicSubnets, "true" ]
    NoPublicSubnetsCondition:
        Fn::Equals: [ !Ref CreatePublicSubnets, "false" ]
```

- `Fn::If`

  This returns one value if the specified condition evaluates to `true`, and another value if the specified condition evaluates to `false`.

- `Fn::Not`

  This returns `true` for a condition that evaluates to `false`, and returns `false` for a condition that evaluates to `true`.

- `Fn::Or`

  This returns `true` if any of the specified conditions evaluate to `true`.

CloudFormation best practices

As you start to work to build out your IaC in CloudFormation templates, there are some best practices and recommendations to follow that have been set forth by AWS. These tips can help you more effectively organize and plan your resource creation, along with minimizing the time that you spend troubleshooting initial runs of your templates.

Don't embed sensitive information in your templates

Rather than placing secrets or sensitive information that could be compromised directly in a CloudFormation template, *store secrets inside of AWS Secrets Manager*. Even better, the best practice is to use a *dynamic reference*. A dynamic reference allows you to cite an external value in `Systems Manager` (SSM) Parameter Store or AWS Secrets Manager. For SSM Parameter Store, it supports both `ssm`, which are plaintext values, and `ssm-secure`, which are encrypted values.

So instead of using template-based parameters for the database username and password that you create, your RDS resource block can look like this:

```
MySQLInstance:
  Type: 'AWS::RDS::DBInstance'
  Properties:
    DBName: MyRDSInstance
    AllocatedStorage: '20'
    DBInstanceClass: db.t2.micro
    Engine: mysql
    MasterUsername:
'{{resolve:secretsmanager:MyRDSSecret:SecretString:username}}'
    MasterUserPassword:
'{{resolve:secretsmanager:MyRDSSecret:SecretString:password}}'
```

Use AWS-specific parameter types

To make AWS specialized parameter input easier on both yourself and your users, especially when asking for items that could need looking up in your account. Rather then use a string for the type of parameter you can define the AWS specific type of parameter. These could be items such as security group IDs, VPC IDs, or Route53 HostedZone IDs. A good example of this would be `AWS::EC2::KeyPair::KeyName`, which would then provide a dropdown of the EC2 keypairs available.

Make use of parameter constraints

If you are not using an AWS-specific parameter and using a string parameter, the use of parameter constraints can help catch user mistakes on parameter entry before the template even starts to create and then has to take the time to roll back. Parameter constraints are constructed like regular expressions and can also have a description for the user so that they have a better understanding of what needs to be entered.

Understand how to use AWS::CloudFormation::Init to deploy software to EC2 instances

When you launch an EC2 instance from a CloudFormation template, software running on that instance can be installed and configured using the `cfn-init` helper script and the `AWS::CloudFormation::Init` resource. These helper scripts can be used to install not only system scripts but also necessary system packages via the OS package manager, such as `apt` or `yum`.

Modularize your templates

Making your templates modular allows you to accomplish a few things at the same time. The first and most important is to make your templates reusable. This becomes especially relevant as we start talking about nested templates. The second is to allow multiple team members to concentrate on the smaller templates where they have more subject matter expertise.

Lint templates before attempting to launch

Just like any other form of code, CloudFormation templates should go through a verification process to make sure that there are no formatting issues with the template before you attempt to launch it. There is a built-in template checker right from the AWS CLI that you can use to find any issues that might be present in your templates.

The AWS CloudFormation Linter (`cfn-lint`) is an open source tool that intergrates into most IDEs and is a tool that can be run either via the command line, inside of the IDE itself, or integrated into a CI/CD pipeline. The tool will validate both JSON and YAML templates and includes additional checks for things in line with CloudFormation best practices.

> **Note**
>
> Although these best practices don't usually appear in the context of the test questions, these are more relevant for day-to-day usage of CloudFormation and making you the best DevOps professional that you can be.

Creating nested stacks with dependencies

As you start to organize your templates logically, you will find that breaking your larger templates up into smaller, more manageable pieces is a smart strategy. It will not only make them easier to work with but also allows for greater reusability with each template having a specific purpose, such as creating a VPC with subnets or a set of reusable IAM roles. Adding outputs to these smaller templates can allow other templates to use the resources other templates have created previously, acting like building blocks stacked on top of each other to create a complete solution. This also allows multiple team members to work on separate sections of the cloud infrastructure as each person can concentrate on their area of expertise. If you have a team member whose strength is networking then they can concentrate on the VPC and networking pieces of the stacks. If you have another team member who has a background in databases, then they can concentrate on the nested piece that creates and configures the database(s), including any specialized parameter groups for that particular database engine.

Nested stacks all start with a root or parent stack, depicted in *Figure 7.9* with the letter **A**. The root stack then has other child stacks as its resources instead of other AWS services. We can also see in the diagram that the stack labeled **B** is providing outputs that are being consumed by level **C**. The stack on level **C** creates another child stack, as well as resources whose outputs are consumed by the stack on level **D**:

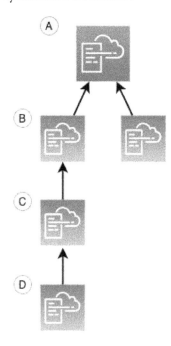

Figure 7.9 – CloudFormation nested stacks

Another benefit that nested stacks can provide is the ability to go beyond the 200 resource limit for CloudFormation templates.

> **Note**
>
> Nested stacks can be more complex to troubleshoot as a whole. If you run into errors while creating and deploying nested CloudFormation stacks, then try deploying just the single template that is causing you issues and then update for your changes.

Packaging up your nested stack for deployment

Once you have all your templates ready to upload to S3, then you can package them all together using the AWS CLI. Since nested stacks need their child templates to be in S3 for deployment, this saves you time in having to upload each template individually. It will then generate a new template for us and give us the command for deploying the full nested stack using that newly generated template. The generic code for creating a CloudFormation package looks like the following:

```
$aws cloudformation package \
  --template-file /path_to_template/template.json \
  --s3-bucket bucket-name \
  --output-template-file packaged-template.json
```

We will look at this command in greater depth when we go through our exercise of creating a nested template with the AWS CLI.

Let's recap what the `package` command does:

- It creates a ZIP file with all of your child template files and extra code such as `lambda` code.

- It uploads those items into the S3 bucket designated by you.

- It generates a new template with your local templates and replaces it with the S3 URIs.

Creating a nested stack using the AWS CLI

In the following exercise, we will create a nested stack using a set of templates that you can find on the GitHub repository under the `Chapter-7/nested` directory. Download all of these YAML templates before you begin the following exercise. Also, make a note of the S3 bucket that you want to deploy the package to. As we noted earlier when we showed the example command, an S3 bucket is necessary as part of the CLI command to upload the files:

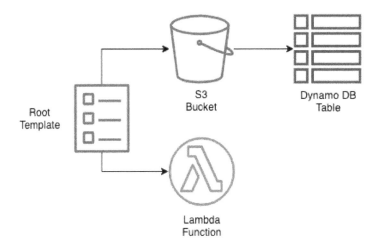

Figure 7.10 – Items created in the nested stack exercise

The diagram in *Figure 7.10* shows what we are about to package and deploy using our nested template:

1. Open up your terminal so that you have access to your AWS CLI.

2. Now navigate to the folder where you have downloaded all of the files to create the nested stack. If you have downloaded the book's entire GitHub repository, then the path will be `Chapter-7/nested`.

3. Because the packaging for nested stacks only works one level deep, we will have to manually upload our `nested_dynamo.yml` template (make sure that you substitute your own bucket name when performing the following command):

```
aws s3 cp nested_dynamo.yml s3://devopspro-beyond/
```

4. Open the `nested_root.yml` file and edit the default value for the S3 URL for the HTTPS value of the bucket where you placed the `nested_dynamo.yml` file.

5. Once in the nested directory, perform the following command:

```
aws cloudformation package \
   --template-file nested_root.yml \
   --s3-bucket devopspro-beyond \
   --output-template-file packaged_template.yml
```

6. Once completed, the `package` command should tell you that it was successful and give you a cut and paste command to run, where you need to substitute the stack name at the end of the command:

    Successfully packaged artifacts and wrote output template to file `packaged_template.yml`.

    Execute the following command to deploy the packaged template:

```
aws cloudformation deploy --template-file /AWS-Certified-
DevOps-Engineer-Professional-Certification-and-Beyond/
Chapter-7/nested/packaged_template.yml --stack-name <YOUR
STACK NAME>
```

7. Run the command to create your nested stack, but be sure to change the stack name and to add the flag for `--capability CAPABILITY_IAM`:

```
aws cloudformation deploy --template-file /AWS-Certified-
DevOps-Engineer-Professional-Certification-and-Beyond/
Chapter-7/nested/packaged_template.yml --stack-name
Chapt7 --capabilities CAPABILITY_IAM
```

8. Log in to the AWS Console and go to the CloudFormation service. At this point, you should be able to see both your root stack and the nested stacks that you have created. By clicking on the **Resources** tab, you can see all of the resources that you just created.

9. If you don't want to incur charges then delete the stack.

Using DependsOn to order resources

Although CloudFormation creates the resources in the order you specify in your template, it does not wait for the completion of any one resource before starting to build the next resource unless specifically instructed. The DependsOn construct allows you to pause the creation of specific resources until other resources have been completed.

There are a number of situations in which you would want to invoke DependsOn. The first set of scenarios involve those resources that need access to the internet, and hence they require an internet gateway to be completed before moving forward.

Adding wait conditions to a template

A WaitCondition adds a pause to the tasks that your template is performing on your stack until a success signal has been received, indicating readiness to move on.

Whenever a WaitCondition is added as a resource to a CloudFormation template, you must couple this with a WaitConditionHandle resource.

> **Note**
>
> Even though the DependsOn construct and WaitCondition resource seem similar in function, they are different in a few ways. First is the fact that DependsOn is a much simpler implementation as it doesn't require a helper script. DependsOn also doesn't check for a success or failure signal for a resource once it has been created and therefore only governs the order of items produced in the template. A WaitCondition, on the other hand, needs to explicitly receive a success signal and will make the template (or change set) pause until this signal is received.

Using curl to signal

When using curl to signal to the WaitCondition that the resource, in most cases an EC2 instance, is done with its configuration, then it can be done dynamically. The first step is to create the WaitHandle, and then to construct the URL that the CloudFormation will use in the curl command. At that point, we are ready to call the curl command from within the UserData section of the template. We can see an example of this in an abbreviated template snippet, as shown next:

```
ServerWaitCondition:
    Type: 'AWS::CloudFormation::WaitCondition'
    DependsOn: Server
```

```
    Properties:
      Handle: !Ref ServerWaitHandle
      Timeout: '1200'
...
      UserData: !Base64
        'Fn::Join':
          - ''
          - - |
              #!/bin/bash -v
            - |
              # Send Wait Condition URL
            - '/usr/local/bin/aws s3 cp s3://'
            - !Ref BucketFolder
            - |
              /templates/success.json /tmp/a
            - SignalURL=
            - !Ref ServerWaitHandle
            - |+

            - |
              echo $SignalURL
            - curl -T /tmp/a "
            - !Ref AdminServerWaitHandle
            - |
              "
```

The WaitCondition is the signal that is used in concert with one of the few helper scripts available in CloudFormation. Next, we will look at the different helper scripts available in CloudFormation templates.

Adding a helper script to a CloudFormation template

In CloudFormation, there are a few different helper scripts, written in Python, that allow you to install software and start services when requesting and configuring EC2 instances as part of a template:

Helper script	Script function
`cfn-init`	This is the script used to install packages and software on an instance, along with creating files, starting services, and retrieving metadata.
`cfn-signal`	This script is used in concert with a `WaitCondition` or `CreatePolicy` in order to coordinate when items are ready.
`cfn-get-metadata`	This script is used to retrieve metadata and especially retrieve a specific key value.
`cfn-hup`	This script is used to check for updates in the metadata and can then execute updates to resources via hooks when changes are detected.

Table 7.1 – CloudFormation helper scripts

As you can see, using these helper scripts provided by CloudFormation can greatly simplify EC2 configuration. Next, we will see how to detect drift in CloudFormation templates if your infrastructure deviates from your template code.

Understanding how to detect drift in CloudFormation templates

CloudFormation templates allow you to create and manage your infrastructure and resources as code in the various AWS accounts that you oversee. Following this method of provisioning items as code that can be checked in and out of version control is a best practice since it is repeatable, rather than trying to build and maintain cloud assets by hand.

Can you stop people from altering the assets you have provisioned in this manner? Unless you have used IAM policies to prevent groups from modifying any resources other than those submitted through CloudFormation templates or via code deployment pipelines, then there is a chance that there could be drift in some of the resources spun up by your CloudFormation templates.

When you initiate a check for drift detection, the CloudFormation service compares the current stack and the resources that are currently provisioned versus what was originally specified in the template used to create or update that particular stack. It then reports any differences that it finds.

With a good understanding of CloudFormation templates under our belt, we will now look at a complimentary service – AWS Service Catalog. Service Catalog is another tool in the AWS management toolbox that allows your users to quickly and easily provision resources that have been pre-created with a set guardrails around them.

Managing templates with Service Catalog

As we continue the conversation about CloudFormation templates, we start to look at other ways to manage the templates in an organization. The Service Catalog product from AWS allows us to use our CloudFormation templates and create a self-service portal for users to provision known patterns of items as long as they have been given the appropriate access. These templates, which now become products in our Service Catalog, can even become parameterized so that our user can choose items such as a vetted, predefined set of EC2 instance sizes and AMIs in the case of EC2.

First, you need to understand a few concepts regarding Service Catalog. The first concept is a product. A **product** is an IT service that you want to make available on AWS. This could be something as simple as a single S3 bucket or as complex as an EC2 instance backed by an RDS database along with predefined CloudWatch alerts. The next concept to understand in Service Catalog is a portfolio. **Portfolios** are groupings of specific products along with configuration information. Portfolios are also tied to particular user groups, and give those user groups access to launch the products.

Looking at *Figure 7.11*, we can see the two different workflows that are available in Service Catalog from both an administrator's and user's perspective. The top view shows how an administrator would load up a template containing an EMR cluster, an S3 bucket, and CloudWatch alarms for repeated use as a product for the data users' group.

Any user in the data user's group can then quickly provision not only the EMR cluster but a corresponding S3 bucket along with relevant S3 alarms after entering a few pieces of relevant information into the Service Catalog screens:

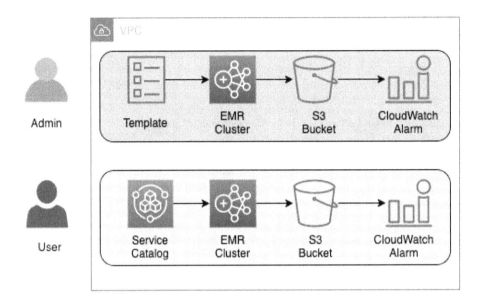

Figure 7.11 – Service Catalog workflows

Service Catalog also allows you to update the templates containing your products and release newer versions with updated features. This can give your users the choice of which version to use: a version they are familiar with, or a version with newer features.

There are quite a number of situations where this can become useful:

- When a solutions architect needs to be able to quickly provision a demo environment for a customer sales call.

- The QA team would like to stand up a QA environment for testing.

- In a marketing department, where they would like a promotional application to run for a specific period of time.

- A data scientist who needs a set of specialized software provisioned, such as an EMR cluster or a server with R Studio, but doesn't have the background in system administration to install and configure all of the necessary software and secure it correctly.

Access control in Service Catalog is handled via IAM. The Service Catalog administrators create IAM roles specific to the products in the catalog so that users have just enough access to run the services provisioned by the catalog.

There are a number of different types of constraints that you can place on products in your Service Catalog in order to apply governance.

There are template-based constraints that can shrink the number of options available to users when launching a product. Examples of this include the sizes of an EC2 instance or RDS instance or the languages allowed for a Lambda product.

Tag-based constraints either enforce specific tags that need to be filled in when launching a product or disallow users to use any additional tagging besides those tags that have already been predefined.

You can specify a particular SNS topic to receive updates about a product using notification constraints.

Defining which IAM roles can be used to run a particular product and be controlled via a launch constraint. This provides Service Catalog administrators an extra set of controls to say which services are allowed to be governed by the Service Catalog product.

We just discovered how Service Catalog can allow even non-developers to quickly spin up predefined patterns of AWS infrastructure for users. These users don't need to worry about how to configure the underlying resources since all of that is handled by the underlying CloudFormation templates.

We will next examine another way of creating CloudFormation templates that allows us much more flexibility then only using JSON or YAML in the Cloud Development Kit.

Using the Cloud Development Kit

Developers are used to creating reusable libraries and using things like loops to take care of repetitive tasks. The **Cloud Development Kit** (**CDK**) allows those with a programming background to use a variety of languages (TypeScript, JavaScript, Python, Java, and C#) to create CloudFormation templates using techniques that they know, such as the following:

- Logic (if statements, for loops)
- Object-oriented techniques
- Organization via logical modules

This is in contrast to CloudFormation templates, which require you, your developers, or DevOps personnel to write them in either JSON or YAML, although both of these options are somewhat universal in nature and can be picked up in a short time with or without a programming background. CloudFormation templates are also quite limiting in the number of programmatic options and logic that you can perform while creating your stacks:

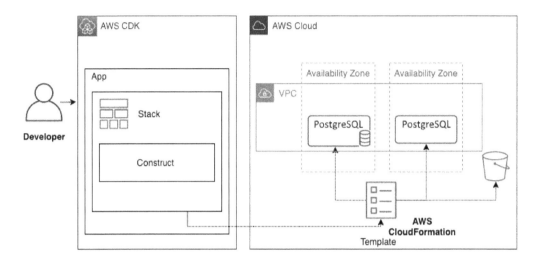

Figure 7.12 – CDK workflow

In *Figure 7.12*, you can see the workflow of how a developer would create an app using the CDK. This app then creates a stack that deploys the CloudFormation template to the AWS cloud environment.

Concepts of the AWS CDK

Inside of the AWS CDK, there are three essential components – apps, stacks, and constructs – that developers can use to create cloud services.

Apps

With the AWS CDK, you are building an application, which in turn is an app and is composed of the CDK App class.

Stacks

Units of deployment in the AWS CDK are stacks, and all resources defined within the scope of a stack are provisioned as a single unit. Since CDK stacks are implemented as CloudFormation stacks, any boundaries or restrictions that CloudFormation stacks need to abide by also are present for CDK stacks.

Constructs

The building blocks of the AWS CDK are constructs. These can be a single AWS resource like a load balancer, or a single component consisting of multiple resources such as a VPC and subnets. Constructs create reusable components that can be shared like any other piece of code.

Advantages of using the AWS CDK

We will examine some of the advantages of using the CDK over normal CloudFormation templates.

Faster development process

With the AWS CDK you have the ability to use the languages that you natively program in, such as Python, TypeScript, .NET, Go and Java. In using these languages you retain the ability to create object, loops, and conditions that you are familiar with rather than having to learn domain-specific functions and work-arounds.

Code completion within your IDE

Using popular IDEs such as Visual Studio Code allows for code completion using the language of your choice when programming with the AWS CDK.

The ability to deploy code and IaC synchronously

Since the AWS CDK uses the same native language that you write your code in, combining pieces of infrastructure along with the code that would run your application becomes easy without any context switching.

Summary

In this chapter, we looked at how to create IaC using both CloudFormation templates and the CDK. We examined some best practices of template construction and organization. We also practiced hands-on implementation of deploying and updating CloudFormation templates into stacks using both the AWS Management Console and the CLI.

In the next chapter, we will begin our examination of some of the AWS code tools that are part of the SDLC process, starting with CodeCommit and CodeBuild. We will create a code repository and commit to it using CodeCommit and then start building out code using AWS CodeBuild. Finally, we will see how we can tie both services together by triggering the CodeBuild job from pushes to our CodeCommit repository.

Review questions

1. What are the two major components of CloudFormation?

2. You have been contracted by a company to help review and optimize their CI/CD processes, especially in the area of IaC. The company currently has a single CloudFormation template, which they have developed over time, that is used to create the IAM account roles, VPCs, subnets, Lambda functions, CloudWatch alarms, along with SNS topics and other resources, including databases and EKS EC2 instances and their corresponding AutoScaling groups. How would you go about creating a recommendation for optimizing their process?

   a. Check the current template to make sure that it has the correct `DependsOn` and `WaitCondition` embedded so that all resources come up without conflicts.

   b. Create a plan on how to break up their large template.

   c. Use the CloudFormation `package` command to package all of the templates together.

   d. Take the smaller templates and launch with a single `deploy` command.

3. After assessing a client's needs for a repeatable three-tier web application, you have decided to build a CloudFormation template that you can hand over at the end of the engagement. The middleware Linux servers have an elaborate user data script that takes a while to run in order to install completely. These servers are sitting behind a Network Load Balancer resource and need to be fully operational before being added. How can you make sure that the servers are fully operational and configured correctly before you attach them to the Network Load Balancer resource? (Choose two.)

   a. Make sure that the Network Load Balancer resource is located directly after the middleware servers in the CloudFormation resources section of the template.

   b. Add a `WaitCondition` that depends on the middleware servers. Once the user data finishes, it uses `cfn-signal` to notify that it is ready.

   c. Launch everything from a nested template where the middleware servers are in a template launched from the Load Balancer template.

   d. Add a `WaitCondition` that depends on the middleware servers. Once the user data finishes configuring the middleware servers, it uses a `curl` command to notify a pre-signed URL that they are ready.

4. You are working at a large enterprise company and have a number of projects in the backlog that are waiting to be deployed to different AWS cloud environments. In order to minimize the amount of time it takes for each deployment, you want to create a reusable set of components that can be easily swapped out based on the architecture of each project. There is already a basic CodePipeline service instance in place to add the test and deployment steps. Whatever solution you choose should be able to be checked into a code version control system and be tested easily. How should you organize your resources for optimal reusability?

a. Use the AWS CDK to create apps made of a shared library of components. Add a testing library and then run the tests as a step in CodePipeline before deploying to the development environment.

b. Create a library of small resource-specific CloudFormation templates that can be easily nested and packaged up according to the architecture guidelines. Use the `yamllint` and `cfn-lint` commands to lint the CloudFormation templates as a test step in CodeBuild to catch any errors.

c. Develop a foundation of 5-10 blueprint patterns that all projects fall into. Using these blueprints, develop a library of CloudFormation templates with parameters and conditions so that they meet the majority of your project's needs. Use the `cfn-lint` command to lint the CloudFormation templates as a test step in CodeBuild to catch any errors.

d. Create CloudFormation templates for each solution that have the necessary parameter values and conditions so that they work seamlessly on any Region and can be used in any environment (DEV, TEST, and PROD). Upload the templates to CodeCommit for version control.

Review answers

1. Templates and stacks

2. b

3. b and d

4. a

8
Creating Workloads with CodeCommit and CodeBuild

AWS has come up with an admirable set of tools to help developers and organizations run their CI/CD operations entirely from the Amazon ecosystem.

Continuous integration starts with the process of source code versioning and then building and testing your code in a repeatable, automated fashion. Two of the tools provided by the **AWS Codes Services Suite**, **CodeCommit** and **CodeBuild**, help engineers achieve these goals.

In this chapter, we're going to cover the following main topics:

- Using CodeCommit for code versioning
- Setting up your CodeCommit repository
- Approvals in CodeCommit
- Using AWS CodeBuild
- Creating a `buildspec` file

Technical requirements

As we start on this path, a few assumptions are going to be made, especially with pursuing the professional DevOps certification. The first is that you have a basic knowledge of using the software versioning system known as Git. This basic knowledge includes creating a bare repository, creating a branch, making a commit to that branch, and then pushing that commit to a remote repository. If you need to get up to speed with Git, then using the tutorial *Getting Started with Git and AWS CodeCommit* is suggested. You can find it at `https://docs.aws.amazon.com/codecommit/latest/userguide/getting-started.html`.

The second assumption is that you already have Git installed on your laptop or workstation. Suppose you don't have Git installed yet or cannot access it due to a lack of administrative permissions. In that case, a perfect alternative is to spin up an Amazon EC2 instance in your AWS account that you can install Git on so that you can follow along with the exercises.

If Git is not installed on your local workstation and you either cannot install it or would not like to install it, AWS provides the Cloud9 development environment, which allows you to create code inside of a browser window.

Using CodeCommit for code versioning

There are plenty of choices regarding where you can store your code, both on-premises and within SaaS solutions. Features and values are what tend to drive decisions on which product to use. AWS CodeCommit provides both a robust set of features and the value of being a pay-per-use service, containing elements such as native connectivity to AWS services, along with them complying with multiple assurance programs such as SOC2, PCI, HIPAA, and others.

In *Chapter 6*, *Understanding CI/CD and the SDLC*, we discussed the four major phases of the software development life cycle: source, build, test, and deploy. The initial phase, source, is what we will be concentrating on in this section. As a quick recap, the source phase allows you to check your code into a central repository and also allows multiple people to collaborate on the same code base.

What is CodeCommit?

CodeCommit is a secure, highly scalable, managed source control service that hosts private Git repositories. This may sound a bit foreign initially, but you have probably used a SaaS service to host code already if you have used BitBucket or GitHub private repositories.

The fundamental object in CodeCommit is the repository. The repository is where users store code and objects of any type and size for their projects. CodeCommit also stores and tracks changes that have been made to uploaded files, along with changes that have been made to files. You can configure the repository to send notifications when events happen, such as branch merges or comments being added to code. It also allows users to work on their local systems and then push changes to the CodeCommit storage system.

Benefits of CodeCommit

As part of the AWS ecosystem, CodeCommit can integrate with services such as KMS for encryption, as well as CloudWatch to incorporate metrics and alarms for our repositories, and this doesn't even touch on some of the development tools that are tightly integrated with the CodeCommit service. Aside from just being tightly coupled with numerous other AWS services, CodeCommit provides many other benefits:

- IAM for CodeCommit provides high service availability and durability.
- CodeCommit repositories are encrypted at rest and in transit, so you know that your code is always secure.
- It scales easily and has no limits on the size of repositories or the types and sizes of the files you can store.
- Seamlessly integrates with several AWS services (Lambda, SNS, CodePipeline).
- Allows you to migrate from other remote Git-based repositories easily.
- Allows you to use the Git tools and commands that you already know.

With this overview and the benefits of AWS CodeCommit under our belt, we will look at the available access control features before creating our repository and our new group specifically for CodeCommit access.

Controlling access to repositories

Before we go through the process of creating the CodeCommit repository, we need to do a little pre-work as far as permissions are concerned. Up until this point, we have been doing most things as an administrative user. We are starting to add developers and development teams into the mix, some of which only need access to the CodeCommit repositories and no other services. We are going to create a new IAM group and then add a developer entity to that group. This will demonstrate good GitFlow practices of the developer contributing code and then asking for their branch to be merged into the main branch.

This works when the developers are all part of the same account where we are creating our CodeCommit repositories. There are other scenarios that we need to be aware of as well. For example, what happens when a developer in another account needs access to CodeCommit repositories in our account? CodeCommit can provide cross-account access to users and roles in another AWS account. This can solve the issue of not having to add external developers to your IAM list of users if they already have an AWS account.

Setting up your CodeCommit repository

A CodeCommit repository can be created via the CLI. However, when using the AWS management console, there are some additional features, such as the info panel, that are only available in that environment.

Before we create the repository, be sure to open your browser to the AWS CodeCommit home page, `https://console.aws.amazon.com/codesuite/codecommit/home`, and sign in if prompted.

Once you have signed in, we can go through the process of creating a repository in CodeCommit:

1. Once you're on the `CodeCommit` main page for the region you are working in (we are doing everything in our examples in the Ohio region/`us-east-2`), click on the orange **Create repository** button in the top right-hand corner to start the process of creating a new repository:

Figure 8.1 – Create repository button

2. On the **Create repository** screen, under **Repository settings**, use `chapter8` for the name of the repository. If you like, you can add a description for the repository, but it is not necessary. You may wish to enable the CodeGuru reviewer at this point; however, since this is a new feature, we will not be going over this:

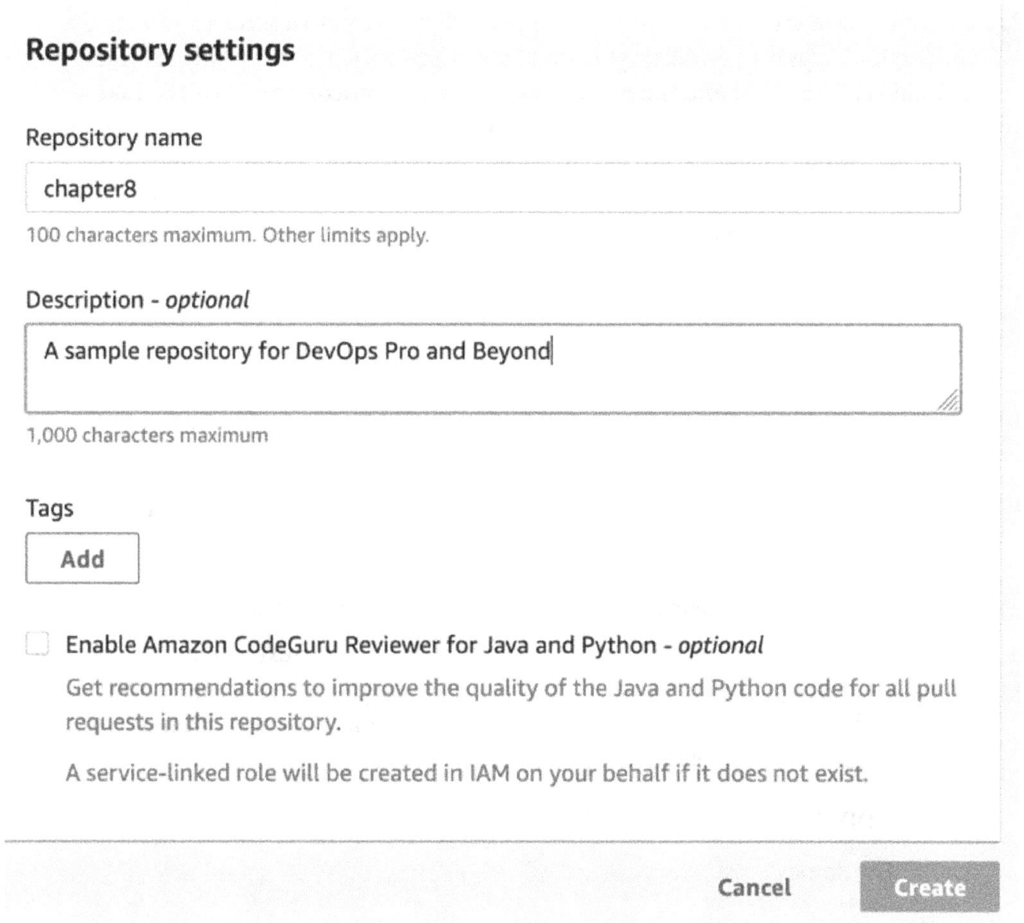

Repository settings

Repository name

chapter8

100 characters maximum. Other limits apply.

Description - *optional*

A sample repository for DevOps Pro and Beyond

1,000 characters maximum

Tags

Add

☐ **Enable Amazon CodeGuru Reviewer for Java and Python** - *optional*

Get recommendations to improve the quality of the Java and Python code for all pull requests in this repository.

A service-linked role will be created in IAM on your behalf if it does not exist.

Cancel   Create

Figure 8.2 – New code commit repository settings

> **Note**
> The Amazon CodeGuru reviewer is an optional feature and an additional service of AWS that, at the time of publishing, only works for the Java and Python languages. More languages may be available at a later time.

3. Once you have created your repository, you will be taken to a screen showing the connection steps for connecting to your new repository in several ways. These include HTTPS, SSH, and the git-remote-connect protocol over HTTPS. Take a moment to look over these. We will be using this section more once our developer has been created and we are using their set of permissions. Scroll down past the connection steps. At the bottom of the page, you will see a button called **Create file**. Click this button so that we can set up our main branch:

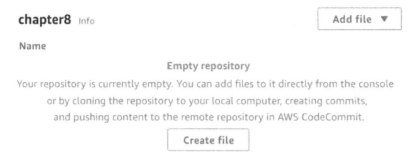

Figure 8.3 – Creating an initial file for our CodeCommit repository

4. On the **Create a file** screen, type some text into the box. Any text will do; for example, Welcome to my repository. Once you have done this, in the **File name** box, name the file sample.txt. You will also have to enter a name and an **Email address** for the commit on the web interface, as shown in the following screenshot:

Commit changes to main

File name
For example, file.txt

sample.txt

chapter8/sample.txt

Author name

Example

Email address

admin@example.com

Commit message - *optional*
A default commit message will be used if you do not provide one.

Figure 8.4 – Committing the change to CodeCommit via the web interface

5. Once you've entered this information, you can press the orange **Commit changes** button. From here, you will be taken to the file in the repository; however, it's more important to note that, in the top right-hand corner of the screen, we have successfully created the **main** branch:

Figure 8.5 – The main branch of our CodeCommit repository

With our repository and main branch now created, we can start creating the permission set that our developers will use to push code up to the repository. Depending on how your organization is structured, you may give all developers the ability to create new repositories and merge pull requests to the main branch.

In our example scenario, as we work through the exercises, we will be separating the duties from what a developer can do versus an administrator or DevOps power user. It's imperative to think about how to divide responsibilities as you prepare for the exam, as well as some of the questions and scenarios that may be presented.

Creating your IAM group for developers

Before we begin, make sure that you download the files from this book's GitHub repository, from the Chapter-8 folder. We will start by creating the IAM policy for the developers. Once our policy has been uploaded, we will create the developer group and then attach our policy to the newly created group:

1. Open your terminal so that you have access to your AWS CLI.

2. We want to create a group just for our code commit developers. You may have made a developers group previously, but we want to make a new group just for this chapter:

```
aws iam create-group --group-name CC_Developers
```

3. Now that you have created the group, we will attach the policy to it. We will need the policy ARN from when we created our CC_Developers policy just a few moments ago:

```
aws iam attach-group-policy
   --policy-arn arn:aws:iam::aws:policy/
AWSCodeCommitPowerUser -group-name CC_developers
```

4. Now that we have created the `CC_Developers` group, we can create our developer and allow them to log in and add their SSH key to start using `CodeCommit`.

Creating your developer

In this section, we are going to create a fictional developer called `Mariel`. If you like, you can change the name of the developer from `Mariel` to something else. If you haven't already, download the `change-password.json` file and follow these steps:

1. If your terminal isn't open still, open it back up so that you can perform commands in the AWS CLI. Use the `create-user` command from `iam`:

    ```
    $aws iam create-user --user-name mariel
    ```

2. Once this command has finished running, it should return a JSON statement that shows, among other things, the User ID, as shown in the following example:

    ```
    {
        "User": {
            "Path": "/",
            "UserName": "mariel",
            "UserId": "AIDAW24Q7QQFVGQHQJM3Y",
            "Arn": "arn:aws:iam::000066100007:user/mariel",
            "CreateDate": "2021-05-02T11:06:47+00:00"
        }
    }
    ```

3. With our user created, we can attach them to the `CC_Developers` group and allow them to inherit all the group's permissions:

    ```
    $aws iam add-user-to-group --user-name mariel --group-
    name CC_Developers
    ```

4. Now that we've created our user and attached them to the `CC_Developers` group, we will need to create an initial password for the user so that they can log into the management console and upload their SSH key. It is also a good idea to require them to reset their password:

    ```
    aws iam create-login-profile --user-name mariel
    --password DevOpsPRO123! --password-reset-required
    ```

With that, we have created an initial password for our user to configure their SSH key settings in the console and view the CodeCommit branches via the console.

Now that we've created our developer using the AWS CLI, we can switch contexts and assume the role of the developer as we continue.

Adding your developer's SSH key

Now that we have created our developer, we will switch from the CLI and move to the **Management Console**. Before we do this, however, we must ensure that we have an SSH key pair in our working environment that we can use to authenticate ourselves when creating code.

> **Tip**
>
> When using a specific repository or set of repositories for an account, client, or specific purpose, it's a good idea to create a particular key just for that project, account, or client. Using distinct keys is part of an overall risk mitigation strategy in case anything were to happen to the servers, SaaS offerings, or your workstation.

Let's start by creating an SSH key pair that's specific for our developer to be able to add to the AWS IAM console.

In your terminal, use the following command to generate a new key pair file for Linux or macOS users:

```
$ ssh-keygen -t rsa -b 4096
```

Running this command will start by asking you a few prompts, the first of which will be where you want to save your key. The default should be in your local .ssh folder, but with an id_rsa name. Instead, we will use a custom name for our key so that we know it is specifically for our code commit project:

```
Enter file in which to save the key (/Users/abook/.ssh/id_rsa):
/Users/abook/.ssh/cc_developer
```

After you have chosen where to save you key with its name, then you will be presented with a prompt for a passphrase:

```
Enter passphrase (empty for no passphrase):
```

We have no requirements for a passphrase, so just hit *Enter* twice to leave this blank. At this point, you should see statements saying where your identity and public key have been saved, along with something that looks like ASCII art.

You will need to cat out your public key and have that available when we log into the AWS console as the developer.

After you have generated your key

Now that we have generated our key, we need to log into the AWS console using the developer's username and credentials that we just created (and not the DevOps user who has administrator privileges, which is what we have been using up to this point.)

> Tip
> You may want to use a different browser or start an incognito/private session, just to be sure that you don't have any lingering cookies.

In your new browser, log in as the developer user that we just created:

1. We will now use the developer account instead of our administrator account. Log into your account at `https://aws.amazon.com/` and use the developer's name (`mariel`, in our case) and password (`DevOpsPRO`) to sign into the AWS Management Console initially.

2. Once signed in, you will be immediately prompted to change your password from the initial password that was set to one of your choosing. If you previously put a password policy in the **IAM Account** settings, you must adhere to these guidelines when changing the password:

Figure 8.6 – Mandatory password change for the developer user

3. Once you have changed the password, you will be brought to the main page of the AWS management console. Now, we need to type **IAM** in the search box to go to the **IAM service**.

4. Once you have reached the IAM service, it will look as if you don't have permission to do anything. However, this is only because the developer role has a limited scope of permissions. This user can list users and update information about their user. Click on the **Users** menu item on the left-hand menu.

5. Once you're on the user's menu, you will see a list of users. Click on the name of the developer that we created (in our case, `mariel`):

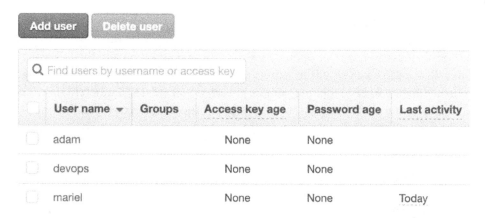

Figure 8.7 – List of IAM users

6. On the **Summary** page for the user, you will be met by a few red boxes, but here, we are interested in the **Security credentials** tab. Click on this tab so that you can start modifying the **Security credentials** page:

Figure 8.8 – IAM user summary top menu

7. Scrolling down on the **Security credentials** page, you will come to SSH **Keys for AWS CodeCommit**. Underneath this header will be a button we need to press labeled **Upload** SSH **public key**. Before pressing this button, make sure that you have the output of your `public key` not only available but in your clipboard, ready to paste. Press the **Upload** SSH **public key** button to continue.

8. When the pop-up window appears, paste your public key into the text box and then click the blue **Upload** SSH **public key** button. Once your key has been uploaded and the pop-up has disappeared, you should see a unique identifier for your key under the SSH **key ID** listing. Take note of this as we need this ID in just a moment for our local setup:

SSH key ID	Uploaded	Status
APKAW24Q7QQFSRFEETDF Show SSH key	2021-05-02 21:55 EDT	Active

Figure 8.9 – The SSH key ID is given in the IAM console once the SSH public key has been uploaded

9. Back in your workstation, we need to either create a .ssh/config file or modify the existing config file by adding a few lines specific to the code commit. Use your editor of choice to create or open ~/.ssh/config and then add the following lines. Be sure to change out the sample user ID for the one that the AWS console returned to you after uploading your key. Also, if you created a different name for your SSH key, you will have to substitute that in the IdentityFile line:

```
Host git-codecommit.*.amazonaws.com
User APKAW24Q7QQFSRFEETDF
IdentityFile ~/.ssh/cc_developer
```

10. Going back to the browser and the AWS Management console, we will now switch over to the CodeCommit service so that we can get the repository information to clone the repository. In the top search box, type CodeCommit and then click on the icon to be taken to the **CodeCommit** service page.

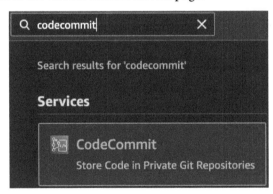

Figure 8.10 – The CodeCommit icon from the search bar in the AWS Management Console

11. Once on the **CodeCommit** page, you should see the **chapter8** repository that we created previously. Click on the SSH link on the right-hand side to copy the **Clone URL** property:

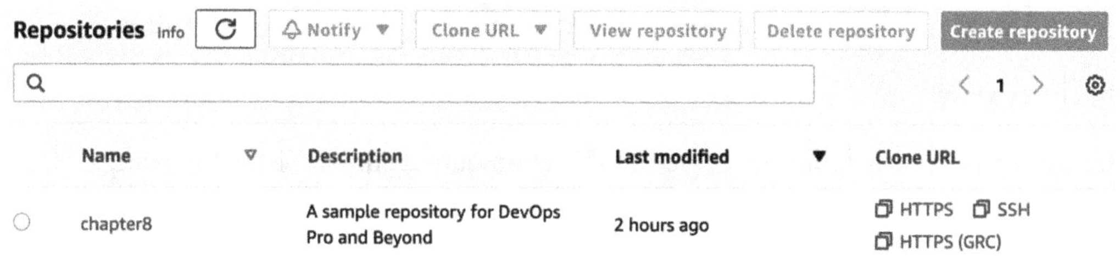

Name	▽	Description	Last modified	▼	Clone URL

Repositories Info ⟳   ⌂ Notify ▼   Clone URL ▼   View repository   Delete repository   **Create repository**

🔍 _____   ‹ 1 › ⚙

Name ▽	Description	Last modified ▼	Clone URL
○ chapter8	A sample repository for DevOps Pro and Beyond	2 hours ago	📋 HTTPS 📋 SSH 📋 HTTPS (GRC)

Figure 8.11 - The CodeCommit repository as it appears to the developer user

12. Switch back to your terminal and navigate to the root of your home folder. Then, cut and paste the copied URL after the `git clone` command, as shown here:

```
$cd ~
$git clone ssh://git-codecommit.us-east-2.amazonaws.com/
v1/repos/chapter8
```

13. Finally, move to the `chapter8` directory. Now, you are ready to start adding some code or objects.

At this point, we have created an SSH key and added that to our development user. Then, we updated our local SSH config to let CodeCommit know that when working with code commit, which user and keypair are connected. Finally, we downloaded the repository that the administrative user had created earlier. We are now ready to create our developer branch and start pushing up some commits.

Creating a developer branch and pushing commits to that branch

Now that we have successfully cloned our repository to our local workstation, it's time to create a branch and submit it. All of these actions will happen in the directory where you cloned your repository. If you followed the same naming convention as in the preceding examples, you should go to your home directory and find the `chapter8` folder.

The first thing that we will do is create a feature branch off of the main branch:

```
$git checkout -b feature1
Switched to a new branch feature 1
```

Now that we're in our branch, we can start adding some files to make the commit. There are two sample files (`hello.py` and `loops.py`) in the `Chapter-8` folder of the GitHub repository. You can view these files and then cut and paste the raw contents into new files of the same name in the `chapter8` CodeCommit repository that we have downloaded.

Before we make our commits, we should check the status of our files in our directory:

```
$git status
```

After using the status command, we should see an output similar to the following:

```
On branch feature1
No commits yet
Untracked files:
   (use "git add <file>..." to include in what will be
committed)
  hello.py
  loops.py
nothing added to commit but untracked files present (use "git
add" to track)
```

Now, we need to add the two files to the commit. Since they are both of the same types, we can use a wildcard in our commit command. As an alternative, you can use just a pure wildcard and add all the untracked files to the commit at once:

```
$git add *.py
```

At this point, we need to commit the files to our feature branch:

```
$git commit -m "adding python files for feature 1"
```

Finally, with the two files committed to the feature branch, it's time to push that feature branch up to CodeCommit:

```
$git push origin feature1
```

Upon successfully pushing the new feature branch, you should receive confirmation from the command line:

```
* [new branch]        feature1 -> feature1
```

Now that we have made the commit to our feature branch and pushed it up to the CodeCommit repository, it's time to make the merge request. This merge request will allow our code to become part of the main branch for everyone else who is working off of the same code repository, along with keeping up to date with the main branch.

We will go back to the AWS console as our developer user to create the pull request. If you still have your developer's separate session open in a web browser, then you can switch back to that browser session; otherwise, log into your account at https://aws. amazon.com/ and then use the developer's username (mariel, in our case) and the new password you changed for your developer user:

1. In the search box at the top, type CodeCommit and then click on the **CodeCommit** icon to be brought to the **Service** page.

2. You should see your repository, as shown in *Figure 8.7*. Click on the name of the repository (chapter8).

3. In the top menu, you will see a dropdown that currently says **main**. Click on this dropdown and select your branch (**feature1**). Once you've done this, the two files that you have uploaded will appear.

4. Now that your branch name is showing in the top menu, you should see a button called **Create pull request**. Click on this button to start the merge request:

Figure 8.12 – The Create pull request button alongside the branch name

5. A green box should appear, showing that we don't currently have any conflicts between our branch and the main branch. Before you create the merge request, you will need to add something to the **Title** box in the **Details** section. It can be as simple as Python files. With the title filled in, click on the orange **Create pull request** button at the bottom of the screen:

Figure 8.13 – The pull request showing that it has no conflicts

6. Once you have created the pull request, you will be taken to a screen where you can view the details of the currently opened pull request.

Now that we have entered the pull request, this is where the developer user stops. The IAM policy that we have allocated will not allow them to merge the branches, and this is by design.

Next, using our administrator, we will learn how to merge the request into the main branch.

Approvals in CodeCommit

As developers update code and then push those updates to the CodeCommit repository, especially if they are using their feature branches, there needs to be a process for incorporating their changes into the main branch. Let's walk through how to take the code that our developer created on their branch and merge it into the main branch.

Merging your developer branch with the main branch

At this point, we haven't enabled any notifications for our repository, but in a real-world scenario, this is an excellent way to add notifications for when someone has created a pull request. Our developer has created a branch on our repository, which is now ready to be merged. As the account user with merge privileges, we will log in and merge the feature branch and the main branch:

1. Open your console and log into **Amazon Web Console** as the primary administrative user that you have been using up to this point, not the developer user.

2. Navigate to the **CodeCommit** service by searching for CodeCommit in the top search bar in the middle of the screen. Once the **CodeCommit** icon becomes visible, click on it:

Figure 8.14 – Searching for CodeCommit in the top search box for services

3. Once on the main CodeCommit screen, so long as you are in the correct region, you should see the repository that we created earlier by the developer user named **chapter8**. Click on this repository name to be taken into the repository:

Name	▽	Description	Last modified	▼	Clone URL
chapter8		A sample repository for DevOps Pro and Beyond	2 hours ago		🗋 HTTPS 🗋 SSH 🗋 HTTPS (GRC)

Repositories Info 〇 | 🔔 Notify ▼ | Clone URL ▼ | View repository | Delete repository | **Create repository**

〈 1 〉 ⚙

Figure 8.15 – The chapter8 repository inside of CodeCommit

4. After clicking the repository's name, a set of menu options will appear on the left-hand side of the screen, under **Repositories**. Here, you will find a sub-heading called **Pull requests** that will show us any outstanding pull requests, including the one that was created by our developer user previously. Click on the **Pull requests** menu item to be taken to the pull requests screen:

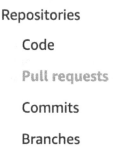

Repositories

Code

Pull requests

Commits

Branches

Figure 8.16 – Pull requests on the Repositories menu of CodeCommit

5. Our pull request name should be **Python Files**, predicated by number one. **Python Files** was the name that our developer user used when creating the pull request originally. We should also see a blue status of **Open** on the right-hand side of the pull request name. Click on the pull request's name, **1:Python files**, to be taken to the **Merge** screen:

Pull request	Author	Destination	Last activity	Status	Approval status
1: Python files	mariel	main	5 days ago	Open	No approval rules

Figure 8.17 – Pull request with status

6. Now, on the pull requests screen, which will be labeled with **1: Python files** at the top of the screen, you should see three colored labels right below the heading: **Open** (in blue), **No approval rule** (in dark gray), and **No merge conflicts** (in green). The latter is telling us that we can easily merge our pull request into the main branch. Do this by clicking on the orange **Merge** button at the top right-hand side of the screen:

Open   No approval rules   No merge conflicts

Figure 8.18 – Notifications about the CodeCommit pull request

7. At this point, we will come to the **Merge pull request** screen. This screen is all about the merge strategy, but for this exercise, we are going to leave the defaults as-is. This includes keeping the box checked at the bottom that will delete the branch that is being merged in and using the fast forward merge strategy. Click the orange **Merge pull request** button at the bottom right of the page to merge the feature branch into the main branch.

8. Finally, you should see a notification at the top of the screen in green, telling you that your `feature1` branch has been merged into the main branch.

We have just taken a thorough look at the CodeCommit service and looked at how to perform commits and merges from the perspective of multiple team members. With our source code in order, we will look at an AWS service that can produce packages along with test software: AWS CodeBuild.

Using AWS CodeBuild

AWS CodeBuild is a fully managed service that compiles source code, produces packages and container images, and runs tests. These steps are part of the continuous integration process, although not every project needs each of these steps. CodeBuild can also scale to meet your needs without any complex provisioning. Once you have configured the `buildspec` file and the job is initiated, CodeBuild will allocate the specified resources.

This is in contrast to other build systems, where you need to manually provision the compute for the workers or perform complex setups to ensure that autoscaling environments are available for heavy build and test periods.

You can run AWS CodeBuild as a standalone service, or it can be integrated with other services, such as **CodeCommit** and **AWS CodePipeline**, to create a repeatable, automated process that is part of your continuous integration life cycle.

Features to know about CodeBuild

The following are some of the features you should know about CodeBuild:

- **It's a fully managed build service**: There is no need to set up, patch, or update any servers or software.

- **It scales to meet your needs**: CodeBuild can scale up and down to meet the demand you place on it.

- **It's securely backed by AWS**: With optional encryption powered by KMS, along with permissions down to the specific job with IAM, you can feel confident in securing your build environment.

- It's one of the AWS developer tools that tightly integrates with CodeCommit, CodeDeploy, and CodePipeline.

- CodeBuild integrates natively with **CloudWatch Events** so that failed builds and items can be triggered, such as sending SNS messages.

- Logs can be set as output to S3 or **CloudWatch Logs**.

- CloudWatch metrics and alarms can be used to monitor CodeBuild thresholds.

Creating a CodeBuild job

CodeBuild is very versatile in the tasks it can perform. For our example, we will create an **Elastic Container Registry (ECR)**, use our `buildspec` file to create a Docker image, and then push that Docker image to the ECR:

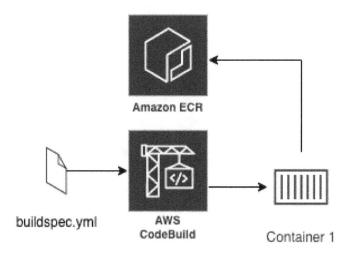

Figure 8.19 – Layout of what the example CodeBuild job does

If you haven't already downloaded the `Chapter-8` files from this book's GitHub repository, then now is the perfect time since we will be using the `docker` directory in our first CodeBuild exercise:

1. Navigate to the directory where you have downloaded the `Chapter-8` files. Do not go into the `docker` directory, as we will upload all of the files at once using a recursive command:

    ```
    $ aws s3 cp docker s3://{yourS3bucket}/docker/
    --recursive
    ```

2. First, we're going to create our ECR repository so that our container has a place to go once it's been built:

    ```
    $aws ecr create-repository \
        --repository-name chapter8
    ```

3. With our ECR repository created, we will open the AWS Console to complete the rest of the CodeBuild project. Make sure that you do this using the administrative user and not the developer user. Once your AWS console is open, search for `CodeBuild` in the top search box and then click on the icon to be taken to the CodeBuild service.

4. On the CodeBuild service, click the orange button at the top right-hand corner that says **Create build project**.

5. To create our project, we will start with the initial section: **Project configuration**. Use the following specified values to fill out this section. We are not going to use the optional build badge since it does not work with source code located in S3 buckets, but if we did, it would be a quick way to display whether our build is passing or failing to both ourselves and others on our team:

    a. **Project name**: `chapter8_docker`

b. **Description**: An example build for chapter 8:

Project configuration

Project name

chapter8_docker

A project name must be 2 to 255 characters. It can include the letters A-Z and a-z,
the numbers 0-9, and the special characters - and _.

Description - *optional*

An example build for chapter 8

Build badge - *optional*

☐ Enable build badge

Enable concurrent build limit - *optional*
Limit the number of allowed concurrent builds for this project.

☐ Restrict number of concurrent builds this project can start

▶ **Additional configuration**
tags

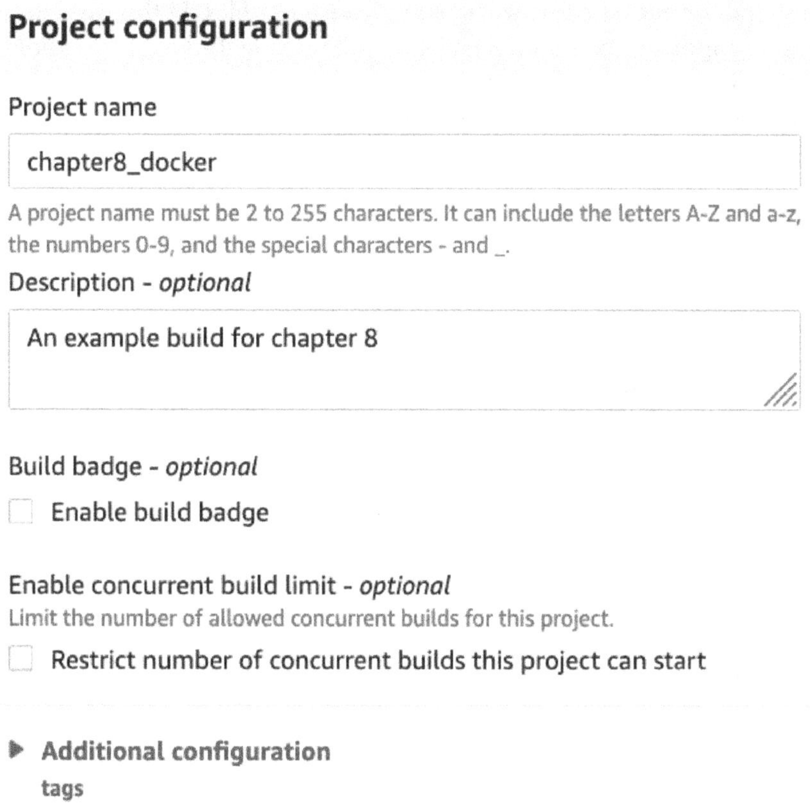

Figure 8.20 – Project configuration on the CodeBuild Create build project page

6. Next, we will move to the **Source** section. First, change the source from
    CodeCommit to **S3** since we have uploaded our source files to our S3 bucket. Then,
    we must find the bucket name where we uploaded our files. In this case, the bucket
    name is `devopspro-beyond`. Finally, we will provide the key that is asking for the
    folder that we uploaded the files in. In our case, this would be `docker/`. We don't
    have any versioning on our bucket, so we can leave this field blank:

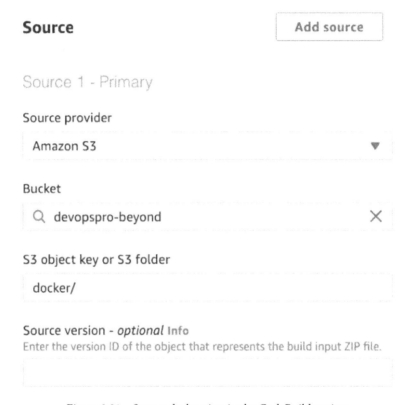

Figure 8.21 – Source declaration in the CodeBuild project

7. Next, we will be taken to the **Environment** section of the CodeBuild project. Use
    the following values to complete the **Environment** section:

    a. **Environment image: Managed image**

    b. **Operating system: Ubuntu**

    c. **Runtime(s): standard**

    d. **Image**: `aws/codebuild/standard:5.0`

    e. **Image version: Always use the latest version for this runtime version**

f. **Environment type: Linux**

g. **Service role: New service role**

> **Note**
>
> For the image in the CodeBuild project, we have specified version 5.0 here. However, you may want to look for the latest version available from AWS.

8. At this point, you can leave the rest of the options as they are, scroll down to the bottom of the page, and click on the orange **Create build project** button.

9. You should now be taken to the screen of your build project. At the top, you will see a green banner stating that your build project has been created:

⊘ **Project created**
You have successfully created the following project: chapter8-docker

Figure 8.22 – A successful project created in AWS CodeBuild

10. Now, we can try to run our job and see how it turns out. If you would like to run the job and start a build, then click the orange **Start build** button at the top right-hand corner of the screen.

This is just a base project and doesn't have to be the stopping point. We can still edit both the job and the `buildspec` file to add enhancements to the job if we so wish.

With our project created via the console using an S3 bucket as the source, we will now take a closer look at the actual file that runs the CodeBuild job: the `buildspec` file.

Constructing the buildspec file

The `buildspec` file is what drives the CodeBuild process and has a few strict requirements:

- It must be named `buildspec.yml`, and it can go by no other names.
- It must be in the `root directory` of your source code folder.
- It must be in the YAML language.

If you open the `buildspec` file to examine it after completing the previous exercise, then you would notice that it contains three major sections:

- **Version**: This tells the CodeBuild service which version of the syntax you are using and is one of the few required sections.

- **Phases**: These are the actual build phases that CodeBuild executes. The **Phases** section is one of the most powerful in CodeBuild as it performs many (if not all) of the following actions that comprise the build process:

  a. `Install`: This is only used for installing packages in the build environment. If you need to install a testing framework such as `pytest` or `Mocha`, this is the step where you would perform this task.

  b. `pre-build`: These are commands that happen before the actual build command is run. Pre-build commands can include signing into an **Elastic Container Repository** or installing packages from a **PyPI server**.

  c. `Build`: At this part of the phase, these are the commands that are run during the build. These can be commands that build the software, create a container, or create commands that test the software.

  d. `post-build`: Once the build has been completed, these are any commands that are performed afterward, which can include packaging up the software in a `.jar` or `.war` file, or creating a Python egg file. It can even involve uploading a container to a repository such as ECR or DockerHub.

- **Artifacts**: This is the build output. This can include the names of the files, the actual files themselves, or both. The artifacts phase also allows you to keep the directory structure where the files have been created or strip that away so that you just have the files themselves.

There are quite a few other options that are available in the `buildspec` file, depending on your use case. You can use environment variables, create reports, and even run as a specific user or a Linux operating system. For a complete list of options, please reference the AWS build spec documentation available at `https://docs.aws.amazon.com/codebuild/latest/userguide/build-spec-ref.html`.

We just took a deeper look at constructing `buildspec` files and what components they comprise. Next, we will concentrate on the artifacts that CodeBuild produces and how to export them and store them elsewhere.

Storing CodeBuild artifacts

When you create a software package, you can have AWS CodeBuild automatically save this artifact into an S3 bucket for use during deployments.

Inside of your `buildspec` file, if you are looking to create artifacts, you can declare just one artifact, or you can declare multiple artifacts. In the *Creating a CodeBuild job* section, the `buildspec` file had one artifact declared at the end.

> **Note**
>
> While the name CodeArtifact may lead you to believe that you can automatically store CodeBuild artifacts after a build, you can't. CodeArtifact can be used in conjunction with CodeBuild by supplying language packages just like a **NuGet**, **PyPi**, or **npm server** would do during the build process so that your team can have standardized packages and forgo download limits from public servers.

Using CodeBuild to test

While the first use of AWS CodeBuild that comes to mind may be creating packages for deployment and building Docker containers, this service also can serve a dual function in the SDLC process, and that is for testing. CodeBuild allows you to run the predefined unit tests and see reports of both a graphical and textual nature in the console regarding the results of those tests.

Reports expire after 30 days of creation and once they have expired, you nor anyone else with access to the report can view the report. If you need to keep the results of a report for more than 30 days, then you have the option to export the raw results of the report to an S3 bucket, where they will not expire and will only be phased out based on the bucket's life cycle policy.

CodeBuild supports the following test report file formats:

- Cucumber JSON
- Junit XML
- NUnit XML
- TestNG XML
- Visual Studio TRX

Specifying a report group name in your `buildspec` file with information about your test cases allows CodeBuild to create a report group for you if one doesn't already exist.

Once tests have been created, AWS CodeBuild can show reports on the console, providing a quick summary of the status of all of our tests.

Triggering CodeBuild jobs via CodeCommit

AWS CodeCommit can be used as the input source for the build. CodeCommit by itself cannot natively signal to CodeBuild that it has received either a new commit or a merge into a branch. However, with the help of the Lambda service, you can add a trigger that will initiate the CodeBuild job.

Having an automated process such as this without any manual intervention helps teams find bugs, along with deployed software, in a more expedient fashion. Creating a trigger that automatically signals the CodeBuild project to kick off the build is a great way to add continuous feedback to your **Software Development Life Cycle** (**SDLC**):

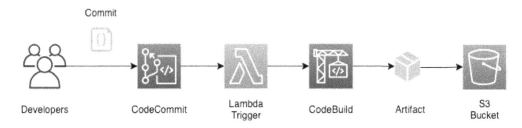

Figure 8.23 – Triggering a CodeBuild job from CodeCommit

Now that we have looked at how to start a CodeBuild job using the CodeCommit service automatically, we know how to automatically start CodeBuild jobs as soon as new code has been successfully merged into the project.

In the next chapter, *Chapter 9, Deploying Workloads with CodeDeploy and CodePipeline*, we will learn how the CodePipeline service can also be used to start a build process from CodeCommit.

Next, we will look at some of the advanced features of AWS CodeBuild.

Advanced features of AWS CodeBuild

Some of the features we will talk about in this section aren't mandatory for the DevOps professional exam. Since this book has *beyond* in its title, the following are some helpful features that can make using the service a bit easier. They are also good tips that you can share with teammates and clients.

Using AWS Session Manager to help troubleshoot builds

If you are trying to troubleshoot builds, then rather than looking up a way to `ssh` into the AWS-based docker images session manager, AWS Session Manager will allow you to jump into the build and try and troubleshoot what is happening. You can pause your build by adding `codebuild-breakpoint` to your `buildspec` file:

```
phases:
  pre_build:
    commands:
      - echo Entered the pre_build phase...
      - echo "Hello World"> /tmp/test.txt
      - codebuild-breakpoint
```

Once you have finished troubleshooting, you can pass `codebuild-resume` via the Command Prompt to pick up where you left off in your `buildspec` file.

Summary

In this chapter, we looked at how we can get our teams to share code using the AWS native source code tool CodeCommit. We also looked at how to package, build, and test our software using the CodeBuild tool.

In the next chapter, we will continue looking at **AWS Developer Tools**. This includes using CodeDeploy to deploy workloads and then tying everything together using AWS CodePipeline.

Review questions

1. You have been asked to set up a CodeCommit repository for your organization for the development team. The developers need the ability to push commits to their branches but not merge commits to the main branch, nor push commits to the main branch. The project manager also needs to be notified whenever a merge or a commit to the main branch occurs. Which combination of steps will safeguard the master branch and send the notification in the shortest time possible?

    a. Attach a resource policy to the CodeCommit repository that denies members of the IAM developer group actions of pushing commits, merging requests, and adding files to the main branch.

b. Affix an IAM policy to the developer IAM developer group that denies the actions of pushing commits, merging requests, and adding files to the main branch.

c. Configure AWS CloudTrail to send log events to send logs events to Amazon CloudWatch logs. Create a CloudWatch alarm based on a defined metric filter to identify CodeCommit Repository events. Use an SNS topic as a target that the project manager is subscribed to for the CloudWatch Alarm.

d. Create an Amazon CloudWatch Events rule that's triggered by a CodeCommit Repository State Change event by the main branch. Use an SNS topic that the project manager is subscribed to for the target.

e. Create a Lambda function to check for repository changes that will be sent to an SNS topic that the project manager is subscribed to if changes are found. Have AWS CloudWatch Events run the Lambda every 15 minutes.

2. You have a CodeCommit repository in your account that your developers have a limited set of actions for. Two new developers from a different Organizational Unit need to access this CodeCommit repository, but their users are based in a different AWS account. What is the most effective way to grant these new developers access?

a. Enable public access for the repository.

b. Create IAM users for each new developer and then grant them access to the repository.

c. Create an IAM group for the external developers, add the IAM users, and then provide access to the repository.

d. Create a cross-account role in your account, assign the necessary privileges to the role. and then provide the role ARN to the developers so that they can assume the role.

3. A client is looking for a new code versioning service and no longer wants the management hassle of running his servers. Currently, his team is small and not looking to grow exponentially over the next four quarters, but he is extremely concerned about the encryption abilities of any service he considers. What can you explain to him regarding AWS CodeCommit's ability to protect objects and code with encryption?

a. All transmissions are encrypted in transit and can only be done through the HTTPS or SSH protocols.

b. Transmissions are encrypted in transit when sent through the SSH protocol.

c. CodeCommit repositories are automatically encrypted using KMS.

d. CodeCommit repositories can be encrypted using KMS.

4.   You are setting up a new build process that is using S3 as the source of the code. You want to have unit tests run against the code that you and other developers have previously created and placed in a directory called `test/`, which is located in the child directory in the main source code directory. You plan to work on this build process for 60 days and want to be able to view the historical test reports, even from the first build. How can you configure this with native AWS services?

a. The test reports will be available in the **Reports** tab of the CodeBuild section of the console if you specify a report section in your `buildspec` file.

b. Export the CodeBuild test data to S3 after every build and use AWS QuickSight to create the test reports.

c. The test reports will be available in the Reports tab of the CodeBuild section of the console if you specify a report section in your `buildspec` file. Create a Lambda function that is triggered by CloudWatch events to export the raw data from the test reports that will be saved to an S3 bucket and available to be reviewed.

d. This is not possible with native AWS services.

Review answers

1.   b, d
2.   d
3.   a, c
4.   c

9
Deploying Workloads with CodeDeploy and CodePipeline

Teams are looking for an easy-to-use tool that allows the automation of the release process and gives a consistent release process. We will look at incorporating **CodePipeline** and the third-party tool Jenkins to automate our deployments' release cycle, which can then integrate into **CodeDeploy** for the actual code deployments.

In this chapter, we're going to cover the following main topics:

- About AWS CodePipeline
- Setting up a code pipeline
- Using Jenkins to build your workloads
- About AWS CodeDeploy
- Use cases for AWS CodeDeploy

Technical requirements

If you plan to follow along with the exercises in this chapter, there are some dependencies in the previous chapter, *Chapter 8*, *Creating Workloads with CodeCommit and CodeBuild*. Just as in the real world, we are building on what we have previously started in earlier chapters. Hence, if you have not created a developer user from the previous chapter, you will need to do so with the corresponding **Identity and Access Management** (**IAM**) permissions in their group.

About AWS CodePipeline

AWS CodePipeline can be thought of as a conductor in an orchestra. Using either code or the AWS console, you can put together your software development life cycle process in a visual representation that is either fully automated or has manual checks along the way for certain stages to be passed. This whole process is then laid out visually for your team members (including developers, testers, and others) to see which deployments happened and which deployments failed.

AWS CodePipeline helps you automate the steps you require to release your software and infrastructure changes in a continuous manner, as illustrated in the following diagram:

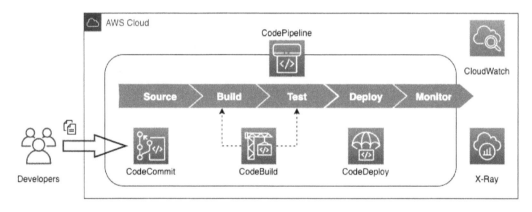

Figure 9.1 – CodePipeline and its integration with other AWS developer tools

The different AWS developer services, which CodePipeline is one of, are depicted in *Figure 9.1* and shown underneath their respected stages in the **Systems Development Life Cycle** (**SDLC**).

CodePipeline structure for actions

CodePipeline is structured into several different categories that allow either native AWS services to act or allow a supported third-party service to be integrated and perform the necessary action.

There are six valid action categories, as listed here:

- Source
- Build
- Test
- Deploy
- Approval
- Invoke

Each of the action categories has its own set of providers that can invoke actions or have actions invoked from that resource, as outlined in the following table:

Action category	Valid action providers
Source	Amazon **Simple Storage Service (S3)**
	Amazon **Elastic Container Registry (ECR)**
	Bitbucket Cloud
	GitHub
	GitHub Enterprise Server
Build	AWS CodeBuild
	CloudBees
	Jenkins
	TeamCity
Test	AWS CodeBuild
	AWS Device Farm
	BlazeMeter
	Ghost Inspector
	Micro Focus StormRunner
	Runscope
Deploy	Amazon S3
	AWS AppConfig
	AWS CloudFormation
	Amazon **Elastic Container Service (ECS)**
	Elastic Beanstalk
	OpsWorks
	Service Catalog

Table 9.1 – CodePipeline source actions and action providers

In addition to the action integrations, a few other AWS services integrate without any specific action category, as follows:

- **Amazon CloudWatch**: CloudWatch can monitor your resources that are built with the pipeline or are being tested and integrated into the pipeline.

- **Amazon CloudWatch Events**: CloudWatch Events can detect changes to your pipeline as a whole or even at certain stages of your pipeline. CloudWatch Events can even listen to other outside services and then trigger a CodePipeline run if specific scenarios happen, such as if someone updated a CloudFormation stack. This would then create a need for the software to be redeployed.

- **AWS Cloud9**: Cloud9 is the cloud-based **integrated development environment (IDE)** that is accessible via a web browser.

- **AWS CloudTrail**: If the CloudTrail service is active for the particular region, any **application programming interface (API)** actions that happen via the AWS console, **software development kits (SDKs)**, or **command-line interface (CLI)** are captured and recorded.

- **AWS KMS**: **Key Management Service (KMS)** can be integrated with AWS CodePipeline for source S3 buckets and artifacts, which either are encrypted or need to be encrypted. Suppose the artifacts are coming from an account other than the one where the CodePipeline is being executed. In that case, the key that is encrypting the bucket and objects will need to be a customer-managed key.

Looking at *Table 9.1*, you can see that there is a combination of both native AWS services and specialized third-party partner tools that you may already be using to integrate into your code pipeline stages.

We have just looked at the different actions along with their corresponding AWS services and third-party services that can help perform those actions. Next, we will look at some of the use cases for AWS CodePipeline, including real-world use cases.

Use cases for AWS CodePipeline

When thinking about what you can do with CodePipeline, there are specific scenarios that have already identified that CodePipeline is an optimal fit as a tool. Let's take a look at some of these next.

Automation of your build and release process

CodePipeline allows your developers to concentrate on the code they are working on and then commit that code to either an Amazon-hosted repository or a third-party repository such as GitHub or Bitbucket Cloud. The push of a new code commit by a developer can then trigger the build process.

Creating a consistent toolset for developers to use

One of the most challenging parts of getting a new developer up and productive quickly in an organization is the onboarding process. CodePipeline helps with this process by presenting a consistent toolset no matter which time zone a team member is in or the operating system that they are using.

Using CodePipeline to integrate with third-party providers

Suppose your team currently uses third-party tools such as Jenkins for the code build or testing process, BlazeMeter for the load testing process, or StormRunner for the testing process. In that case, CodePipeline can help present a unified orchestration front for all of these tools.

There can also be a cost-savings aspect in using only one service account called from the CodePipeline service rather than each individual or individual team requesting their own license.

Continuously deploy your web applications with Elastic Beanstalk and CodePipeline

Although Elastic Beanstalk is many times looked upon as a service which helps developers who minimal exposure to AWS infrastructure items to get their code up and running quickly, those items of code can become successful items in an organization. The incorporation of CodePipeline allows to take the reliance away from the Elastic Beanstalk CLI or try to track down deployments in the AWS Management Console and allow for a more structured and trackable approach.

Now that we have examined a few of the various use cases where CodePipeline can be used successfully, we can now go on to our hands-on example. This is where we will go through the steps of setting up our own AWS CodePipeline that utilizes a CodeCommit repository.

Setting up a code pipeline

One of the best ways to get a feel for AWS CodePipeline is to go through the exercise of setting up a code pipeline. Many times, it will be those designated tools' team members who will set up the pipelines. These **tools** team members have a unique set of permissions that differ from that of developers.

We will need to set up our tools team group and assign them the correct permission set. After this, we can create a tools team member and associate them with the tools team IAM group. Then, we can log in as that tools team member and have them build out the pipeline.

Creating our code base prior to setting up the pipeline

We are going to set up a brand new CodeCommit repository before setting up our code pipeline. Creating the repository beforehand will allow us to have a fresh set of code to use to run through the steps of our pipeline.

In the `chapter9` section of our GitHub repository, there will be a folder named `code`. This folder will hold the source code that we will need to upload to the CodeCommit repository we are about to create using the following steps:

1. Open your browser to the AWS CodeCommit home page, `https://console.aws.amazon.com/codesuite/codecommit/home`, and sign in if prompted.

2. Click the orange **Create repository** button on the top right-hand side of the screen.

3. On the **Create repository** screen, under **Repository settings**, use `chapt9` for the name of the repository. We will add a description that this repository is for `CodePipeline` to distinguish it from the `chapt8` repository, which we created as a test in the previous chapter, *Chapter 8*, *Creating Workloads with CodeCommit and CodeBuild*, as illustrated in the following screenshot:

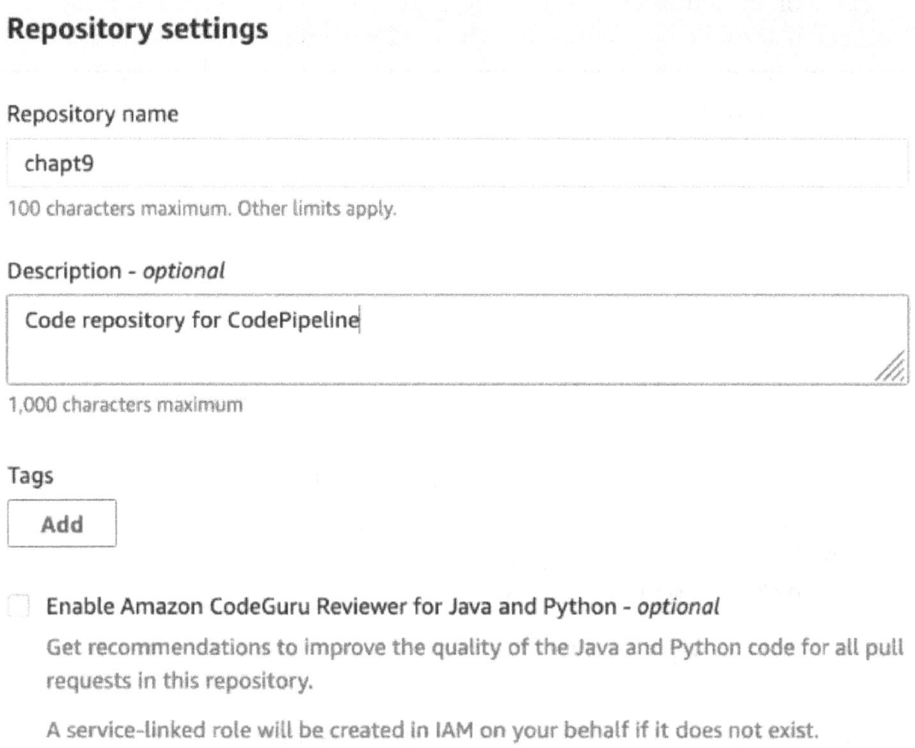

Repository settings

Repository name

chapt9

100 characters maximum. Other limits apply.

Description - *optional*

Code repository for CodePipeline

1,000 characters maximum

Tags

Add

☐ Enable Amazon CodeGuru Reviewer for Java and Python - *optional*

Get recommendations to improve the quality of the Java and Python code for all pull requests in this repository.

A service-linked role will be created in IAM on your behalf if it does not exist.

Figure 9.2 – CodeCommit repository settings for chapt9 repository

4. Click the orange **Create** button.

5. Since we have already created a user that can upload files in *Chapter 8, Creating Workloads with CodeCommit and CodeBuild*, simply click on the **Repositories** menu item in the left-hand menu, as illustrated in the following screenshot. This will bring up the names of all the repositories that we have in CodeCommit and will allow us to copy the link we need to clone the repository locally to our workstation:

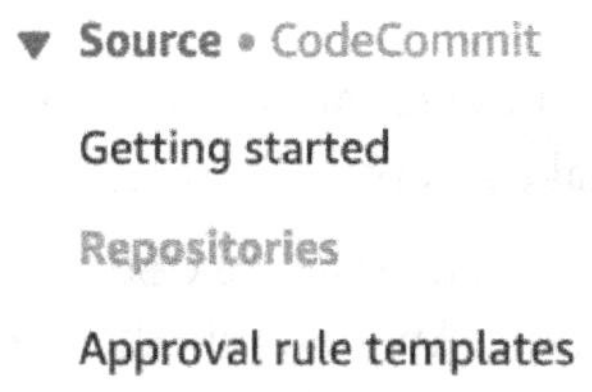

▼ **Source** • CodeCommit

Getting started

Repositories

Approval rule templates

Figure 9.3 – The side menu on CodeCommit with Repositories highlighted

6. Now, click on the link labeled SSH, to the right of the chapt9 repository, as illustrated in the following screenshot. This will pop up a small dialog box confirming that the link has been copied and you are ready to clone the repository to your local workstation:

Clone U   ⊘ Copied

⬚ HTTPS   ⬚ SSH   ⬚ HTTPS (GRC)

⬚ HTTPS   ⬚ SSH   ⬚ HTTPS (GRC)

Figure 9.4 – SSH cloning Uniform Resource Locator (URL) copied to clipboard

At this point, we are done with the AWS console for the moment, and we will move on for the next set of commands to the terminal on your local workstation.

7. Now, with your terminal open, go to the root of your home directory. In Linux, you can get there quickly with the $cd ~ command. Next, perform the following command to clone the repository locally. Now, we can use the URL on our clipboard to clone into the repository:

```
$ git clone ssh://git-codecommit.us-east-2.amazonaws.com/
v1/repos/chapt9
```

Once you have successfully cloned into the repository, then you should get a confirmation message stating that you have cloned into an empty repository.

8. Even though we have an empty repository, we are going to take the example code from the GitHub repository in the /code directory and then copy (or move it if you don't want two copies of the code on your local machine) to this new local CodeCommit repository so that we can push it up to CodeCommit.

Make sure that you are starting from the directory where you have cloned the example code from GitHub, as illustrated in the following code snippet:

```
$ cp -R * ~/chapt9/
```

If you don't want to copy the files via the command line you could also use either File Explorer or Finder to make a copy of the files into your new git repository.

9. Now that we have copied the files over to the CodeCommit local directory, we need to add all of them to the commit. After adding them and the commit message, we will push the files up to the repository. We can do all of this in just a few simple commands. First, we will need to change directories from where we cloned the example code to our local CodeCommit repository, as follows:

```
$ cd ~/chapt9
```

10. Since we are in our local directory, we can now add all of the files and then push them up to the remote `CodeCommit` repository. We do this by using the `git add`, `git commit`, and `git push` commands, like this:

```
$ git add *
$ git commit -m "adding sample code to CodeCommit"
[master (root-commit) f85e8f2] adding sample code to
CodeCommit
 2 files changed, 52 insertions(+)
 create mode 100644 buildspec.yml
 create mode 100644 src/app.py
$ git push
Enumerating objects: 5, done.
Counting objects: 100% (5/5), done.
Delta compression using up to 8 threads
Compressing objects: 100% (4/4), done.
Writing objects: 100% (5/5), 825 bytes | 825.00 KiB/s,
done.
Total 5 (delta 0), reused 0 (delta 0), pack-reused 0
To ssh://git-codecommit.us-east-2.amazonaws.com/v1/repos/
chapt9
 * [new branch]      master -> master
```

Now, with our sample code ready, we are prepared to move on to the next step, which is creating our tools team member. Our tools team member is the person on our team who focuses on creating and managing the pipelines and processes more than the code itself.

Creating our tools team member

Just as we had previously created our development group and member, we will need to do the same thing for our tools team member. It is important to separate the duties of our team members and give each person only the privileges they need to perform their job duties. After creating our tools member, we will then log in as that tools teams member and then run the CloudFormation template named `pipeline1.yml` as this tools team member to construct the pipeline.

Let's go ahead and set up the group for our tools team member, as follows:

1. Open up your terminal and type the following commands so that we can create our new group:

    ```
    $aws iam create-group --group-name Tools
    ```

    After running the command, you should see a confirmation like the one shown next:

    ```
    {
        "Group": {
            "Path": "/",
            "GroupName": "tools",
            "GroupId": "AGPAW24Q7QQFSHYZHE6N6",
            "Arn": "arn:aws:iam::470066103307:group/tools",
            "CreateDate": "2021-05-14T01:25:58+00:00"
        }
    }
    ```

    Take note of the ARN which is returned for the group name. You will either need to save this to a note pad or scroll up to copy and paste this when we attach the policy which we create next to this group

2. Next, we will need to create our policy using the pre-made policy document that you should have downloaded from the GitHub repository in the Chapter9 folder named IAM_Tools.json, as follows:

    ```
    aws iam create-policy --policy-name Tools_Members
    --policy-document file://IAM_Tools.json
    ```

    This should return the **Amazon Resource Name (ARN)** of our policy in our success JSON after creation, as in the following statement:

    ```
    {
        "Policy": {
            "PolicyName": "Tools_Members",
            "PolicyId": "ANPAW24Q7QQF6FPOCHV5V",
            "Arn": "arn:aws:iam::470066103307:policy/Tools_
    Members",
            "Path": "/",
            "DefaultVersionId": "v1",
            "AttachmentCount": 0,
    ```

```
        "PermissionsBoundaryUsageCount": 0,
        "IsAttachable": true,
        "CreateDate": "2021-05-16T19:32:41+00:00",
        "UpdateDate": "2021-05-16T19:32:41+00:00"
    }
}
```

3. After creating our policy, we will then attach our policy to our group, like this:

```
aws iam attach-group-policy \
```

In the `policy-arn` field use the value which was returned to you in step 1:

```
    --policy-arn arn:aws:iam::470066103307:policy/Tools_
Members \
    --group-name tools
```

If the policy attachment is successful, then there will be no confirmation message.

4. Now that we have created our `tools` group, we can create a user to be part of the tools team. This tools team member will be the user that we will use to create the actual code pipeline. Let's go ahead and make our new user, as follows:

```
aws iam create-user --user-name peter
```

5. Now, we can take our new user and add them to the `tools` group so that they have the group permissions, as follows:

```
aws iam add-user-to-group --user-name peter --group-name
Tools
```

6. And just like the developer user we created in *Chapter 8*, *Creating Workloads with CodeCommit and CodeBuild*, we will need to set an initial password that the tools user will need to reset on the first login, as follows:

```
aws iam create-login-profile --user-name peter --password
DevOpsPRO --password-reset-required
```

At this point, we have created our tools user and are ready to log in to the AWS console as the tools user and build out our AWS pipeline.

Creating a pipeline

With our tools user (Peter, in our case) created, we will make a context switch from using the AWS CLI to the browser and the management console and act as if we were performing these acts as the tools user. Before we start, you need to be sure to download the `pipeline1.yml` file from the `Chapter9` folder in the GitHub repository for the book.

Let's open up a fresh web browser that doesn't have cookies or a session attached so that we can log in to the AWS console as Peter. You may need to either open up an incognito window or a different browser than the one you have been using. For example, if you have been running the administrator in Chrome, then you would open up a new session in Firefox or Edge using the following steps:

1. Sign in to the AWS console using your account number or account alias and the user's name, `peter`, and the `DevOpsPRO` password, as we created previously.

2. You will instantly be prompted to change the password for the user `peter`. Change this password to anything you like, but either write it down or keep it something you can remember as you may need to access this user later. The process is shown in the following screenshot:

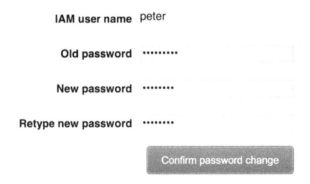

Figure 9.5 – Password change for tools user peter

3. Once you have changed the password, the AWS console will bring you to the main console page. Use the top unified search box to search for the **CloudFormation** service. Once the CloudFormation service appears, click on the service icon to be taken to the main CloudFormation page, as illustrated in the following screenshot:

Figure 9.6 – CloudFormation service icon

4. Once on the main **CloudFormation** service page, click the orange **Create stack** button.

5. On the **Create stack** page, use the following options:

    a. **Prepare stack**—Keep this selected on **Template is ready**

    b. **Specify template**—Choose the checkbox that is labeled **Upload a template file**

6. Then, click the **Choose file** button and find the `pipeline1.yml` file and open this file. Once this file has been selected, click the orange **Next** button at the bottom of the screen, as illustrated in the following screenshot:

Prerequisite - Prepare template

Prepare template
Every stack is based on a template. A template is a JSON or YAML file that contains configuration information about the AWS resources you want to include in the stack.

◉ Template is ready	○ Use a sample template	○ Create template in Designer

Specify template
A template is a JSON or YAML file that describes your stack's resources and properties.

Template source
Selecting a template generates an Amazon S3 URL where it will be stored.

○ Amazon S3 URL	◉ Upload a template file

Upload a template file

Choose file 🔺	*pipeline1.yml*

JSON or YAML formatted file

S3 URL: https://s3.us-east-2.amazonaws.com/cf-templates-14e6hhrdu34vu-us-east-2/2021142wWx-pipeline1.yml   View in Designer

Cancel   Next

Figure 9.7 – Creating a CodePipeline stack

7. Once you click the **Next** button, it will take you to the **Specify stack details** screen. Everything should already be filled out for you using CloudFormation default parameter values except for **CloudFormation Stack name**—use C9-demo and your email address where you would like to receive notifications about the stack. You can leave this blank for the email address, or you can put in your own email address. If the branch you created was named main, then you do not need to make any changes; however, if the branch you created was named master as in our example, then you will need to change the default value in the **Branch name** field to master before moving on or the pipeline will not execute right away. Once you have filled in these values, click the orange **Next** button at the bottom of the screen, as illustrated in the following screenshot:

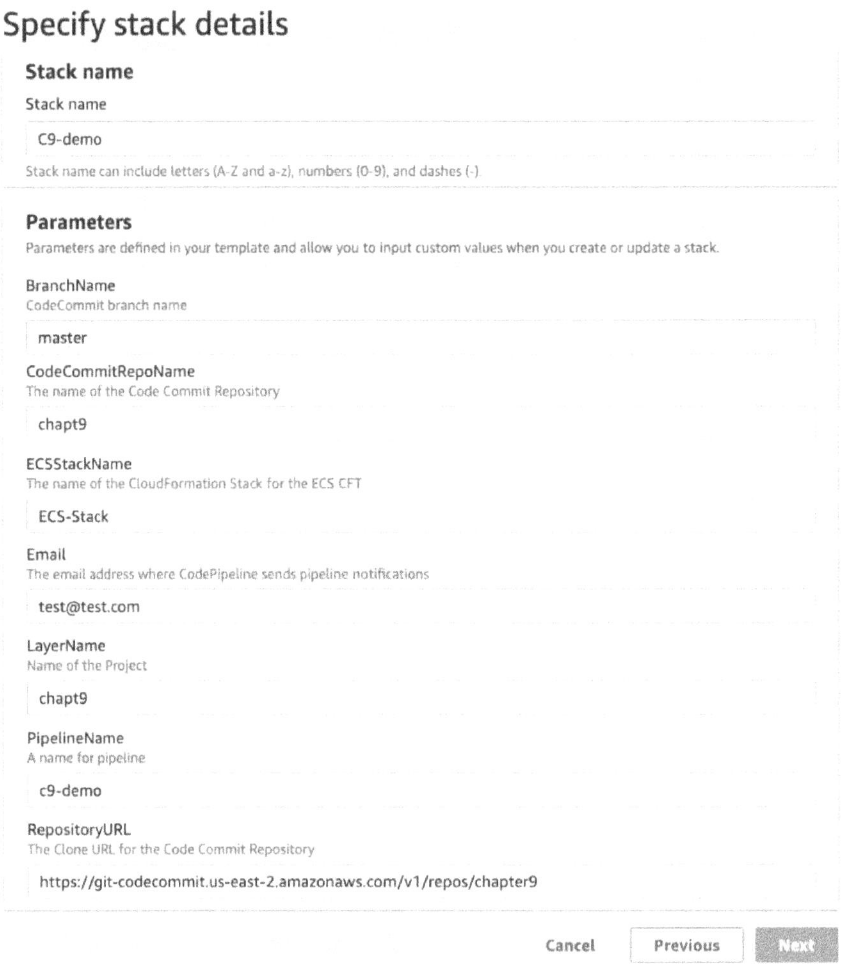

Specify stack details

Stack name

Stack name

C9-demo

Stack name can include letters (A-Z and a-z), numbers (0-9), and dashes (-)

Parameters

Parameters are defined in your template and allow you to input custom values when you create or update a stack.

BranchName
CodeCommit branch name

master

CodeCommitRepoName
The name of the Code Commit Repository

chapt9

ECSStackName
The name of the CloudFormation Stack for the ECS CFT

ECS-Stack

Email
The email address where CodePipeline sends pipeline notifications

test@test.com

LayerName
Name of the Project

chapt9

PipelineName
A name for pipeline

c9-demo

RepositoryURL
The Clone URL for the Code Commit Repository

https://git-codecommit.us-east-2.amazonaws.com/v1/repos/chapter9

Cancel   Previous   Next

Figure 9.8 – Adding values to the Specify stack details screen

8. On the **Configure stack options** screen, we will not be adding any tags. Simply go to the bottom of the page and click the orange **Next** button.

9. Now, on the **Review C9-demo** page, scroll down to the bottom of the page and check the box in the blue area that acknowledges that this stack will create an IAM role under the heading of **Capabilities**. After you have checked the box, you can press the orange button that is labeled **Create stack**, as illustrated in the following screenshot:

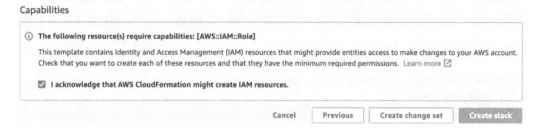

Figure 9.9 – Checking the acknowledgment in the Capabilities section on the review

10. Once pressed, you will be taken to the **Stacks** page in CloudFormation. The stack itself will show the status of CREATE_IN_PROGRESS until it has finished creating our code pipeline. Once the status has been completed, then we can move on to the next step.

11. After the CloudFormation process completes, we will then be able to go back up to the unified search and search for the CodePipeline service. Once the icon for **CodePipeline** appears, as illustrated in the following screenshot, click on it to be taken to the current pipelines:

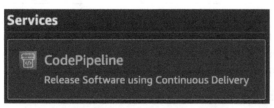

Figure 9.10 – CodePipeline from the unified search bar

12. You should see the pipeline that you just created, named C9-demo. Click on this name to view the details of the pipeline.

We can see from the pipeline we created that we now have a pipeline created with three stages: **Source**, **Build**, and **Deploy**. There are integrated processes including manual approval processes on the deploy stage along with the CloudFormation creation..

> **Note**
> If you failed to complete the previous exercise of creating the CodeCommit repository named `chapt9`, you would run into problems with the pipeline. This `chapt9` repository is the code source for our AWS code pipeline. If you named your repository something different, you would need to either modify the CloudFormation template or go into the AWS console and modify your source stage.

Now, with our AWS code pipeline built, we can move forward with our developer testing a commit and seeing the stages of the pipeline in action, just as it would work in real life. Before we do that, however, we need to give our developer users more IAM permissions since when they were initially created, they only had permissions for CodeCommit and to modify their own password.

Updating our developer users

In the previous chapter, *Chapter 8*, *Creating Workloads with CodeCommit and CodeBuild*, we created a group for developers. We gave them a scoped set of permissions that was limited to AWS CodeCommit and CodeBuild. Since we are now adding CodePipeline and CodeBuild to the mix, we will now have to expand their permissions so that they can use these services as well. In the GitHub `Chapter9` repository, there is a file named `IAM_Developers.json`—be sure to download this file so that you can update the permissions for your developer. While updating the commands, you will need to use your administrative user. We'll proceed as follows:

> **Note**
> Before you start to perform these commands, make sure that you have downloaded the `IAM_Developers.json` file from the `Chapter9` folder in the GitHub repository and are in the same directory where you downloaded the file.

1. The first thing that we need to do is find the ARN of the policy attached to our developers. We created this policy in the exercise in *Chapter 8*, *Creating Workloads with CodeCommit and CodeBuild*. Log in to your terminal and run the following command to extract the ARN for the developer group:

```
aws iam list-policies --query 'Policies[?PolicyName==`CC_
Developers`].Arn' --output text
```

After we run this command, we should get a return of the current ARN being used for our Code Commit developers' group. This ARN is going to be needed in the following step so that we can update our policy.

2. After we have the ARN, we can then create a new policy version and set that version as `default`, as follows:

```
aws iam create-policy-version \
--policy-arn arn:aws:iam::470066103307:policy/CC_
Developers \
--policy-document file://IAM_Developers.json --set-as-
default
```

If this is successful, we will be returned a **JavaScript Object Notation (JSON)** message showing that we are now on the second version of our policy, as illustrated in the following code snippet:

```
{
    "PolicyVersion": {
        "VersionId": "v2",
        "IsDefaultVersion": true,
        "CreateDate": "2021-05-16T19:01:09+00:00"
    }
}
```

Updating our developer user group to allow for CodePipeline access permits the developers to see the pipelines previously created when running, along with any errors that may have been encountered during individual steps of the pipeline. These permissions have been scoped in such a manner so as not to allow developers to create or modify any new or existing pipelines.

CodePipeline concepts

There are several basic concepts and terms to understand when using AWS CodePipeline, as depicted in the following diagram:

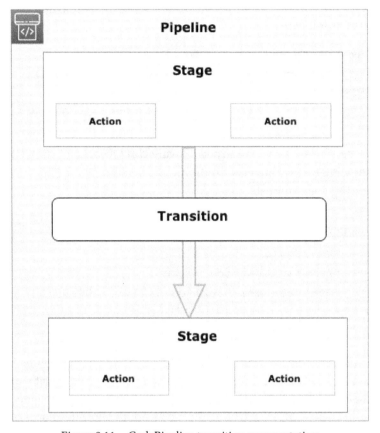

Figure 9.11 – CodePipeline transition representation

Understanding the CodePipeline and stage requirements

As you start to build pipelines, there are some rules and requirements that should be known as this can help you avoid headaches in trying to debug errors. Look through the list stated next for best practices as you build your AWS code pipelines along with the knowledge to remember during exam time:

- All stage names in a pipeline must be unique.
- Only the first stage in a pipeline can contain source actions.
- A pipeline must have at least two stages.
- All actions within a single stage must be unique.

- The input action of a stage must match the output action of the preceding stage exactly.

- Output artifact names must be unique to a pipeline. If a stage has an output artifact named `TestPackage`, no other stage may have another output artifact named `TestPackage` in that pipeline.

- For all supported action types, the only valid owner strings are `AWS`, `ThirdParty`, or `Custom`.

There are other more complex requirements; however, if you understand these requirements, then you have a solid foundation for using the CodePipeline service along with an understanding of the nuances of stages for the DevOps professional exam.

Now that we have looked at the stages as a whole in an AWS code pipeline, let's look at one aspect that is not automated: approval actions.

Approval actions in a code pipeline

In a code pipeline, there is an opportunity to pause between stages using approval actions. Approval actions allow for the manual review of an action before proceeding to the next stage of the pipeline.

Actions that are approved by the reviewer move on to the next stage of the pipeline. If an approval is rejected, then the pipeline does not continue to the next stage. You also have 7 days to approve a pipeline action, or it will result in the pipeline failing.

There are a few common reasons why you would use approval actions in an AWS code pipeline, outlined as follows:

- To perform manual **Quality Assurance (QA)** testing before moving on to the next stage

- To allow for a code review or change management review before proceeding

- To allow for manual review of a web page before publishing to production

Using Jenkins to build your workloads

The developer tools provided by AWS can give you just about all the functionality that you need without any extra configuration or setup. There are instances where teams have already built parts of their **Continuous Integration/Continuous Deployment (CI/CD)** process around existing tooling and may want to retain some of the work that they have already poured time and effort into refining.

Teams that use the Jenkins server can be one of those cases. Along with its vast ecosystem of plugins, Jenkins can provide an extreme amount of functionality in the CI/CD process.

Many teams use Jenkins for the build stage of the CI process since, in Jenkins, it can feel easier to build the steps in shell scripts without a need to create additional `buildspec` files.

The following diagram depicts Jenkins being used in conjunction with CodePipeline:

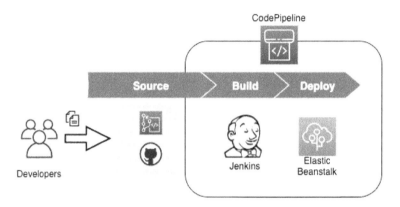

Figure 9.12 – Using Jenkins in conjunction with CodePipeline

Jenkins itself has plugins that will work with many of the AWS services, including the developers' suite of services.

Now that we have looked at the CodePipeline service, including how to incorporate third-party tools such as the Jenkins build server, let's look at how we can use AWS CodeDeploy to deploy our workloads either by themselves or as part of a pipeline.

About AWS CodeDeploy

AWS CodeDeploy is a service that assists in automating deployments of application code and files to EC2 instances, on-premises servers, Lambda functions, and containers running on AWS ECS.

CodeDeploy is a service that makes the following things easier:

- Update Lambda functions and create new versions
- Avoid downtime during application deployments
- Swiftly release new features
- Promptly roll back to a known good version in case of deployment failure

AWS CodeDeploy is focused on the following three platforms for deployments:

- EC2/on-premises instances
- Lambda functions
- ECS containerized applications

Using these three platforms as a basis, you describe your deployments in a file named `appspec`. This file can be written in either JSON or **YAML Ain't Markup Language (YAML)** format.

There are multiple types of files and components that CodeDeploy can distribute, including the following:

- Executables
- Packages
- Scripts
- Multimedia files

Now that we have a grasp of the basics of CodeDeploy, let's look at the basic concepts we need to understand when using the CodeDeploy service.

Basic CodeDeploy concepts to understand

As we start to employ the CodeDeploy service, there are some components that should be understood first and foremost.

Application

This is a name that is unique, and CodeDeploy uses it to identify the application which you want to deploy. It uses this unique name to make sure that the correct version of deployment, deployment group, revision, or rollback is referenced during a particular deployment.

Compute platform

This is the actual platform where CodeDeploy deploys an application. There are quite a few choices for CodeDeploy deployments. Deployments can happen to EC2 instances or on-premises servers as long as they meet operating system requirements. These deployments can consist of configuration files, images, and executable files, along with other types of files.

CodeDeploy can also be used to update Lambda functions. CodeDeploy also has the ability to shift traffic from one version of a Lambda function to a newer function in a multitude of deployment options, including **canary**, **linear**, and **all-at-once** deployments.

If you are using containers on the ECS platform, then CodeDeploy can be used to update the tasks using a **blue/green deployment** strategy by first installing a new version of the task set and then shifting the traffic over to the latest version of the task set. As with Lambda deployments, these deployments can be done in a canary, linear, or all-at-once fashion.

Deployment configuration

Using a set of success or failure criteria along with deployment rules, CodeDeploy is guided by a deployment configuration for each deployment. Inside of the deployment configuration, especially in the case of EC2 or on-premises deployments, you can set the minimum number of healthy instances that need to succeed for the deployment to be a success. If you are targeting either a Lambda application or ECS tasks, then your deployment configuration can specify how the traffic is routed during the deployment. The following deployments are available:

- **Canary**: Traffic is shifted in two separate increments, and you have the ability to specify the percentage of the traffic shifted initially before the rest of the traffic is shifted at a time specified in minutes.

- **Linear**: Traffic is shifted in equal increments. You have predefined options for percentages that can be shifted and for the intervals between shifts.

- **All-at-once**: All the traffic is shifted from the original Lambda function or ECS task at the same time.

Deployment groups

Deployment groups pertain to EC2 instances, either individually or in **Auto Scaling Groups** (**ASGs**), and are explicitly tagged for targeting a deployment.

Deployment groups can be as straightforward as using a single tag to designate the deployment group, or you can become elaborate as using up to 10 different tags in a deployment group.

Deployment type

The deployment type is the technique used by CodeDeploy to place the most up-to-date version of an application into a deployment group. There are two different types of deployment types: **in-place** and **blue/green**.

With in-place deployments, although possibly more cost-effective, the application on the current instances in the deployment group is stopped while the latest version of the application is installed. The new version of the application is restarted and then validated. In-place deployments are only available for EC2 instances or on-premises deployments.

Blue/green deployments provision a new set of resources. This can consist of creating a new version of a Lambda function, a new task set in the case of ECS, or a whole new instance in the case of EC2.

Blue/green deployments do not work with on-premises deployments.

Revision

A revision for an AWS Lambda deployment is either a file in YAML or JSON format that states information about the Lambda function to deploy. Revisions for Lambda are stored in an S3 bucket.

In the case of an EC2 or on-premises deployment revision, this is not just one file but a collection of files that contain the components (such as web pages, executable files, source code, and deployment scripts) along with a specification for the application, which is then packaged up into archive format. Revisions for EC2 or on-premises instances can be stored in an S3 bucket or GitHub repositories.

Target revision

This is the most recent version of an application that has been uploaded to the code repository, which is going to be targeted for deployment.

Installing the CodeDeploy agent file

When using CodeDeploy with EC2 instances, an agent file is placed on those instances, making it viable for the instance to achieve deployments from the CodeDeploy service. A configuration file is also placed on the instance with an agent, and this file specifies how the agent will work. These instances not only have to be in the AWS cloud but can also be on a specific operating system of an on-premises data center.

Take a glance at the following lists to see which operating systems have been tested using the AWS CodeDeploy agent:

On-premises operating systems	AWS EC2 operating systems
Ubuntu Server 20.04 **Long-Term Support** (**LTS**), 18.04 LTS, 16.04 LTS, 14.04 LTS	Ubuntu Server 20.04 LTS, 18.04 LTS, 16.04 LTS, 14.04 LTS
Microsoft Windows Server 2019, 2016, 2012 R2, and 2008 R2	Microsoft Windows Server 2019, 2016, 2012 R2, and 2008 R2
Red Hat Enterprise Linux (RHEL) 7.x	RHEL 7.x

Table 9.2 – Tested operating systems to use the AWS CodeDeploy agent

Any EC2 instance that you wish to use with CodeDeploy will also need to have a service role attached that gives it enough permissions so that the CodeDeploy service can perform its duties.

Understanding the appspec file

The application specification file, or `appspec` file, as it is most commonly called, is a YAML file in the majority of cases (although it may be formatted in JSON format) that has a specific number of sections along with some optional sections depending on which type of deployment you are trying to perform.

The `appspec` file details the deployment actions that you want to take during the deployment.

There are different types of `appspec` files if you are doing an ECS deployment for containers, for EC2, or on-premises instances, and then if you are trying to deploy to Lambda instances.

The following is an example `appspec` file:

```
version: 0.0
Resources:
  - myLambdaFunction:
      Type: AWS::Lambda::Function
      Properties:
        Name: "myTestFunction"
        Alias: "myTestFunctionAlias"
```

```
          CurrentVersion: "1"
          TargetVersion: "2"
  Hooks:
    - BeforeAllowTraffic:
  "LambdaFunctionToValidateBeforeTrafficShift"
    - AfterAllowTraffic:
  "LambdaFunctionToValidateAfterTrafficShift"
```

Important items to note in our example `appspec` file are the version, the resources, and the hooks.

Deployment rollbacks and content redeployment

AWS CodeDeploy can roll back a deployment that has either been stopped manually or that has failed during the deployment process. Properly speaking, these rollbacks are new deployments and receive new deployment **identifiers** (**IDs**). The rollbacks do restore previous versions of a set of code. There are two different ways that rollbacks can happen, via automated rollbacks or via a manual process.

Automated rollbacks

Your deployment group can be configured to automatically roll back either on failure during a deployment or if certain monitoring thresholds are met during a deployment. If one of these limits has been set and then triggered during the deployment for a case of automated rollback, then the deployment will go back to the last known good configuration.

You may also choose to override the automated rollback option if it has previously been put in place when starting a new deployment by configuring one of the advanced configuration options for the deployment group.

Manual rollbacks

Even if you have not set up your deployment to roll back automatically, using AWS CodeDeploy, you can push out a previous version of a deployment. This would create a new deployment version. You could do this if your deployment failed or if your instances have got into an unknown state, and they might be fixed by pushing updates to the application and configuration.

Rollback and redeployment workflow

If a rollback has been induced either automatically or manually, then CodeDeploy will start to try to remove all of the files that were successfully installed during the deployment.

The cleanup file, if it exists, is a type of log file that CodeDeploy keeps so that it knows which files have been installed and can remove these files before starting a new deployment.

During deployment, the CodeDeploy agent will write out the filenames that are being deployed so that it has a record in case a rollback is necessary.

In the case of a rollback, CodeDeploy will refer to the cleanup file so that it knows which files to remove. It can then proceed with the previous version of the deployment in case of automatic rollback.

Knowing now how deployments and rollbacks work, let's take a look at some of the use cases for AWS CodeDeploy.

Use cases for AWS CodeDeploy

As we have looked at the AWS CodeDeploy service, let's think about some of the situations where this service would serve us best.

Deploying application updates to servers in an on-premises data center

If you are running a compatible operating system on your on-premises hardware, then you use AWS CodeDeploy to coordinate the deployments as well as have a single pane of glass to see successes and failures. Installing the AWS CodeDeploy agent is a required prerequisite. These operating systems include Windows Server, Ubuntu Server, or RHEL.

Deploying application updates to Windows or Linux servers in the AWS cloud

If you are deploying to an NGINX or Apache server on a Linux EC2 instance or an **Internet Information Services (IIS)** instance on a Windows server, then you can use the CodeDeploy agent to choreograph the placement of files and restart any necessary services needed to update files on these instances.

Deploying application updates to multiple regions with one deployment push

When you are looking for a way to build a solution that will create an automated **End-to-End (E2E)** release flow for deployments in multiple regions, then AWS CodeDeploy can help do this with help from AWS CodePipeline. This is especially true in the case of trying to keep Lambda deployments in sync for either **High-Availability (HA)** purposes or a **Disaster Recovery (DR)** strategy. In the following diagram, you can see CodeDeploy being used to deploy to multiple regions:

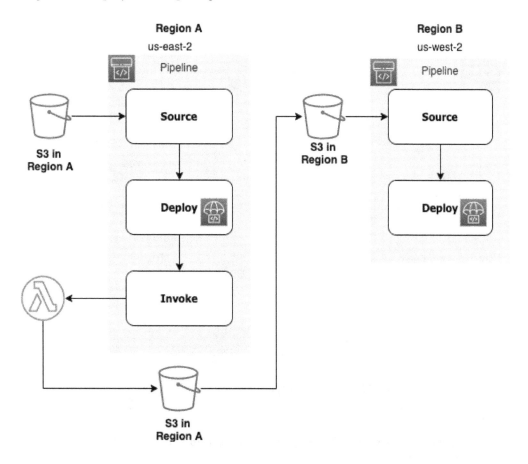

Figure 9.13 – Using CodePipeline and CodeDeploy to deploy to multiple regions

Using S3 as the origin for the source code will kick off the deployment via the AWS code pipeline. If the pipeline is successful, then it will move on to an invoke stage where a Lambda function will copy the source code to an S3 bucket in the replicated region.

This replication of the source code to the S3 bucket in the **B** region will start the process again on another pipeline that has been set up in the second region.

Deploying a new task to ECS in blue/green fashion

CodeDeploy can give you the ability to seamlessly switch between task sets that are behind a network or application load balancer. It does this by deploying the new version of the task set and then switching the listener to the new version at the load balancer level, as illustrated in the following diagram:

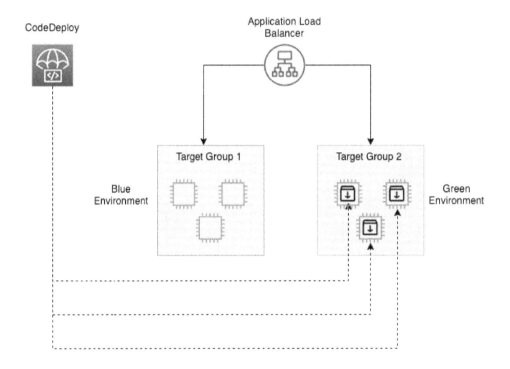

Figure 9.14 – Blue/green deployment using CodeDeploy

> **Note**
>
> We will go more in-depth into blue/green deployments in *Chapter 13, Blue Green Deployments*.

Deploying a task to Amazon ECS and using Lambda to validate before switching traffic over

With the use of CodeDeploy deployment groups, along with the assistance of a Lambda function, you can create a test to ensure that your new task is up and running successfully. This Lambda function can be specified in the deployment `appspec` file, and if the validation fails, then the deployment is stopped and rolled back. If the validation succeeds, then the deployment continues.

Monitoring CodeDeploy jobs

Whether your CodeDeploy jobs are running standalone with the CodeDeploy service or as part of a deployment pipeline, monitoring their status and being notified promptly when there is a failure can be critical for solving issues in a timely fashion.

Partnering CodeDeploy with the monitoring abilities of CloudWatch allows you to monitor specific metrics during deployments. Using these metrics, you can create CloudWatch alarms. Up to 10 CloudWatch alarms can be associated with a CodeDeploy deployment group. Triggering any one of the alarms will cause the deployment to stop and the status of the deployment to be reflected as `Stopped`. In order to monitor CodeDeploy with the CloudWatch service, you must grant your CodeDeploy service role permissions to use the CloudWatch service.

CloudWatch events can be used to help not only detect but also react to failures of CodeDeploy jobs based on rules that you create. Once these rules are created, CloudWatch events will initiate actions on specific targets. The following targets will work from CloudWatch events with rules from CodeDeploy jobs:

- AWS **Lambda** functions
- Kinesis streams
- Amazon **Simple Queue Service (SQS)** queues
- **CloudWatch** alarm actions
- Amazon **Simple Notification Service (SNS)** topics (and notifications)

CodeDeploy monitoring use cases

If your team uses Slack as a communication channel, then you can integrate a Slack notification via a Lambda function whenever a CodeDeploy deployment fails.

CloudWatch alarm actions can be used to programmatically reboot, stop, or terminate EC2 instances if a specific event occurs during deployment.

Now that we have looked at the abilities to monitor our CodeDeploy deployments, let's recap what we have learned in this chapter.

Summary

In this chapter, we looked at the other pieces of the AWS developer tools we are going to cover in depth in this book. We learned about deploying our software to different environments, both in the cloud and on-premises, using AWS CodeDeploy. We examined how CodeDeploy can be used to not only push out new versions of an application but also to control traffic during deployments. We also studied the AWS CodePipeline orchestration tool and how it as a service can incorporate not only all of the other three services we have been looking at from a development perspective but also other third-party partner tools as well.

In the next chapter, we will look at the AWS OpsWorks service and how this can be helpful for managing infrastructure and application services using stacks, especially if your team members are well versed in Chef or Puppet.

Review questions

1. One of the team of developers at a company has made multiple deployments. The last deployment with overwrite content failed. You have been tasked with rolling back to the previously working version with all the files necessary for the application. Which option would you choose to fulfill this requirement in the most expedient way possible?

   a. Manually roll back to the last known application version, which will add files required for application revision.

   b. Manually roll back to the previous deployment and then manually add files for the application revision.

   c. Automatically roll back to the last known version, which will add files required for application revision.

   d. Automatically roll back to the last known application version and manually add files for application revision.

2. You have constructed an AWS code pipeline that carries out a code release process. There are two stages to this pipeline: a source stage and a deploy stage. The source stage is using the third-party provider GitHub to source the code for the deployments. AWS CodeDeploy is being used to deploy the new versions of the application to multiple EC2 instances in a target group. The last few deployments have not gone successfully, and failures appear during the CodeDeploy stage. You need the ability to increase monitoring and notifications for deployments in order to cut down your Mean Time to Resolution (MTTR). How can you create notifications as soon as an issue is detected?

    a. Set up CloudWatch events for both CodeDeploy and CodePipeline. Use Amazon Inspector to create an assessment target to assess code deployment issues and create an SNS topic so that you can be notified of deployment issues.

    b. Set up CloudWatch events for both CodeDeploy and CodePipeline. Use an AWS Lambda function to help assess code deployment issues and create an SNS topic so that you can be notified of deployment issues.

    c. Set up a new AWS CloudTrail trail for the region that the pipeline is running in. Use AWS Config to assess any code deployment issues and create an SNS topic so that you can be notified of any deployment issues.

    d. Set up a new AWS CloudTrail to capture API events from CodeDeploy and CodePipeline. Use an AWS Lambda function to help assess code deployment issues and create an SNS topic so that you can be notified of deployment issues.

Review answers

1. d
2. b

10
Using AWS Opsworks to Manage and Deploy your Application Stack

If your operations team is already managing infrastructure and applications with either Puppet or Chef, then you should have already heard of OpsWorks. We'll look at how stacks can use pre-defined code to create the infrastructure applications and even the data tier for you. Need to perform updates to your apps? We will look into that as well.

In this chapter, we're going to cover the following main topics:

- OpsWorks essentials
- Available OpsWorks platforms

- OpsWorks and Chef Automate
- Creating and deploying a recipe

Technical requirements

Although not necessary, familiarity with either the Chef or Puppet platform will be helpful when reading this chapter. It is required that you understand the different components of a full-stack application and how they interact with each other.

OpsWorks essentials

In *Chapter 7, Using CloudFormation Templates to Deploy Workloads,* we took a look at the **Infrastructure as a Service (IaaS)** offering from **Amazon Web Services (AWS)**. Now we will look at one of the **Platform as a Service (PaaS)** offerings from AWS, OpsWorks.

OpsWorks is, in essence, a wrapper around the automation packages Chef and Puppet. Why would you want to use OpsWorks? The following bullet points outline why you would choose to implement OpsWorks as a solution for your client or enterprise:

- **Simple**: OpsWorks is easy to use.
- **Productive**: OpsWorks helps reduce errors by using scripted configuration.
- **Flexible**: Deployments of any size can be simplified by using OpsWorks.
- **Powerful**: OpsWorks helps reduce both the time of deployment and the costs to deploy.
- **Secure**: The OpsWorks service allows for fine-grained access control.

> **Important note**
> OpsWorks is one of the few services that doesn't require a region. It is a global service that shows all stacks created, no matter what region your cloud resources are currently residing in.

OpsWorks can be especially helpful if you have a lean team that is responsible for all or most components:

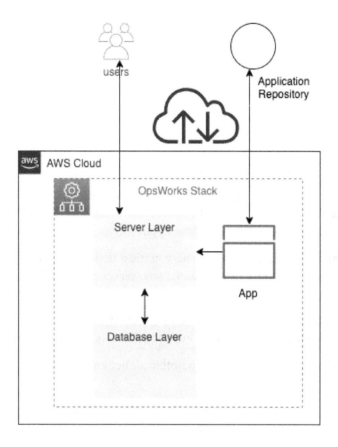

Figure 10.1 – OpsWorks serving users from an app

Through a series of events, you can both construct and manage your applications using AWS OpsWorks:

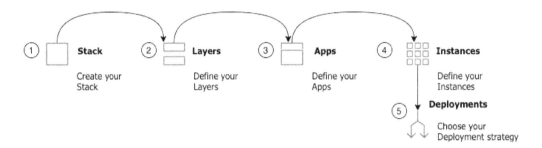

Figure 10.2 – Overview of how OpsWorks operates

First, you create a stack for your application. Multiple applications can reside on the same stack. They should all be components of the same system, such as a LAMP stack, a Rails deployment, or even a single- or double-tiered application.

Second, after your stack has been provisioned, you must define your layers. In the case of a LAMP stack, those layers would be an Apache web server with PHP pages, along with a MySQL data store.

Third, you will have your applications defined in a combination of code versioning repositories and deployment code, such as Puppet manifests or Chef recipes.

Fourth, EC2 instances need to be defined in both number and size so that our application has a spot to be deployed to.

And finally, for the fifth step, we need to choose a deployment strategy for our different layers. This strategy can vary based on whether you are performing manual or automated deployments.

Understanding the components of OpsWorks

As we start to talk about OpsWorks, it will be essential to understand specific keywords. Let's now look at the components that make up the parts of OpsWorks.

Stacks

A **stack** represents the cloud infrastructure and applications that you want to manage.

You cannot mix and match Windows and Linux instances in a stack. You can, however, run different distributions of Linux in the same stack.

Your stack can comprise custom **Amazon Machine Images** (**AMIs**) as long as the base operating system from which the image was made is one of the supported operating systems. OpsWorks does not support custom Windows AMIs.

Layers

A **layer** defines how to set up and configure a set of instances, as well as any related resources. Every stack in OpsWorks contains one or more layers. You can think of a layer as a blueprint for OpsWorks to use for a set of one or more EC2 instances.

At the layer level, you can define how to configure the instances, what packages are installed, and also define critical events that will happen during the lifecycle.

There are several different types of layers that you can define in OpsWorks:

- Load balancer layers.
- Database layers.
- Application server layers.
- Custom OpsWorks layers (this is when no pre-defined layer will meet your requirements).
- Other layers: Linux stacks have the ability to create Ganglia layers for monitoring or Memcached layers for caching.

Instances

An instance in OpsWorks comprises a computing resource, such as an EC2 instance or on-premises server.

Users can add instances to an OpsWorks stack in one of two ways:

- You can use OpsWorks to provision the instance.
- In the case of Linux-based stacks, you can register a previously created EC2 instance or on-premises server. This would then allow OpsWorks to manage the instance.

AWS OpsWorks supports 64-bit versions of a few of the operating systems offered on the AWS EC2 platform, including Ubuntu and Amazon Linux, along with Windows Server.

Apps

In OpsWorks, an **app** represents code that you would like to apply to an application server. The code itself will reside in an application repository, such as GitHub or S3. When you deploy an application, OpsWorks triggers a deploy event.

Lifecycle events in OpsWorks

There are five series of events that happen during the lifecycle of an instance when managed by OpsWorks. The events are as follows:

- Setup
- Configure
- Deploy
- Undeploy
- Shutdown

Let's take a look at some of the events in a bit more detail:

- **Setup**: This is triggered every time an instance boots up. It is also used for the initial installation of packages.

- **Deploy**: This event is triggered when deploying a new software version. The Deploy event also allows for customizable software deployments.

- **Configure**: The Configure event is triggered on all instances when there is a state change. This event is used to ensure that the configuration is up to date on all instances.

Access management in OpsWorks

If you have a requirement to provide individual user-level access to instances, be they Linux or Windows instances, then OpsWorks might be the solution you are looking for. The AWS OpsWorks service provides an easy way to allow individual users to log in to the server with their own credentials without the use of custom scripts.

If you have ever stood up EC2 instances and tried to get in via SSH, then you are most likely going to use the EC2 user or Ubuntu user, depending on the flavor of the operating system that you have used to provision your instance. If you would like to have custom users be able to access individual EC2 instances then this can become a complex setup, provisioning individual users and having them authenticate with either LDAP or an Active Directory Server.

The same is true if you are allocating Windows servers on AWS. You need to take the key pair that you used to stand up the instance and then, using that key pair, decrypt the administrator password.

> **Important note**
> All users for AWS OpsWorks Stacks must originate from **Identity and Access Management (IAM)** users.

Opsworks lets you provision access for users to access to the instances in the stack quickly and easily without the need to set up an Active Directory or LDAP server.

Types of users in OpsWorks

There are two types of users inside of the AWS OpsWorks service: administrative users and regular OpsWorks users.

Regular users in OpsWorks do not require an attached policy or any special OpsWorks permissions in their IAM user permissions. On a stack-by-stack basis, you can designate one or more of the following four permissions:

- **Show**: This permission allows users to view the stack only.

- **Deploy**: The deploy permission also includes the show permission, and adds on extra permissions for users to update and deploy apps.

- **Manage**: The manage permission includes the deploy permission set along with adding stack management permissions, such as adding layers or instances.

- **Deny**: The deny permission is used to deny specific permissions across a stack.

Importing users into OpsWorks

Throughout the course of the exercises we have performed so far, we have created a number of IAM users in the account we have been using. We are going to use these users in our next exercise to practice importing users into OpsWorks:

1. Log on to the AWS terminal as the administrative user.

2. Once logged in, navigate to the **OpsWorks** service by searching for OpsWorks in the top search box. Once **OpsWorks** appears, click on the **OpsWorks** icon to be taken to the **OpsWorks** service:

Figure 10.3 – The OpsWorks service from the top search bar in the AWS Console

3. Even though once you import users, they appear in any one of the three sections of the OpsWorks service, we are going to import our users into **OpsWorks Stacks**. Click on the blue button on the bottom right, which is labeled **Go to OpsWorks Stacks**.

4. Now, in the OpsWorks Stacks service, look to the left-hand menu and find the menu item named **Users**. Click on **Users** to be taken to the **Users** management screen:

Figure 10.4 – The Users menu item on the OpsWorks Stacks side menu bar

5. In the main window pane, under any current users, find the link labeled **Import IAM users to US East (Ohio)**. Click on this link to be brought to the import page.

6. Once you have clicked on this link, a new menu will appear, allowing you to select the users that you want to import into OpsWorks. Select one or more users that you want to import into OpsWorks as OpsWorks users, and then click the blue **Import to OpsWorks** button:

The following users have IAM-only permissions for all AWS OpsWorks resources. To grant per-stack permissions and instance access, select the user(s), and then choose **Import to OpsWorks**.

☐ **Select all users**

              devops
             ☑ mariel
             ☑ peter

Cancel   **Import to OpsWorks**

Figure 10.5 – Selecting the users to import inside of the OpsWorks service

7. Now we should see a list of our users that we have imported into OpsWorks. If we wanted to add further users or restrict their permissions (such as giving a user self-management access to add their own SSH key, or the ability to change their own password), we could do that by editing the user. Let's edit one user. Click on the edit link next to our developer, Mariel.

8. In another terminal window, print out the contents of your developer's public SSH key that you created previously for the exercise in *Chapter 8, Creating Workloads with CodeCommit and CodeBuild*. Copy this public key to your clipboard for use in the next step:

```
$cat ~/.ssh/cc_developer.pub
```

9. Switching back to the AWS Console, and now on the user Mariel's screen, scroll down until you see the multi-line text box to the left of **Public SSH key**. Paste the public key that you have previously copied into this box:

SSH User Name mariel

Public SSH key

```
/JSUJGULE+Ji941RiFeGDShBSKEemd6StQaM9dUyREpMGN1NIUK0hQnhP2xqmNCautpMGi4U4v+j5V7jzgAnucSXjcr
y+PxqPAEPAxWmMKE2Rejadfasfasdfsdfsdfd++S
/O6SqCd+RlnyMZRadsfsdfasfdJ2OPanahCVdLIPz33JxbpD3asdfsdfsdfsadfaslasdfdsffd5MKNdhNFd
/rhXAHZ0T20Q44prELFPY11XUw==
```

A user is created on **Linux-based instances** if the user has a **public SSH key**.

Figure 10.6 – Adding the developer's public SSH key to the OpsWorks user

10. With the public key added, our developer can log in to an instance with the following:

```
$ ssh -i ~/.ssh/cc_developer mariel@INSTANCE-DNS
```

We have now gone through the exercise of importing previously created IAM users into OpsWorks, along with setting up our developer's previously created SSH key so that they can easily log into an OpsWorks instance with their own username.

Next, we will look at some of the scenarios where you would want to use the OpsWorks service.

Use cases for AWS OpsWorks

After getting this far, you may be wondering what the best cases to put this service into practical use are. We will look at a few scenarios detailing when it would make sense to use the OpsWorks service. You need to understand that when evaluating either scenarios presented by test questions or real-life situations, OpsWorks is one of a number of deployment and management tools that AWS offers. Elastic Beanstalk is another tool that can be used to provision, deploy, and manage instances and code. We will cover this in much more depth in the next chapter, *Chapter 11, Using Elastic Beanstalk to Deploy your Application*.

Available OpsWorks platforms

OpsWorks, by its nature, is a configuration management service that provides managed instances of Chef and Puppet. Both of these software packages are industry-recognized automation platforms that enable you to use code and configure your servers via automation.

Now let's take a look at the three platforms that OpsWorks offers in a bit more depth.

OpsWorks Stacks

The original service offering from OpsWorks was OpsWorks Stacks. By using layers, it helps you manage and organize your EC2 instances.

OpsWorks for Chef Automate

If you have a team that is well versed in the creation of Chef cookbooks and knows how to use some of the more advanced features of Chef, such as Knife, yet doesn't want to deal with the hassles and headaches of maintaining their own Chef servers, then Chef Automate can be a logical choice for your team.

If you are currently using Chef to manage some of your on-premises infrastructure, OpsWorks for Chef Automate can be set up to manage hybrid environments from a single source.

OpsWorks for Chef Automate also has the ability to perform compliance scans. These compliance scans allow you to audit your applications on a regular basis to detect exposures and non-compliant configurations.

There is no need to worry about patching, updating, or backing up your Chef server, since these are some of the tasks that OpsWorks takes care of for you. OpsWorks for Chef Automate also allows you to manage the instance of Chef through the Chef Console, or via the command line with Knife.

OpsWorks for Puppet Enterprise

Puppet helps enforce the desired state of the infrastructure with its set of tools.

If you are an existing Puppet user, then OpsWorks for Puppet Enterprise would be the most logical choice for you. Using OpsWorks for Puppet Enterprise, you can launch a Puppet master server in minutes, and then allow the OpsWorks service to handle operations tasks such as backup, software upgrades, and restorations.

When using the Puppet Enterprise version of OpsWorks, the Puppet master houses and provisions all of the Puppet modules for the instances that need to run **Puppet Agent**.

Now that we have looked at the various different options available for OpsWorks, including the two options that use Chef recipes as the basis, and the one option that uses Puppet manifests, we will now examine a quick example. This example will give us a better understanding of how all the different components of OpsWorks come together when trying to pick out the best solution, both for our clients and when looking at possible solutions for the *DevOps Professional Exam*.

Creating and deploying a recipe

In this example, we will use OpsWorks to set up a stack and then create a layer. After our layer has been created, we will then use a recipe that is publicly available to deploy to our app:

1. Log in to the AWS Console as your administrative user and use the unified search bar to search for the OpsWorks service. Once you see the OpsWorks icon appear, click on the icon to be taken to the **OpsWorks** main page:

Figure 10.7 – The OpsWorks service from the unified search bar

2. On the OpsWorks Stacks page, click on the blue **Go to OpsWorks Stacks** button, which should be at the bottom left. This is the original offering from OpsWorks:

OpsWorks Stacks

Define, group, provision, deploy, and operate your
applications in AWS by using Chef in local mode.

Go to OpsWorks Stacks

Figure 10.8 – The OpsWorks Stacks option on the OpsWorks service page

3. Once on the OpsWorks Stacks page, click on the blue **Create stack** button at the top right-hand side of the page. This will start the process of creating a new stack in OpsWorks.

4. Next, we are going to create a `Chef-11` stack with the following options:

    *   **Stack name** – `chapt-10`

    *   **Region** – `US East (Ohio)`

    *   **VPC** – (choose your default VPC)

    *   **Default subnet** – (leave the default subnet)

    *   **Chef version** – `11.10`

    *   **Use Custom Cookbooks** – `YES`

    *   **Repository Type** – `Git`

    *   **Repository URL** – `https://github.com/amazonwebservices/opsworks-example-cookbooks.git`

5. Leave everything else as it is.

6. Click on the blue **Add Stack** button.

7. Refer to the following screenshot:

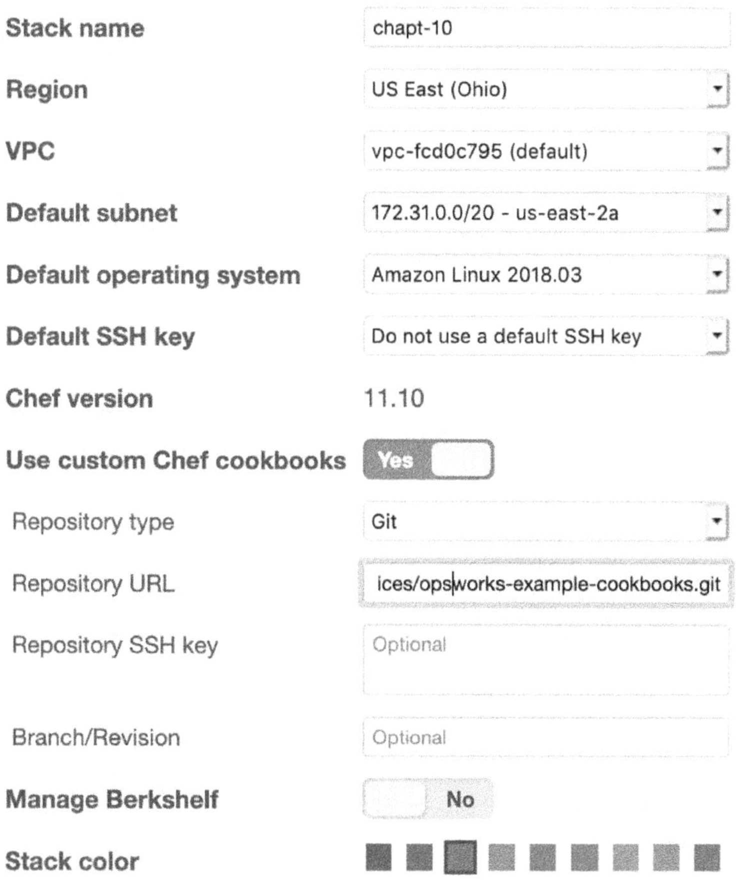

Stack name	chapt-10
Region	US East (Ohio)
VPC	vpc-fcd0c795 (default)
Default subnet	172.31.0.0/20 - us-east-2a
Default operating system	Amazon Linux 2018.03
Default SSH key	Do not use a default SSH key
Chef version	11.10
Use custom Chef cookbooks	Yes
Repository type	Git
Repository URL	ices/opsworks-example-cookbooks.git
Repository SSH key	Optional
Branch/Revision	Optional
Manage Berkshelf	No
Stack color	

Figure 10.9 – Adding the options to create the stack

8. Now we should have seen a **Congratulations! Your stack was created** notification. Our next step is to add a layer to our stack. We can click the **Add a layer** link and click add a layer:

Congratulations! Your stack was created.

Next step: Add a layer.

Figure 10.10 – The congratulations banner showing the link to add a layer

9. Now on the layer screen, we will select **PHP App Server** from the drop-down list. Once that is selected, then we can click the blue button that says **Add Layer**:

Figure 10.11 – Adding the layer in OpsWorks

10. With our layer available, we need to add an instance to our layer. With our **PHP App Server** page showing, we need to click the **Add instance** link at the very right-hand side:

Figure 10.12 – Showing our created layer in OpsWorks

11. Now add the instance with the following details:

Hostname – `php-app1`

Size – `c5.large`

Subnet – (leave the default subnet that is selected)

When you have filled in all of these options, click the blue **Add Instance** button at the bottom right-hand side of the selection window:

New	Existing OpsWorks	EC2 instances and own servers
Hostname		php-app1
Size		c4.large
Subnet		172.31.0.0/20 - us-east-2a
Advanced »		

Cancel   **Add Instance**

Figure 10.13 – Adding an instance to the OpsWorks layer

12. With your instance added, you can now start your instance. You will notice that on the screen this instance type is designated as a 24/7 instance. Press the **start** link under the **Actions** column to start your instance.

13. After about a minute or so, the instance status should change from `stopped` or `booting` to `online`.

Congratulations, you now have your stack up and running!

Now that we have seen how to create an OpsWorks stack, including its layers, instances, and app, we will take a closer look at deployments in OpsWorks. We will continue with our current stack that we just created, using that stack to create an app and then deploy it to the stack.

Deployments in OpsWorks

When you deploy an app in OpsWorks, its primary purpose is to deploy the application code and all of the files to the server instances. We previously created a stack, a layer, and an instance in OpsWorks. Now we will add an app and make a deployment to the stack:

1. Return to the OpsWorks Console and enter back into the stack we had previously created – named `chapt-10`. You can enter the stack by clicking on the stack name.

2. Once inside the stack, on the left-hand menu, click on the item named **Apps** to be taken to the apps page. At the top of this page, click on the blue button labeled **Add App** to create a new app:

 Apps

⋏ Deployments

Figure 10.14 – Apps and Deployments in the left-hand menu in OpsWorks

3. Use the following values to populate your app:

   - **Name** – `SampleAPP`
   - **Type** – `PHP`
   - **Repository Type** – `Git`
   - **Repository URL** – `https://github.com/awslabs/opsworks-demo-php-simple-app.git`

All other values can be left blank. After you have filled out these values click the blue **Add App** button at the bottom of the screen:

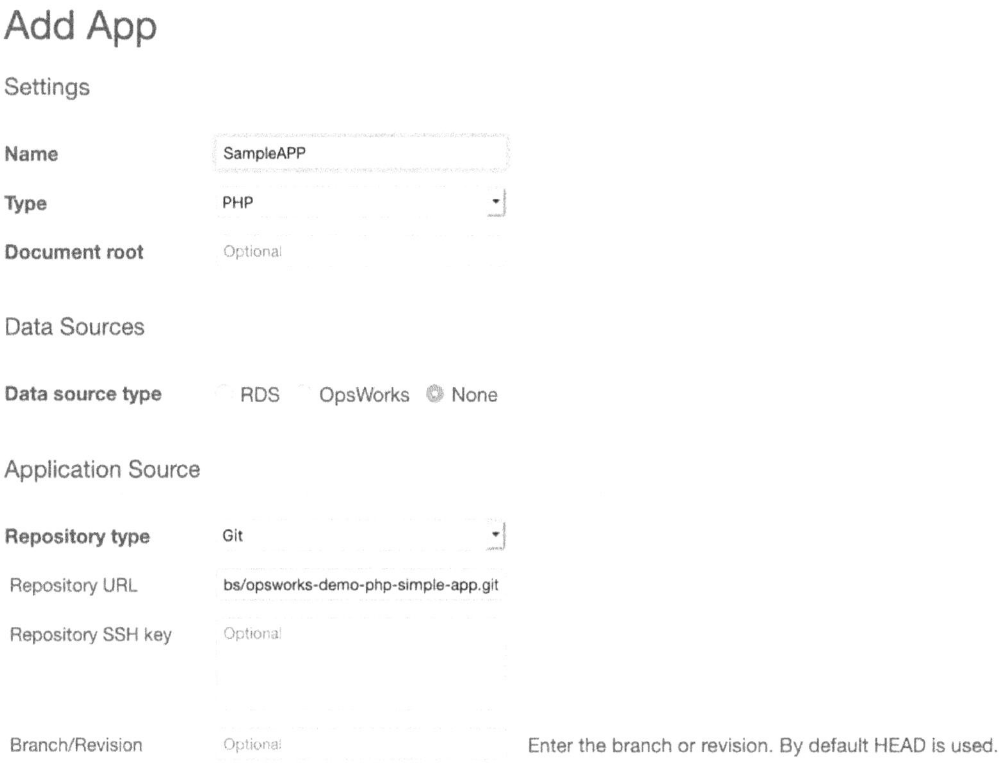

Add App

Settings

Name	SampleAPP
Type	PHP
Document root	Optional

Data Sources

Data source type	○ RDS ○ OpsWorks ◉ None

Application Source

Repository type	Git
Repository URL	bs/opsworks-demo-php-simple-app.git
Repository SSH key	Optional
Branch/Revision	Optional	Enter the branch or revision. By default HEAD is used.

Figure 10.15 – Adding the app settings

4. You should now see a screen showing our app. It's now time to deploy the app using OpsWorks. OpsWorks makes this easy with the **deploy** link being directly under the **Actions** column at the right-hand side of the app. Click the **deploy** link to start the deployment process:

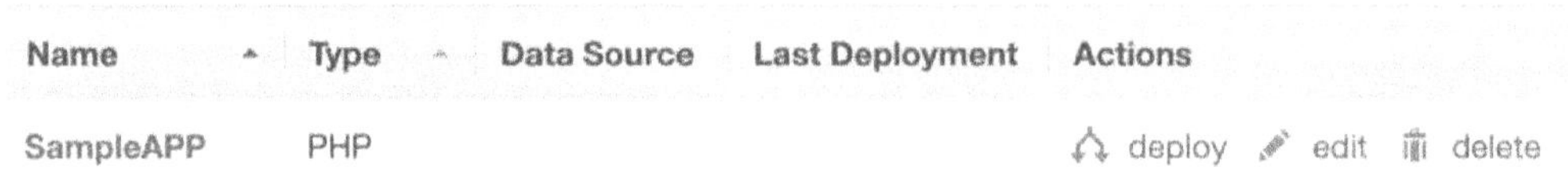

Name	Type	Data Source	Last Deployment	Actions
SampleAPP	PHP			⚡ deploy ✏ edit 🗑 delete

Figure 10.16 – After the app has been created, it is ready to be deployed

5. On the deploy screen, before you deploy, click on the drop-down menu that currently has **Deploy** selected to see all of the different options for deployment. In this section, you can not only deploy the app, but also allow OpsWorks to undeploy, roll back, start, or stop a web server. Make sure that you retain the setting on deployment, and simply keep all of the defaults. Then click the blue **Deploy** button at the bottom of the screen.

We have just completed provisioning an app using a custom recipe. We also deployed that app to the instance that we previously created.

Monitoring OpsWorks

You can monitor your AWS OpsWorks stacks in a number of ways:

- Using Amazon CloudWatch
- Using Amazon CloudWatch Logs
- Using Amazon CloudWatch Events
- Using Amazon CloudTrail

With Amazon CloudWatch, there are 13 custom metrics provided for detailed monitoring of each of the instances in the stack. There is also a custom monitoring page created in OpsWorks that summarizes the data and makes it easy to understand these metrics. If you are running Windows instances, the monitoring page will not display the metrics for those Windows instances. The metrics are displayed for the entire stack, but you can also concentrate on a particular layer or a particular instance.

If you still have the deployment running from the previous exercise, you can click the **Monitoring** menu icon to be taken to the dashboard. This can give you quick insights into the state of your application:

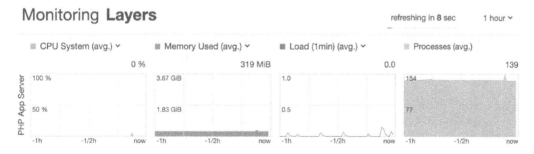

Figure 10.17 – Monitoring dashboard in OpsWorks

We have just covered the full OpsWorks process from creating a stack to monitoring the stack. Now let's review what we learned.

Summary

In this chapter we looked at how the OpsWorks service can help us provision and deploy both our infrastructure and application. It does this by using stacks and layers. We also learned that OpsWorks is a good way to quickly transition a team that is currently using either Chef or Puppet to provision their servers and/or applications in their current environment.

In the next chapter, we will learn about the Elastic Beanstalk service and how it can help you deploy your applications into environments, create revisions for those applications, and help you monitor those applications using a number of different programming languages.

Review questions

1. You have been brought into a company that is starting its move to the AWS cloud. They have a sizable library of Chef recipes that they currently use to manage their on-premises systems. The recipes have already been moved to a private GitHub repository. The individual who managed and administrated the Chef servers left around one month ago. The company is not prepared to re-write all of the recipes to another means in the current fiscal year with their current available resources. What AWS service or services would be best for their migration?

    a) Use AWS Server Migration Service to move the current Chef server to the AWS cloud.

    b) Stand up the latest version of Chef on EC2 with autoscaling for redundancy. Add a Route 53 record in a locally hosted zone for the Chef server. Create a base AMI that already has the Chef agent pre-installed and connects to the DNS entry configured in Route 53.

    c) Set up the applications as stacks in OpsWorks. Create all the different users as IAM users, and then import them into OpsWorks. Then assign the OpsWorks users to individual stacks as needed.

    d) Create a CodePipeline instance to install and configure the applications as needed, using headless Chef in the user `init` script.

2. A company has developed a PHP shopping cart platform. The platform is currently
    being managed and deployed by AWS OpsWorks with separate stacks for dev, QA,
    and production. With most of the PHP developers who originally developed the
    platform no longer with the company, management has given the go-ahead to start
    re-developing the platform in Python. How should the company manage
    the deployment?

    a) Create a new OpsWorks stack that contains a new layer with the new Python
       code. To make the transition over to the new stack, the organization should use a
       blue/green deployment.

    b) Create a new OpsWorks stack that has the new Python code manage separate
       deployments of the application from the secondary stack.

    c) Update the existing stack with the new Python application code and deploy the
       app using the Deploy lifecycle event.

    d) Create a new layer on the OpsWorks stack with the new Python code. To make
       the transition over to the new stack, the organization should use a rolling
       deployment.

3. An enterprise oversees a combination of Windows servers and Linux servers (Red
    Hat Enterprise Linux and Amazon Linux) in their on-premises data center, along
    with their AWS account. After recently being audited, the CTO was told that there
    was no process in place for core application patches, nor for updating the operating
    systems. There was no baseline level of patching for any of the servers, either in
    the on-premises data centers or in the AWS cloud. You have been called in to help
    remedy the situation. What solution would you recommend that would be not only
    the most consistent for providing and maintaining OS and core application patch
    levels but would also prove the most reliable?

    a) Use AWS Systems Manager to store the credentials in the parameter store,
       and then create resource groups for the instances. Allow the Systems Manager
       Run command to deploy patches remotely using the credentials stored in the
       parameter store.

    b) Install the OpsWorks agent on all the servers, both on-premises and in the AWS
       account. Using a single stack, create different layers for the different operating
       systems. Create maintenance windows whenever an operating system or core
       application patch needs to be applied.

c) Configure the AWS Systems Manager agent on all of the servers, both on-premises and in the AWS account. Create resource groups for the instances and then allow Systems Manager Patch Manager to run with a pre-configured baseline, using set maintenance windows.

d) Install the OpsWorks agent on all the servers, both on-premises and in the AWS account. Create two different stacks, one for Windows and the other one for the Red Hat Linux instances. Create a Lambda job that will curl an RSS feed two times a day, checking for new patch updates. If it finds one, then it will trigger the deployment on the OpsWorks layer.

Review answers

1. c

2. a

3. c

11
Using Elastic Beanstalk to Deploy your Application

Knowing and understanding Elastic Beanstalk is essential to the DevOps Pro exam. How to use the service's nuances, what the service's limitations are, and especially the way Beanstalk interacts with containers can be vital knowledge for both passing the exam and getting application teams up and running quickly in the real world.

In this chapter, we're going to cover the following main topics:

- Understanding the built-in functionality of Elastic Beanstalk
- Creating a service role in the IAM console
- Installing and using the Elastic Beanstalk command-line interface (EB CLI)
- Understanding advanced configuration options with .ebextensions
- Using Elastic Beanstalk to deploy an application
- Elastic Beanstalk use cases

Technical requirements

Since Elastic Beanstalk requires some local development and we will be writing our example program in Python, you will need to have Python 3.x locally installed on your workstation. At the time of writing, the most up-to-date version of Python, which Elastic Beanstalk supported, was 3.8.5, and hence this would be the version of Python that we will be using in this chapter. It is the suggestion of the author that you install the `pyenv` Python runtime manager to help you to be able to develop and manage multiple versions of Python on your local workstation. You can find the project's information and how to install it on its GitHub site, which can be found at `https://github.com/pyenv/pyenv`.

Understanding the built-in functionality of Elastic Beanstalk

Elastic Beanstalk is a flexible platform that allows developers who develop in languages such as Python, Java, Ruby, .NET, PHP: Hypertext pre-processor (PHP), Go, or even Docker containers to develop and scale out their applications swiftly. Elastic Beanstalk is a **Platform as a Service** (**PaaS**), which means that it is serving as an abstraction layer above all of the other services that it is helping orchestrate and manage. This makes it easy for developers who are eager to get going with the **Amazon Web Services** (**AWS**) cloud but might not have a vast amount of knowledge of items such as setting up infrastructure or monitoring, the ability to get started, and concentrating on their application in the language of their choice.

Elastic Beanstalk is constantly updating the languages and platforms that it supports. While it does not always support the latest version of a language, it does try to keep up to date with newer revisions and then sunset older versions. This is important to remember since legacy applications cannot just be pushed to the cloud and then left to run forever. There are times when applications will need to be upgraded in order to retain the functionality of the Elastic Beanstalk service. Here are some of the services that Elastic Beanstalk incorporates:

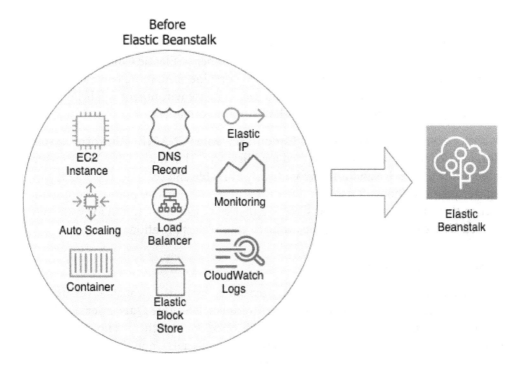

Figure 11.1 – A look at some of the services that Elastic Beanstalk incorporates

Before the Elastic Beanstalk service was available for use, developers had to figure out which services to use to work together, and not only for computing, load balancing, and routing. They also needed to—either manually or with the help of a service such as CloudFormation templates—incorporate security, storage, and monitoring into their application. For developers who wanted to get up and running quickly on the AWS cloud, this could present a challenge, as they were used to creating application code rather than configuring environments.

Different environments in Elastic Beanstalk

Any type of application that you build and deploy using Elastic Beanstalk allows you to manage all of the different components contained for that application as a single environment. There are two major classifications of environments that are run using the Beanstalk framework: web server environments and worker environments.

Web server environments

In a web server environment, Elastic Beanstalk manages three things: a load balancer, an **Auto Scaling Group** (**ASG**), and the requested number of **Elastic Compute Cloud** (**EC2**) instances. Even if your application has been containerized, those containers still need an EC2 instance to run on, and hence the web server environment will bring up a corresponding set of EC2 instances in which to run the containers.

A web server environment also creates a **Canonical Name** (**CNAME**) (**Uniform Resource Locator**, or **URL**) that points to the created load balancer. That initial CNAME is created using the Route 53 service and is therefore highly available and scalable. If desired, you can register your own domain name and use the Route 53 service to point to your load balancer so that instead of the generic CNAME that was originally created for you, your customized URL can be what drives the traffic to your web application.

Worker environments

The big difference in a worker environment provisioned by Elastic Beanstalk versus a web server environment is that in a worker environment, a **Simple Queue Service** (**SQS**) queue is provisioned if you have not already provisioned one and there is no provisioning of a CNAME via Route 53. Beanstalk automatically installs a daemon that allows for autoscaling, as illustrated in the following diagram:

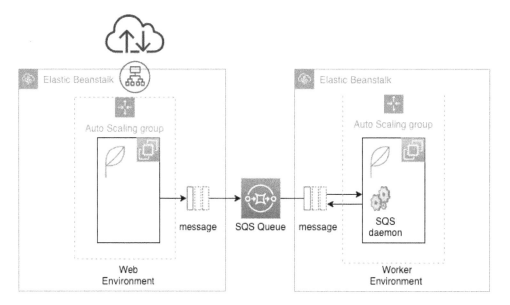

Figure 11.2 – Web server and worker environments in Elastic Beanstalk

Next, we are going to cover the different components that make up Elastic Beanstalk.

The different components that make up Elastic Beanstalk

No matter which of the different types of environments you are trying to build using the Beanstalk service, it will be composed of multiple components.

Application

In Elastic Beanstalk, a logical grouping of components is called an **application**. This includes environments and environment configurations, along with versions. All of these items are grouped into a folder-like structure, and this folder is the application.

Application version

Elastic Beanstalk allows for the packaging and labeling of specific versions of an application. Each specific packaged version of an application is known as an **application version**. These versions are packaged up and sent to the **Simple Storage Service (S3)** service behind the scenes for deployment to one or more of your environments, as illustrated in the following diagram:

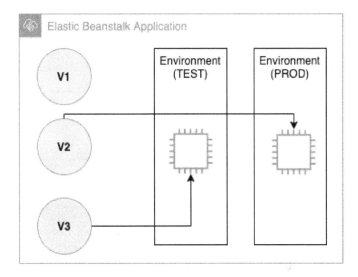

Figure 11.3 – Application versions can only be deployed to a single environment

You can have multiple versions running in various environments using Elastic Beanstalk, such as your current stable version running in your production environment and your latest version running in your test environment. Only one version can be deployed to a single environment at a time.

Environment

A collection of running resources in Elastic Beanstalk running a specific application version is known as an **environment**.

Environment configuration

As you define the parameters and settings for your environment, these become the **environment configuration**. As you iterate on your application and change the settings to the underlying configuration of the environment, then Beanstalk deploys those changes automatically and adds or removes items based on your configuration, as illustrated in the following diagram:

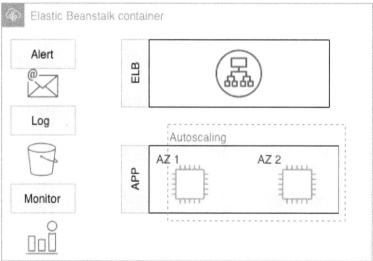

Figure 11.4 – A closer look at an Elastic Beanstalk configuration

Configuration template

The configuration template in Elastic Beanstalk is a starting point for creating customized environment configurations.

Saved configuration

Saved configurations are **YAML Ain't Markup Language** (YAML) files stored in S3 that can be put to use for other running environments or during environment creation. Saved configurations define the following values:

- Platform version
- Tier
- Configuration option settings
- Tags

Platform

When you incorporate the operating system of your choice and the preferred programming language that you will use, your web and application servers all become your **platform**.

Now that we have an understanding of the Elastic Beanstalk components and platform, let's go through an exercise of creating an application in Elastic Beanstalk, starting with creating the necessary permissions.

Creating a service role in the IAM console

Before we begin, we will need to create a service role for Beanstalk to use when pushing out the environments and applications. A failure to do so, or even a failure to update a previously created service role that was created a few years ago, can result in errors and frustration when moving further along in the chapter.

Start by logging in to the **AWS Management Console** with your administrative user, and then follow these next steps:

1. Once you have logged in, navigate to **IAM service**.
2. In the left-hand menu, click on **Roles**.
3. Once on the **Create Role** page, keep the trusted entity as **AWS Service**, and then, in the middle of the page, choose **Elastic Beanstalk** as the service for which you would like to create a service role. Click on the **Elastic Beanstalk** service.

4. After you have selected the **Elastic Beanstalk** service, another set of selections will appear at the bottom of the screen so that you can select your use case. Choose the use case of **Elastic Beanstalk**. After you have clicked on **Elastic Beanstalk** in the use case and it is highlighted in blue, as shown in the following screenshot, then you can click on the blue button on the bottom right-hand side of the screen labeled **Next: Tags**:

Select your use case

Elastic Beanstalk

Allows Elastic Beanstalk to create and manage AWS resources on your behalf.

Elastic Beanstalk - Customizable

Allows Elastic Beanstalk to create and manage AWS resources on your behalf.

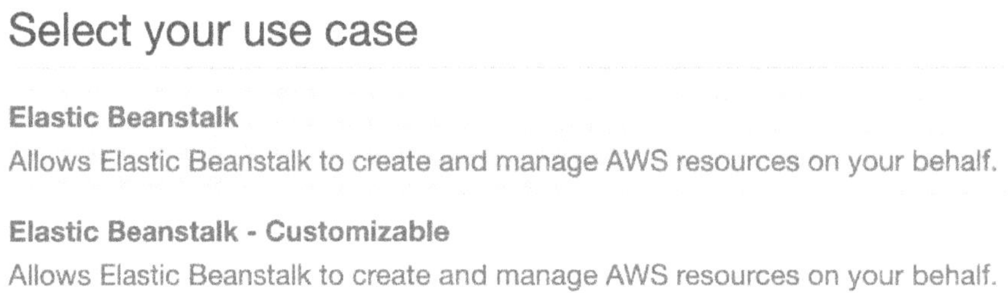

Figure 11.5 – Selecting the Elastic Beanstalk use case in Identity and Access Management (IAM)

5. After clicking **Next** and landing on the **Roles** screen, the default Beanstalk policy should already be populated for you. Click on the blue button at the bottom of the screen labeled **Next:Tags** to move on to the next screen.

6. Tags are optional and you can add them if you wish; otherwise, click the blue **Next:Review** button at the bottom of the screen.

7. On the **Review** page, you should be shown that you are about to create a service role named `AWSServiceRoleForElasticBeanstalk`. Click the blue **Create role** button at the bottom of the page to create this role.

Now that we have our service role created, we can proceed with using the Elastic Beanstalk service.

Installing and using the Elastic Beanstalk command-line interface (EB CLI)

There are a number of commands available from the AWS CLI that allow you, as a user, to take advantage of Elastic Beanstalk.

Basic commands supplied by the EB CLI include those that will do things all the way from creating the correct project structure in your local environment to quickly and efficiently pulling down the log files from your instances for review. We will now review the basic commands that the EB CLI provides, along with a short description of their nature, as follows:

- `eb create`: This command will create a new local environment for Elastic Beanstalk and deploys an initial application version to that directory structure.

- `eb status`: This command returns the status of your environment, including items such as application name, region, CNAME, and health status.

- `eb health`: This command returns the health status of the instances in your environment, updating every 10 seconds.

- `eb events`: This command returns a list of log statements that state the most recent events for the current Elastic Beanstalk environment. Examples of events can include the creation of resources such as instances or load balancers or the changing of the environment status to different health levels.

- `eb ssh`: This command will temporarily open port 22 on your security groups for all incoming traffic if you have not configured port 22 for a specific **classless inter-domain routing (CIDR)** range of addresses. It will then prompt you to connect to your running instance or allow you to select which running instance you would like to connect to via **Secure Shell (SSH)**.

- `eb logs`: This command can perform two explicit commands regarding the logging of files, as follows:

    a. It can toggle the streaming of logs to the CloudWatch Logs service.

    b. It can fetch instance logs for you to review locally.

- `eb open`: This command will use your default browser to open the public URL of your application.

- `eb deploy`: This command will deploy your application to the Elastic Beanstalk service using the current source bundle.

- `eb config`: This command will allow you to change the environment configuration settings.

- `eb terminate`: This command shuts down and stops your instances and environment so that you do not incur additional charges.

Installing the EB CLI

The simplest way to get the **EB CLI** installed on your local workstation, no matter which operating system you are using, is to use the `eb-cli-setup` script, which is available on GitHub at `https://github.com/aws/aws-elastic-beanstalk-cli-setup`).

If you are using a Mac, however, and prefer to install software via the `brew` package manager, then the EB CLI is available through this method as well. Just make sure that your current version of `brew` is up to date before attempting to install the EB CLI. Here's the code you'll need:

```
$ brew update
$ brew install awsebcli
```

No matter which method you choose, once installed, you should then be able to get directly to the EB CLI from a terminal prompt using just a simple `$  eb` command.

> **Note**
> If you have previously installed the EB CLI on your machine in the past, it is a good idea before proceeding to update the package using the `pip3 install --upgrade awsebcli` command.

Now, with the CLI installed, let's look at creating and deploying a project using the EB CLI.

Using the EB CLI to create a project

We will be using our terminal to invoke the `eb  cli` command and then create our Elastic Beanstalk project. Start by opening up your terminal window. Then, follow these steps:

1. With your terminal window open, use the EB CLI to create a new folder to start our project. It's best to start from the root of your home directory so that your folder will be easier to find if you need to come back to it later. We will name our local folder `11-beanstalk`. Here's the code you'll need:

```
$ cd ~
$ mkdir 11-beanstalk
```

2. We need to make sure that the `virtualenv` program has been installed on our system for the Python 3.x set of programming languages. We will do this using `pip`, as follows:

```
$ pip3 install -U virtualenv
```

3. Now, with our folder created, we can move into that folder so that we can create our virtual environment for Python. Once you have enacted the virtual environment, you will see your prompt change slightly to be prepended with 11 before the command prompt, as illustrated in the following code snippet. This lets you know that you are inside the virtual environment:

```
$ cd 11-beanstalk
$ virtualenv eleven
```

4. If the virtual environment has been created successfully, then you should see an output like the one shown next:

```
created virtual environment CPython3.9.1.final.0-64 in
767ms
  creator CPython3Posix(dest=/Users/abook/11-beanstalk/
eleven, clear=False, no_vcs_ignore=False, global=False)
  seeder FromAppData(download=False, pip=bundle,
setuptools=bundle, wheel=bundle, via=copy, app_data_dir=/
Users/abook/Library/Application Support/virtualenv)
    added seed packages: pip==21.1.2, setuptools==57.0.0,
wheel==0.36.2
  activators BashActivator,CShellActivator,FishActivator,
PowerShellActivator,PythonActivator,XonshActivator
```

5. Since we have created our virtual environment, we now need to activate it, as follows:

```
$ source eleven/bin/activate
```

6. With the virtual environment prompt showing, we will now install `flask` locally to make a quick web application using Python, as follows:

```
(eleven) ~/11-beanstalk $ pip3 install flask flask_
bootstrap template-render
```

Installing the `flask` package will also install a number of dependent packages that `flask` needs, including **MarkupSafe**, **Werkzeug**, **Jinja**, **ItsDangerous**, and **Click**.

> **Note**
>
> If you're not familiar with it, `flask` is a lightweight web application framework written in Python. It uses Jinja templates for rendering pages but can also just parse text or **HyperText Markup Language** (**HTML**). It can also be combined with other Python libraries to do more complex calculations and then render the results.

7. With our environment being set up, we can then move the files that we took from the `Chapter-11` GitHub repository and copy them over to our `11-beanstalk` directory. Start by going into the directory where you have cloned the files from the GitHub repository so that the copy will be a shorter command, as illustrated here:

    ```
    $ cp -r * ~/11-beanstalk/.
    ```

    This recursive copy command will bring over not only the `application.py` file that runs the application but also the `templates` folder (this holds the Jinja templates) along with the static folder (this holds the **Cascading Style Sheets** (**CSS**) style sheet).

8. Now, let's capture all of the requirements that we will need to install once we go to our remote environment using the `pip freeze` command and then push that out to a file called `requirements.txt`, as illustrated in the following code snippet:

    ```
    $ pip freeze > requirements.txt
    ```

9. At this point, we should test our `flask` application locally to ensure that it runs and is ready for deployment. Use the following command to test your application:

    ```
    $ python3 application.py
    ```

    Assuming that the application starts successfully, you should see a return like the one shown next:

    ```
    * Serving Flask app 'application' (lazy loading)
     * Environment: production
       WARNING: This is a development server. Do not use it
    in a production deployment.
       Use a production WSGI server instead.
     * Debug mode: on
     * Running on http://127.0.0.1:5000/ (Press CTRL+C to
    quit)
     * Restarting with stat
    ```

```
* Debugger is active!
* Debugger PIN: 688-856-306
```

With the local URL displayed, you can open up a web browser and see the application in action.

Once you are done testing the local application, press *Ctrl + C* in the terminal window to stop running the local server.

10. With our application tested locally, we can now create the environment with the EB CLI. Using the command prompt that you still have open, we will initiate the Elastic Beanstalk environment. Make sure that you are still inside the `11-beanstalk` folder before you run the next command, shown here:

```
$ eb init
```

As you initiate your Beanstalk environment, use the following values:

- **Default region**: `us-east-2`
- **Application**: `flask-quiz`
- **Python**: `(Y)es`
- **Python version**: `Python 3.8`
- **SSH for instances**: `(Y)es`
- **Select a keypair**: Create a new keypair
- **Keypair name**: `DevOps-pro` (no passphrase)

Now that we have created our environment, we will move on to looking at how we can customize our environment using `.ebextensions`.

Understanding advanced configuration options with .ebextensions

Elastic Beanstalk allows you to add configuration files to your application's source code to customize and configure the AWS resources in your environment.

We can create a *hidden* folder in our directory to create and store our custom configurations.

From inside our `11-beanstalk` directory, we will create a `.ebextensions` directory, as follows:

```
$ mkdir .ebextensions
```

We can now run a `tree` command with a level of `1` and ask it to show all files (in order to show the hidden `.ebextensions` folder) and see our current working environment structure, as follows:

```
$ tree -a -L 1
```

This should show a current structure like this:

```
.
├── .ebextensions
├── .elasticbeanstalk
├── .gitignore
├── application.py
├── eleven
├── requirements.txt
├── static
└── templates
```

Next, we will move the file that we copied over with all of the other files named `cloudwatch.config` inside the `.ebextensions` folder, like this:

```
$ mv cloudwatch.config .ebextensions/.
```

This file is also available in the GitHub repository in the `Chapter-11` folder if needed.

With our extra resources created, we are now ready to look at the deployment types available for applications in Elastic Beanstalk before we deploy our application.

Deployment types with Elastic Beanstalk

Elastic Beanstalk allows for multiple deployment types for your application. Let's take a look at each one in detail.

All-at-once deployments

Using an **all-at-once deployment** strategy, Elastic Beanstalk will take the specified version of the application that has been marked for deployment and then simultaneously deploy this version to all instances in your environment.

This deployment type takes the least amount of time to implement and has the lowest associated cost; however, it also comes with the most risk. If something were to go awry during the deployment, then you would be facing downtime with your application. This is a deployment strategy that is best served for development and test environments but it should seldom or never be used in a production environment.

Rolling deployments

A **rolling deployment** will push out a newer version of your application to your existing EC2 instances, but instead of pushing out the application to all instances at the same time, it uses batches to govern how many instances are updated at any single time.

While this deployment type is not as fast as the all-at-once method, it also allows for the safeguard of not allowing all your instances to be down at the same time in case of an application issue. A rolling instance is also less expensive than one of the blue/green types of deployment options since you are utilizing EC2 instances that you already have up and running and don't need to create a whole new environment.

Rolling with additional batch

Although extremely similar to a rolling deployment, **rolling with additional batch** has a very distinct difference. That difference is that this deployment ensures that your full capacity is maintained throughout the deployment process.

This deployment type will launch an additional set of instances before beginning the actual deployment. This is a good option if you are serving a constant flow of traffic and need to ensure capacity even while performing application updates.

Immutable

When you think of immutable infrastructure, you should think of not making an update to a current instance. This is the same concept when using an **immutable deployment**, since we are setting up a whole new set of instances and deploying to those instances, waiting for them to become healthy before switching the **Domain Name System** (**DNS**) of Elastic Beanstalk over from the previous environment, as illustrated in the following diagram:

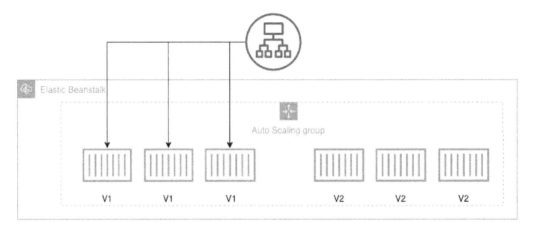

Figure 11.6 – Immutable deployment during deployment

This is one of the safest ways to deploy your application, but it can also be one of the most expensive since you are setting up double the number of instances, as illustrated in the following diagram:

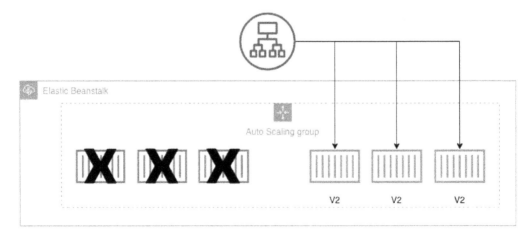

Figure 11.7 – Immutable deployment after deployment

After the health check has been completed on the new version of the instances, then the older version of the instances is taken down and the traffic is routed to the new version of the application. If there were any problems with any of the instances, then the traffic would never get routed to those instances in the first place.

Traffic splitting

A **traffic-splitting** deployment is somewhat like an immutable deployment in the fact that it creates a whole new set of instances. A major difference is that with a traffic-splitting deployment, the older instances don't terminate as soon as the new version is up and healthy. Instead, the traffic is shifted at a controlled pace, one that you set in the console or via the CLI, to direct a portion of the traffic from one version to the next version, as illustrated in the following diagram:

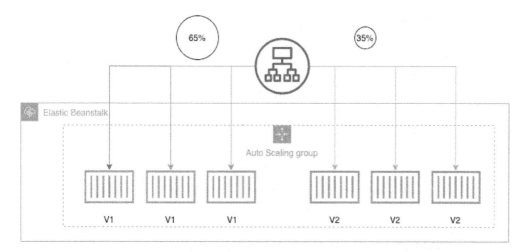

Figure 11.8 – Traffic-splitting deployment showing the percent of traffic

Now that we have looked at different deployment strategies, we will use the EB CLI and our terminal to create and deploy a Beanstalk application.

Using Elastic Beanstalk to deploy an application

The best way to understand some of these services is to test them out with a hands-on example. This is most certainly the case with Elastic Beanstalk, as you need to be able to think about the capabilities of the service when reading through both the questions and answers that the DevOps professional exam presents you with.

We are going to continue with our example code from before that we started earlier in this chapter. If you didn't do the first part of the exercise and you want to deploy the application, then it's suggested that you go back to the part of the chapter entitled *Using the EB CLI to create a project*.

If you have previously closed your terminal window, then you will need to open it up again and navigate back to the `11-beanstalk` folder. We already initiated our environment locally previously, and can now proceed as follows:

1. Since our environment has been initiated previously, we can create our initial environment—`development`, as follows:

   ```
   $ eb create development
   ```

   The terminal should show you a return like the one shown next:

   ```
   Creating application version archive "app-210606_181258".
   Uploading: [#########################################
   #####] 100% Done...
   Environment details for: development
     Application name: flask-quiz
     Region: us-east-2
     Deployed Version: app-210606_181258
     Environment ID: e-mscggmggrw
     Platform: arn:aws:elasticbeanstalk:us-east-2::platform/
   Python 3.8 running on 64bit Amazon Linux 2/3.3.0
     Tier: WebServer-Standard-1.0
     CNAME: UNKNOWN
     Updated: 2021-06-06 22:12:41.224000+00:00
   Printing Status:
   2021-06-06 22:12:40      INFO      createEnvironment is
   starting.
   2021-06-06 22:12:41      INFO      Using elasticbeanstalk-
   us-east-2-182968331794 as Amazon S3 storage bucket for
   environment data.
   ```

2. Although it will most likely take a few minutes to finish the setup of your new environment, once it has been completed, you can see the web application up and running by using the `eb open` command, as follows:

   ```
   $ eb open
   ```

3. You can check on all the events of your application by using the `events` command from within the EB CLI, like this:

```
$ eb events
```

4. If you wanted to check on the health of your environment, then you could log in to the AWS console to see the visual dashboard, or you could simply use the `eb health` command to get a quick look at the statistics for your environment, as follows:

```
$ eb health
```

Alternatively, if you look at the health of the Beanstalk application from the console, then you will be presented with a very easy-to-read **user interface** (**UI**), as illustrated in the following screenshot:

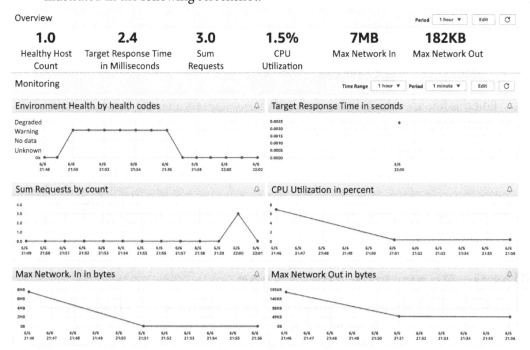

Figure 11.9 – A look at the Elastic Beanstalk monitoring dashboard

5. If you like, you can check the log files directly from the command line using the `eb logs` command.

6. Once you are done, then you should deactivate your virtual environment by running the following command:

```
$ deactivate
```

> **Note**
>
> If you are running into errors from the deployment, please look at the following troubleshooting section regarding the IAM role that the EB CLI is using. As an alternative, you can create an application in the AWS console and then zip up the entire directory using the `zip ../eleven.zip -r *` `.[^.]*` command to use as the source bundle to upload.

Follow the guidance in the next section to help troubleshoot your deployment if you are having issues.

Troubleshooting the deployment with the EB CLI

If you are running into issues during the deployment of your environment, perform the following steps:

1. Find the line that states the **Web Server Gateway Interface** (**WSGI**) path. An example of this is shown next:

   ```
   WSGIPath: application
   ```

2. Edit this line after the colon to ensure that `.py` appears after the word `application`.

3. Save and close the file.

4. Try deploying your application again with the `eb deploy` command.

Now that we have gone through the exercise of deploying our Beanstalk application and looking at the logs and events of the application, we will now look both at use cases and anti-patterns of using Elastic Beanstalk.

Elastic Beanstalk use cases

Elastic Beanstalk makes it easy for developers to get up and running on the cloud without having to worry about the underlying infrastructure or management of underlying components. Next, we will look at some of the optimal use cases for using Elastic Beanstalk.

You have a smaller development team that needs to get going quickly

If you have a smaller team that needs to get up and running quickly on AWS but is not as familiar with all of the other components, services, and interconnectivity, then Elastic Beanstalk can be a good choice to meet deadlines.

You don't have any in-house DevOps expertise

Somewhat like the previous example, if a company or development team has an application that is needing a rapid deployment to AWS, then Elastic Beanstalk presents a very viable solution without the need for any advanced DevOps expertise. Since the product itself can be used with Git, a tool that most developers are familiar with, getting up and running in a minimal amount of time is a simple task. Beanstalk then takes care of tasks such as DNS, autoscaling, easy access to application logs, and even pre-built dashboards for metrics on your environment without any additional setup.

Elastic Beanstalk anti-patterns

As we just looked at some of the cases where the use of Elastic Beanstalk would make sense, there are also cases where Beanstalk would not be a good choice. Here, we will examine some of the known anti-patterns for using Elastic Beanstalk, where you would want to look at a different service offering from AWS.

Applications or projects that need a lot of environment variables

One of Elastic Beanstalk's limitations is the fact that it only has 4 **Kilobytes** (**KB**) to store all key-value pairs. Taking this fact into account, then, if you have numerous environment variables such as different database strings, usernames, and passwords for each environment that you create, then you could run up against this hard limit, and Elastic Beanstalk would not be a good choice for your application.

Applications that are very large

Elastic Beanstalk deployment bundles can only be at a maximum of 512 **Megabytes** (**MB**) in size. Hence, if your application is large and complex, it might not be a good application for deployment and management via the Beanstalk platform. There are some possible workarounds since this only applies to the source code bundle that Beanstalk itself is dealing with and managing. For instance, if you have lots of data, then you can offload that onto a **Relational Database Service** (**RDS**) database and store those configurations in your `.ebextensions` directory. Also, if you have a lot of static assets, such as pictures or media, then those could be stored in S3 with pointers in either your application or in a dynamic store such as DynamoDB, and this can keep your source bundle size down.

Summary

In this chapter, we covered the Elastic Beanstalk service, another one of the deployment services offered by AWS. We covered the different deployment options available with Elastic Beanstalk, along with creating and deploying a Beanstalk application using the EB CLI tool. We also took a look at some of the best-case scenarios for using Elastic Beanstalk, as well as some anti-patterns whereby the Elastic Beanstalk service would not be a good fit.

In the next chapter, we will look at using Lambda functions and step functions in order to deploy our code in a serverless manner.

Review questions

1. You have been brought into a company that has an application team whose application has two tiers—a web layer and a database layer. The application team needs a method to provision and deploy an environment to AWS quickly. Which of the following options would be the fastest and most ideal way to get this team set up?

    a. Use the Elastic Beanstalk service to provision an environment and then push the application to the environment.

    b. Create an OpsWorks stack with two layers, one for the application and another for the database. Deploy the application to the application layer.

    c. Use CloudFormation to create an S3 bucket, an RDS database, and an EC2 instance. Use user-data scripting to load the application from the S3 bucket and parameters for the RDS connection strings, username, and password.

    d. Create an RDS database and then use Lambda to deploy the application.

2. You have been brought into a company that is trying to automate its move to the AWS cloud as quickly as possible. The applications that it is trying to move are built using a multitude of different programming languages. How can you get these applications moved and deployed to the cloud as swiftly as possible?

    a. Create a master CloudFormation template that utilizes a nested stack design for the deployment of the applications. Create a child template for each of the applications after creating a Docker container for the applications.

    b. Develop each application in one Docker container and use Elastic Beanstalk to deploy.

    c. Develop each application in a separate Docker container and use Elastic Beanstalk to deploy.

d. Create a stack using OpsWorks. Create a separate layer for each application and then deploy each application to its layer.

3. There is a service that Elastic Beanstalk can perform for you if your application takes a while to complete its activities or workflows. What is that service?

a. Beanstalk can manage the Elastic Load Balancing (ELB) service and run a daemon process on each of the instances.

b. Beanstalk can manage a Simple Notification Service (SNS) topic and run a daemon process on each of the instances.

c. Beanstalk can manage Lambda functions and run a daemon process on each of the instances.

d. Beanstalk can manage an SQS queue and run a daemon process on each of the instances.

4. You have been brought into a company in order to help automate a recovery strategy in case an outage hits the primary region where resources are hosted in AWS. The company has a priority on minimizing costs for this recovery strategy but also needs to be able to spin up the complete infrastructure if needed. How would you suggest that it does this with the least cost?

a. Create a pilot light infrastructure in another region and automatically resize based on CloudWatch events.

b. Create a duplicate infrastructure by creating a whole new environment in the disaster recovery (DR) region using Elastic Beanstalk. Switch the Route 53 record to the DR region's load balancer in case of an outage.

c. Use Elastic Beanstalk to create another environment as a copy of the application in case of an outage.

d. Use CloudFormation to spin up resources in another region in case of an outage.

Review answers

1. a.
2. c.
3. d.
4. d.

12
Lambda Deployments and Versioning

As more and more application architectures go serverless and more and more cloud budgets are scrutinized, AWS Lambda becomes more and more of a viable option in both the Developer and Operations professions' tool belts. Knowing how to harness the flexibility and power of Lambda and Step functions is key to success in today's AWS environments.

In this chapter, we're going to cover the following main topics:

- AWS Lambda overview
- Lambda functions
- Lambda triggers and event source mappings
- Deploying versions using Lambda
- Working with Lambda layers
- Monitoring Lambda functions
- Optimal Lambda use cases and anti-patterns
- Orchestrating Step Functions with Lambda

Technical requirements

As in the previous chapters, we will be using the Python programming language to create our Lambda functions. Basic working knowledge of Python is suggested to follow the examples and debug and troubleshoot if necessary.

AWS Lambda overview

AWS Lambda is a service that allows you to run your code as a function, without the need to stand up any servers or provision or orchestrate containers. It automatically scales to the number of requests that it receives. One of the most attractive items about Lambda functions is that they are only charged for the time they run. This means that you can have your platform provisioned in one or multiple regions, waiting for requests, without worrying how much the bill will accumulate from idle resources.

Lambda lets you concentrate on the code instead of servers as it is a serverless Platform as a Service offering. Being a PaaS also means that you have no access to the underlying compute platform to make adjustments other than those that AWS exposes to you, such as the runtime (programming language) or your environment, the amount of memory that your function needs to use, and the amount of CPU allocated:

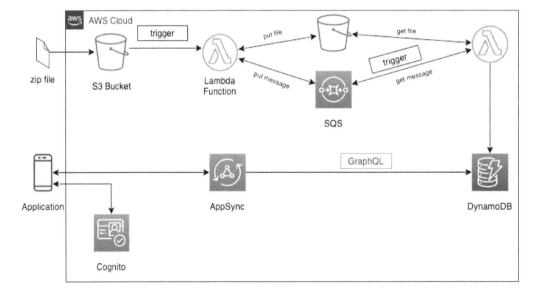

Figure 12.1 – Lambda architecture with triggers

Many Lambda functions are events that are triggered by other AWS services. This is another fact that makes the service so appealing. Lambda functions can be used to do backend processing based on S3 bucket events. Then, you can put them into a decoupled architecture using a message queue, such as **Simple Queue Service** or **Amazon MQ**, for processing by one or more concurrent Lambda functions to put the data into a backend data store.

The data can then be retrieved using either RESTful calls from a service such as **API Gateway** by utilizing GraphQL with **AppSync**.

Serverless instead of servers

The term **serverless** should instantly bring a few critical thoughts to mind. In the world of AWS, this boils down to four fundamental principles:

- **There are no servers to provision**: There should be no actual servers or instances to configure or maintain. There should also be no container orchestration needed on your part.

- **The system and architecture scale with usage**: As requests, data, or events come in, the underlying platform and infrastructure should automatically scale out and scale in to meet the necessary demand.

- **You pay for value**: Resources that are sitting idle, waiting to be used, should not incur charges. Charges should only occur when the resources are actively being utilized.

- **The system is built for availability and fault tolerance**: Once you're launching your platform, it should automatically span multiple availability zones, increasing your availability and fault tolerance.

As you talk to people who have used a serverless platform in the real world, they feel like they have derived numerous benefits from making the switch. One such benefit is **greater agility** and the ability to move faster since they are not spending as much time setting up and provisioning infrastructure. Separating themselves from dealing with instance and container configuration allows them to **better focus** on their business and the features that their customers want and find valuable. They also feel as though they have **increased scale** compared to the previous capacity that customers used to have in their data centers. This is because Lambda will scale up automatically based on the number of invocations to a particular function.

Synchronous versus asynchronous invocation

When you invoke a Lambda function, it can be done in one of two ways: either synchronously or asynchronously.

Lambda will run the function on a synchronous invocation, wait for the response, and then return the response code with any data included in the function's return call. You can use the `invoke` command from the AWS CLI to invoke a function synchronously.

With asynchronous invocations, you are pushing the event to Lambda but not waiting for an immediate response. Lambda will queue up the events before sending them to the function:

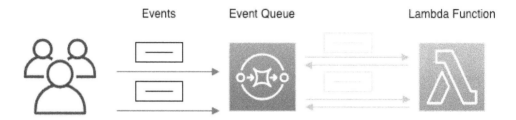

Figure 12.2 – Asynchronous invocation in Lambda

Lambda manages the functions event queue and, upon receiving an error, will attempt to retry the event. If it fails, it will try twice more with a longer time in-between each additional attempt.

Now that we have taken an initial look at how we can invoke a Lambda function, let's take a deeper look at Lambda functions themselves.

Lambda functions

The code you write becomes the function that runs without the need for provisioning or managing any servers. The function itself is the resource and can take in events that have been passed to it, either by you or other AWS services.

You can use a few different languages that are supported to create your Lambda functions. These include Python, Node.js, Ruby, Java, Go, and .NET. You can even create custom runtimes by using containers.

The basic concepts of Lambda

When you start a Lambda function, that process is called **invoking** the function. Lambda functions process events. Events are sent to your functions in a few different ways: you can manually invoke the function, such as with a test event, or you can configure an AWS resource or service to invoke it and start the process.

Since we're talking about Lambda functions, there are a few key concepts that need to be understood.

Function

A function is the code where you process your events. It is what is invoked when you call the Lambda process, either by an event, schedule, or manually.

Qualifier

Lambda functions can have versions and aliases. Once a version has been created, it is a fixed version of the code and contains a numerical qualifier at the end. Although the $LATEST version of your Lambda code is the one that you can constantly be updating, if you wanted to call that version that you snapshotted in time, such as `test-function:1`, you would need to append the version number to the end.

Runtime

The runtime in Lambda allows you to choose the language, along with the language version, that your function will execute. This runtime is not contained within the function itself; instead, it sits between the Lambda service and the function code. Not only can you use the runtimes that the Lambda service provides, but you can also build your own if there is a language and version that you desire that the service itself does not support.

Event

An event in Lambda is a JSON document that consists of data for the function to process. Events can be simple and single-tiered, as shown in the following example:

```
{
    "URL": https://packtpub.com
}
```

They can be also complex, coming from AWS services with nested keys and values that require complex parsing. These complex events can hold valuable data that can automate processes and make your life, as a DevOps professional, easier if you learn how to harness their power.

The Lambda handler

A handler can be any name, but the default name, especially when creating a function in the AWS Console, is `Lambda_function.Lambda_handler`:

```
def Lambda_handler(event, context):
    greeting = 'I am a DevOps Pro and my name is {} {}'.
format(event['firstname'], event['lastname'])
    print(message)
    return {
        'greeting': greeting
    }
```

In our example Lambda code, we can see two arguments being passed to the Lambda handler: the event and the context.

The `event` argument is a JSON formatted document that includes the data for the Lambda function to process. While it is often a dictionary object, it can also be a list, string, integer, or floating-point number.

Using a combination of the event handler and a parser, you can grab information about which specific resources invoked the Lambda function and then perform the necessary actions from there. By using `returns`, such as the greeting in the simple example shown previously, you can call other methods inside your function based on the information you find. You could even have the entire function return a value. This is especially useful in the case of **Step Functions**, which we will look at later in this chapter.

The `context` argument is passed to the Lambda function at runtime. This argument contains information regarding the invocation, the runtime environment, and the function itself.

Limits of Lambda

As you go about creating your Lambda functions, it's helpful to understand some of the constraints that the Lambda service has put into place. The minimum amount of memory for a function is 128 MB, while the maximum is 3,008 MB. The longest execution time allowed for a Lambda function is 15 minutes or 900 seconds. You can only have a maximum of 4 KB in environment variables. There is a concurrency limit of 1,000 concurrent executions per function. If you are extracting data or using the /tmp disk space, you have a limit of 512 MB.

Creating a Lambda function

With an understanding of how Lambda functions work, we will now start creating our Lambda function. The function that we will create will take a URL that's been passed to it and then count the number of words on the web page. Since we will need external packages, we will need to create a ZIP package first and then upload that to the Lambda service.

There are times when you can just write a simple function in the AWS Management Console itself using the built-in editor in the Lambda service. This includes the Python language and the boto and botocore modules, which allow you to take advantage of the Python **software development kit (SDK)**. There are also helpful modules that you would not have to download, such as random to help you generate random numbers and values, OS to allow you to call operating system functionality, and math, gzip, and logging, along with quite a few others.

Follow these steps to build our Lambda package so that it can be uploaded:

1. Let's start by opening our terminal and creating a role for our Lambda function to use. With our terminal open, we will navigate to the beginning of our directory and then create a new directory for our Lambda. Once created, go into that directory:

    ```
    $ cd ~/
    $ mkdir my-wc-function
    $ cd my-wc-function
    ```

2. We will use the following JSON saved to a file named Lambda-role-policy.json; you can also find this file in the Chapter-12 folder of the GitHub repository for this book:

    ```
    {
        "Version": "2012-10-17",
    ```

```
"Statement": [
    { "Effect": "Allow",
      "Principal": {
          "Service": "Lambda.amazonaws.com"
          },
      "Action": "sts:AssumeRole"
    }
  ]
}
```

3. Use this file in the following command to create the role that our Lambda function will use:

```
$ aws iam create-role --role-name Lambda-12 --assume-role-policy-document file://Lambda-role-policy.json
```

4. This should come back with a JSON return showing the successful creation of the role, similar to the following output:

```
{
    "Role": {
        "Path": "/",
        "RoleName": "Lambda-12",
        "RoleId": "AROAW24Q7QQF5NLSQX3L5",
        "Arn": "arn:aws:iam::470066103307:role/Lambda-12",
        "CreateDate": "2021-06-15T01:06:06+00:00",
        "AssumeRolePolicyDocument": {
            "Version": "2012-10-17",
            "Statement": [
                {
                    "Effect": "Allow",
                    "Principal": {
                        "Service": "Lambda.amazonaws.com"
                    },
                    "Action": "sts:AssumeRole"
                }
            ]
        }
```

```
        }
    }
```

5. Our role has now been created for our Lambda to use; however, it cannot do much since there is currently no policy attached. Instead of writing a custom policy, we will use a predefined AWS policy created for Lambdas:

```
$ aws iam attach-role-policy --role-name Lambda-12
--policy-arn arn:aws:iam::aws:policy/service-role/
AWSLambdaBasicExecutionRole
```

6. With our role created and ready to use, let's remove the current file in the directory so that it doesn't get deployed with our zip package later. We are going to use the remove command for the file. However, if you would like to save it, I would suggest using the copy or move command to your /tmp directory or Downloads so that you can access it later:

```
$ rm Lambda-role-policy.json
```

7. Now, we will create a new file called Lambda_function.py. In this function, we will cut and paste (or type, if you're brave) the following code. Alternatively, once again, you can find the full file in the Chapter-12 directory of this book's GitHub repository:

```python
import requests
from bs4 import BeautifulSoup
from collections import Counter
from string import punctuation # already included in
Lambda modules


def Lambda_handler(event, context):

# get the URL from the event
    r = requests.get("https://aws.amazon.com/blogs/compute/
using-Lambda-layers-to-simplify-your-development-
process/") #demo
    bs = BeautifulSoup(r.content)


    # gather all the words within the paragraphs
    p_txt = (''.join(s.findAll(text=True))for s in
bs.findAll('p'))
```

```
    count_p = Counter((x.rstrip(punctuation).lower() for y
in p_txt for x in y.split()))

    # gather all the text in the divs
    d_txt = (''.join(s.findAll(text=True)) for s in soup.
findAll('div'))
    count_div = Counter((x.rstrip(punctuation).lower() for
y in d_txt for x in y.split()))

    # create a sum total of the words
    word_sum = p_txt + d_txt
    # return the number of words
    return word_sum
```

8. Now, the directory structure of your my-wc-function directory should look like this:

    ```
    my-wc-function$
    | Lambda_function.py
    ```

9. At this point, we can start installing our dependent modules locally using the **Python package installer**, **PIP**, along with an extra flag called –target:

    ```
    :pip install --target ./package requests
    pip install --target ./package bs4
    ```

10. Now, let's make the deployment package. First, we will navigate into the package directory that we just created and then create the initial zip file. Take note of the two dots in the zip command; these are telling the zip file to be created in the directory and not in the package directory:

    ```
    $ cd package
    $ zip -r ../my-wc-package.zip .
    ```

11. With our initial zip file created, we can add our Python file to the zip file:

    ```
    $ cd ../
    $ zip -g my-wc-package.zip Lambda_function.py
    ```

12. After running this command, you should see that the Lambda_function.py file has been added to the previously created zip package.

13. We now have our deployment package created and are ready to move on to the AWS Management Console so that we can upload and test our function. This could all be continued from the CLI, but the console has some features that aren't available from the CLI.

14. Open your web browser and navigate to the Lambda service after logging in using your administrative account. You can get to the service directly by going to the `https://console.aws.amazon.com/Lambda`.

15. Find and click on the orange **Create function** button near the top right-hand side of the main screen.

16. Once you're on the **Create function** screen, make sure that the **Author from Scratch** value is selected for creating your function. Under the **Basic Information** section, use the following values:

- **Function name**: `my-word-count_python`.

- **Runtime**: `python 3.8`.

- **Permissions**: You will need to click on the **Change default execution role** to expose the option to use an existing role. Then, you can choose the `Lambda-12` role that we created earlier:

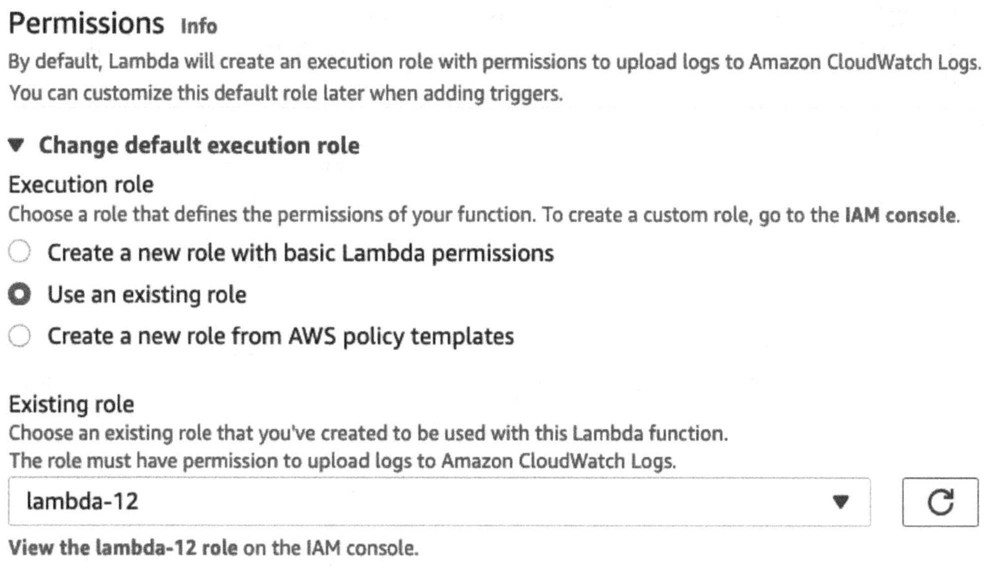

Permissions Info

By default, Lambda will create an execution role with permissions to upload logs to Amazon CloudWatch Logs. You can customize this default role later when adding triggers.

▼ **Change default execution role**

Execution role
Choose a role that defines the permissions of your function. To create a custom role, go to the **IAM console**.

○ Create a new role with basic Lambda permissions

◉ Use an existing role

○ Create a new role from AWS policy templates

Existing role
Choose an existing role that you've created to be used with this Lambda function.
The role must have permission to upload logs to Amazon CloudWatch Logs.

| lambda-12 | ▼ | C |

View the lambda-12 role on the IAM console.

Figure 12.3 – Choosing the existing role we created on the Lambda Create Function screen

17. Once you have filled in all the values, click on the orange **Create Function** button.

18. Once you have created your function, which should take you to the main screen for this Lambda function, we need to upload the ZIP file that we created. So, under the **Code Source** section on the right-hand side, click on the **Upload from** button. Once clicked, you will be presented with two options: **.zip file** or **Amazon S3 location**. Choose **.zip file** and then locate my-wc-package.zip on your local system. Press the **Save** button to send the ZIP file and its code to AWS Lambda:

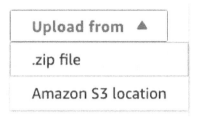

Figure 12.4 – Uploading the .zip file we previously created to our Lambda function

19. Once you have uploaded the ZIP file, multiple folders should appear on the left-hand side of the **Code Source** window. Our function code is available to view if we double-click on the Lambda_function.py file.

20. To see our new function work, we will need to create a test event. Click on the orange **Test** button underneath the **Code Source** heading. This will bring up the dialog to create a test event. We will use the default values for our test event. Set **Event name** to Test1 and then click on the orange **Create** button at the bottom of the dialog.

21. With our test event created, we can run the test. Click on the arrow next to the **Test** button and choose the **Test1** option when it appears. With the correct test event selected, we can click on the orange **Test** button once more to start the test.

After running the test, you should see a count of all the words in our demo URL from the **Execution result** page.

This exercise took us through creating and testing a Lambda function that had dependencies on third-party libraries from scratch. Next, we will look at triggers and source mappings for our functions so that they can run automatically when specific events occur.

Lambda triggers and event source mappings

Lambda triggers are especially useful for kicking off numerous applications when a piece of data is uploaded to a specific S3 bucket. AWS provides examples of images being uploaded to buckets in many of its talks and presentations. This image then triggers a Lambda function, which will resize the image so that it's more compressed and then place it in a folder for GIFs. Many times, this same function will place a pointer for the newly resized image in a DynamoDB table. These resized images are more accessible and quicker for end users to download, and this all happens automatically once a source image has been uploaded:

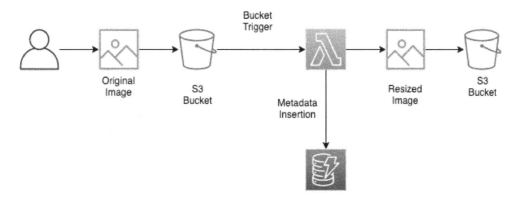

Figure 12.5 – The flow of an image being uploaded to an S3 bucket that triggers a Lambda function for image resizing

There are even more things we can do with bucket triggers than just image resizing, especially in an enterprise and DevOps context. Remember that S3 can be used as source code storage since it has versioning capabilities if they've been turned on. Suppose a new file has been uploaded to a particular folder with a specific file extension (for example, *.py). In that case, that upload could kick off our CodePipeline job to build a new container and push that container through as many steps as we have until any manual gates are encountered.

Now that we've looked at how triggers can invoke Lambda, let's look at how streaming data can be pushed to different queuing services so that Lambda can asynchronously process the data.

Looking at event source mappings

Since Lambda is an automatically scalable service, it can take the information from certain other services that are sending data to it and then process that information. This information may or may not go directly to the Lambda function. These are intermediary services such as message queues, which take the instream of data and then invoke the Lambda function.

Services that Lambda can read from event sources

The following services can provide event source mappings for the Lambda service:

- DynamoDB
- Kinesis
- Amazon MQ
- Amazon Managed Streaming for Apache Kafka
- Amazon SQS

Now that we have seen how Lambda functions can be invoked by different AWS services and even sources, let's learn how to update our functions using versions.

Deploying versions using Lambda

If you have a known good state for your Lambda function, you can freeze it from future changes by publishing a version of the function. Once published, this version will be called and used by users and services independently of any changes or updates made to future versions, including iterations done on the `$Latest` function.

When you publish a version of a Lambda function, it contains the following information:

- The function code, as well as all the dependencies that are associated with it.
- The Lambda runtime that invokes the function.
- All of the function settings, such as memory, VPC association, and IAM role.
- Any environment variables that have been added to the function.
- A unique **Amazon Resource Name** (**ARN**), so that the distinct version of the function can be identified:

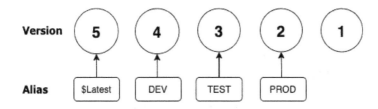

Figure 12.6 – Lambda versions and how they map to aliases

Versions can be referenced in one of two ways: either with a qualified ARN or with an unqualified ARN.

A qualified ARN will have a version suffix at the end of the function ARN:

```
arn:aws:Lambda:us-east-2:470066103307:function:my-word-count_
python:5
```

An unqualified ARN will not have a suffix at the end of the function ARN:

```
arn:aws:Lambda:us-east-2:470066103307:function:my-word-count_
python
```

Now that we have learned about versions for our Lambda functions, let's look at how we can use aliases in conjunction with our versions.

Using aliases in Lambda

Aliases allow you to create named pointers to specific versions of your Lambda deployments. This is especially helpful if you have passed the ARN of your Lambda function to a user or other application.

Unless you were in a testing environment, you wouldn't want to pass along the base Lambda ARN since that would point to the $Latest version. This $Latest version can constantly be changing and be subject to errors and bugs while new features and fixes are being tested and deployed. Instead, by using an alias, you can point users to a specific, stable version of the function and seamlessly transition to a new version of the function once published, tested, and ready for release.

Working with Lambda layers

A Lambda layer is a pre-published collection of code dependencies and libraries. If you find that you (or you, as a team) are using the same snippets of code in multiple functions, you can create a layer to get the benefits of Lambda layers. First, when there is a change to one of the dependencies, you no longer have to update each of the functions; instead, you only maintain a single layer that all the functions can take advantage of and only needs to be maintained one time:

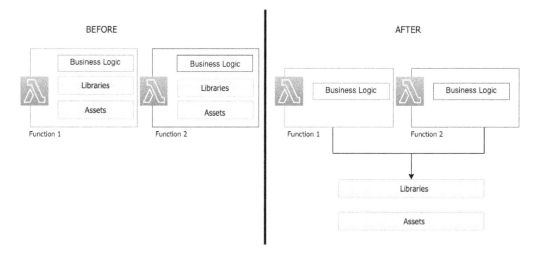

Figure 12.7 – Lambda functions with and without layers

Layers are a great way to speed up development between teams. Shared functionality such as logging, networking, or even database connectivity can be programmed and packaged into a layer once and then called from individual functions.

Adding a Lambda layer to our function

We already have a Lambda function from our previous exercise. Now, let's expand the functionality of our function by adding a layer that works not only for this particular function, but could also be used across numerous functions in our code base and organization.

We have gone through the process of adding a layer to our Lambda function. Next, we will learn how to use native AWS to monitor our functions, as well as what metrics are the most effective to monitor.

Monitoring Lambda functions

Your job is not done once you have developed and deployed your Lambda function. To ensure that it is functioning correctly as it runs, specific metrics should be monitored. Fortunately, Lambda integrates with several other AWS services to help you not only monitor your functions but also troubleshoot them when the need arises.

If you go to the AWS Console in the region where your Lambda function has been deployed, you will find a menu item that you can click on in the vertical menu bar named **Monitor**:

Figure 12.8 – The Monitor menu item from the Lambda function console

Once you enter the **Monitor** section of the Lambda console, you are instantly presented with a pre-built dashboard that allows you to see basic but vital metrics graphically for your Lambda function at a glance. The following metrics are presented:

- **Invocations**
- **Duration**
- **Error count and success rate**
- **Throttles**
- **Async delivery failure**
- **Iterator Age**
- **Concurrent Executions**

The default timeframe for the dashboard is 3 hours; however, there are preset iterations from 1 hour to 1 week. You can also select a custom time range for the dashboard to show monitoring.

Another extremely handy feature regarding the **Monitoring** section in the AWS Management Console is that it has buttons that allow you to jump right to either CloudWatch Logs for the function, X-Ray traces for the function using ServiceLens, or look at Lambda Insights via CloudWatch if you have enabled this extra feature.

Using X-Ray to help troubleshoot your functions

When you try to troubleshoot an application, especially to find where bottlenecks are occurring during the execution of that application or the other services that it's calling, you may need more information than your current metrics and logs are providing you with. This is when the AWS service can become especially helpful.

X-Ray collects data about your application's requests and then provides you with detailed information, including a step-by-step timing of each process. Lambda runs the X-Ray daemon any time a function is invoked. This X-Ray daemon listens for traffic on UDP port 2000 and gathers the segment data. That segmented data is then relayed to the X-Ray API for presentation.

> **Note**
> We will look at the X-Ray service and how it can be used to monitor and watch Lambda functions in more detail in *Chapter 14, CloudWatch and X-Ray's Role in DevOps.*

Now that we have looked at how to monitor our Lambda functions, we will look at both the optimal use cases and the anti-patterns for AWS Lambda.

Optimal Lambda use cases and anti-patterns

Now that we've looked at the AWS Lambda function as code service in detail, let's look at when it is best used and where Lambda is not the best choice. This is essential information to understand both as an AWS professional and while pursuing your AWS DevOps professional certification.

The Lambda service from AWS is very flexible, affordable, and interacts with a vast number of AWS services. It also allows you to write your code in many different languages that you feel comfortable with and use them as your runtime environment.

So, let's move on and look at some of the scenarios where Lambda would serve us best.

AWS Lambda optimal use cases

As more and more teams start to look at serverless solutions and patterns, let's look at where using the Lambda service makes sense.

You want to concentrate on the code and not the underlying infrastructure

If you or your team wants to concentrate on the features and code base instead of provisioning and patching servers, serverless Lambda functions are an excellent choice. You concentrate on the code and simply choose your runtime, amount of memory, how long the function is allowed to run, along with a few other options, and the Lambda service takes care of the rest. This includes scaling to meet demand and managing any underlying hardware.

In return for just focusing on the code, you do give up a few things. You can no longer log into the underlying compute instances, which, in reality, are containers, to check processes or gather and watch logs as they are being generated. Instead, you need to add either logging statements or a logging library so that you can debug your code using CloudWatch Logs.

You need a cost-effective solution

Cost optimization is one of the five original service pillars that we spoke about in *Chapter 1, Amazon Web Service Pillars*. Organizations are constantly looking at how to make their solutions more cost-effective, and using Lambda is usually a great way to do that. There are no idle charges for the Lambda service since the charges are calculated on a pay-per-use basis. The service also has a perpetual free tier of one million invocations per month.

One of the ways that this becomes especially useful is if you're crafting your DevOps pipeline to deploy to both your primary region and your secondary (or disaster recovery) region during each deployment without the worry of incurring extra costs. There would be no charge for the secondary deployment since it would not be invoked unless a regional or service failure occurred in the primary region. If there was an issue, however, you will be steps ahead of others since you would not need to stand up any infrastructure or resources. Instead, all you would need to do would be repoint to which region's Lambdas were being invoked.

Lambda anti-patterns

Since we just looked at where Lambda was the best choice for our platform, we will also look at where Lambda does not make the most sense.

You don't want to update the runtime once the application has been developed

With the AWS Lambda service, a specific number of runtimes are supported. The use of these languages and versions are not set in stone, however. As language versions are deprecated due to lack of support, AWS will no longer support these runtimes, nor will they allow any new Lambda functions to be deployed using these runtimes.

Many times, the fix can be as easy as switching the runtime version in the Lambda console from the previous version to the new supported version of the runtime. If there are dependent packages, libraries, or modules of the older version of the runtime, then updates and replacements might need to be found. This can be a problem if the original developers or contractors are no longer around.

You need to call an asynchronous call from your function

When you need to call an outside service such as an external API to perform your task, this is an asynchronous call. This can be done successfully with Lambda. However, the call it makes to the initial service may not always return an immediate response. Having the function wait for the response from the service is not an optimal pattern since the Lambda service is billed based on the number of resources consumed.

Now that we have looked at when to use and not to use Lambda functions, next, we will learn how to orchestrate multiple Lambda functions together using Step Functions.

Orchestrating Step Functions with Lambda

There may be times when you need to coordinate multiple Lambda functions together to perform a larger task. Step Functions can make decisions based on input received in the state machine, run steps in parallel, and even be connected to other services such as SNS to request human input for tasks.

Step Functions also create a visual workflow that allows you to see the individual steps in the process. As the Step Functions run, you can see the success or failure of your state machine.

Understanding state machines inside of Step Functions

The state machine is the orchestration feature of the Step Function. It defines the order in which the steps are run, along with any data that's been received from previous states and passed out to be used by other states.

State machine commands are always written in JSON format. Even if you have written your CloudFormation template in YAML format, you will still need to create your state machine and its various states in JSON format.

The following is an example of a state machine:

```
{
  "Comment": "A Sample State Machine",
  "StartAt": "StepOne",
  "States": {
    "StepOne": {
      "Type": "Pass",
      "Result": "Hello World!",
      "End": true
    }
  }
}
```

Now that we understand what state machines are and how they differ from Step Functions, let's look at how Step Functions work.

How do Step Functions work?

Step Functions work by using a three-step process:

1. Define the steps in your application.
2. Verify any state changes.
3. Run your application and scale it if needed.

> **Note**
> Did you know that, unlike a Lambda function, you cannot trigger a Step Function from an S3 bucket event? Instead, you need to have the bucket event call a separate Lambda function that can then invoke the Step Function.

Now that we know the basics of how Step Functions work, let's take a closer look at the different states available in Step Functions.

States available in Step Functions

As you start to configure the states for your step functions, you begin to create a **state machine**. You do this by using the **Amazon States Language**, which is a JSON-based language for defining the different states in your state machine. There are states that can do things, transitional states, and then states that stop when the task fails or succeeds:

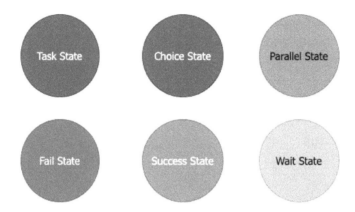

Figure 12.9 – The different states that are available in Step Functions

Let's take a closer look at each of these states and what functions they perform.

Task state

When you create a single unit of work that is to be executed by a state machine, then you are creating a `task` state. The `task` state is where you can invoke Lambda functions.

Choice state

The `choice` state allows the state machine to choose between different branches to follow based on logical evaluations. You provide a set of choice rules, which are rules that evaluate either input or output variables. Based on if the values render as true or not, the next step in the state machine is defined.

Although a default choice is not required, it is recommended just in case none of your choices match any of the logic. Otherwise, your state machine will stop and error out.

Parallel state

As the name implies, a `parallel` state can be used to create multiple branches of a state machine that execute at the same time. This can significantly speed up the execution time for tasks that are not dependent on each other.

Fail state

A `fail` state will stop the execution of your state machine but optionally give you the ability to enter either a `Cause` in a field, an `Error` in a field, or both. These fields can be especially useful when you're trying to debug larger and more complex state machines, and rather than hard-coding in an error code, it is better to pass on some of the information from the system.

Succeed state

The `Succeed` state is a termination state and has no next field.

Wait state

If you need to put a pause in your Step Function, then you can add a `wait` state. The `wait` state can be defined by the number of seconds to pause or by a timestamp on when to resume.

Creating a Step Function

One of the best ways to understand step functions is to create one and then watch it execute. We have created a **CloudFormation** template that will quickly spin up a Step Function state machine. This template also uses an S3 bucket, which is used to store the Lambda functions that have been invoked by the Step Function. If you completed the initial Lambda function exercise, you will already have the Lambda in ZIP format, ready to upload to this new bucket. The template file can be found in this book's GitHub repository, in the `Chapter-12` folder, called `step_function.yml`:

1. Before we run the template, we need to upload the `my-wc-package.zip` file. This can be the S3 bucket that we have been using up to this point. We will do this on the command line using the S3 copy command. Make sure that you are in the same directory where you have the ZIP file, or that you have moved the ZIP file to your current working directory:

    ```
    $aws s3 cp my-wc-package.zip s3://devopspro-beyond/
    ```

2. Open your terminal and go to the directory where you have downloaded the CloudFormation template from the `Chapter-12` folder. Use the following CLI command to create the Step Function from the `step_function.yml` template. We will need to know the bucket that we uploaded the files to so that we can pass that in as a parameter. We will also name our stack `stepTest`:

    ```
    $aws cloudformation create-stack --stack-name stepTest
    --template-body file://step_function.yml --parameters
    ```

```
ParameterKey=LambdaFunctionBucket,ParameterValue=devops
pro-beyond --capabilities CAPABILITY_IAM
```

3. With the Step Function and Lambda function in place, we will log into the AWS Management Console using our administrative user. Once logged in, navigate to the **Step Function** service.

4. Once on the **Step Functions service** page, you should see a state machine named **Chapter-Twelve**. Click on the name of the state machine to be taken to it:

Figure 12.10 – The Chapter-Twelve state machine created

5. Now, on the **Chapter-Twelve** state machine, click on the **Start execution** button. This will open up a dialog box where you can name the execution. Just leave the default values as-is and click the orange **Start Execution** button at the bottom of the dialog window.

6. Once you start the execution, you should see the mapping of the state machine and the flow of the steps. You can click on any individual step to see both the input and output values:

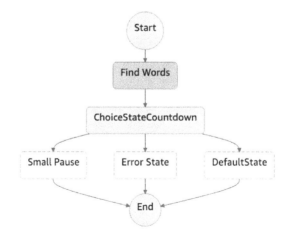

Figure 12.11 – The graphical view of the state machine, as generated by AWS

With that, you have gone through not only the process of creating a Step Function but also incorporating the Lambda function we created earlier into our state machine. Now, let's quickly recap what we learned in this chapter.

Summary

In this chapter, we learned about the Lambda service and how it is used in serverless architectures. We looked at the different components of the Lambda function, from its events to its runtimes. We looked at the different ways that Lambda can be invoked and even used to build a function from scratch. After that, we learned how Step Functions can be used to orchestrate multiple Lambda functions.

In the next chapter, we will do a deep dive into Blue/Green deployments, along with their variations. We will also talk about how they play an important role in the DevOps process since this can be brought up in a variety of ways in different testing questions.

Questions

1. A developer on your team has created a Lambda function that is triggered by an S3 bucket event. The function is supposed to be invoked when an object is placed in the bucket. However, the function is not working correctly. This issue needs to be debugged and fixed. How can this be accomplished simply and quickly by the developer?

    a. Use the Lambda Monitoring Console to help debug the issue.

    b. Use AWS CloudTrail Logs.

    c. Open an AWS support case, noting both the ARN of the Lambda function and the S3 bucket.

    d. Use AWS CloudWatch Logs.

2. You have been asked to help construct a serverless application using AWS Lambda. The application needs to be monitored; however, the company does not want to bring in any third-party services for monitoring. Although logging and tracking functions that call other functions can be tricky, what services from AWS can help you perform this task? (Choose 3)

    a. AWS CloudTrail

    b. AWS CloudWatch

    c. AWS Inspector

    d. AWS X-Ray

3. You have been brought into a company to help with their serverless architecture. They currently have an architecture that has multiple Lambda functions. These functions are invoking other functions and are acting as a state machine. The company has used older obsolete coding patterns to coordinate the state machine and is finding that the code is difficult to repair when it breaks. What service could you suggest to them that will help refactor their application and manage the state machine?

    a. AWS Data Pipeline

    b. AWS Step Functions

    c. AWS Cognito

    d. AWS Beanstalk

Answers

1. d
2. a, b, d
3. c

13
Blue Green Deployments

Releasing applications by shifting traffic between two identical environments is better known as using blue/green deployments. Knowing how to use blue/green deployments through the different AWS service offerings can help mitigate risk when releasing new versions of an application. Walking through the various techniques for the various services and understanding the usage of each is essential to passing the AWS DevOps Professional exam.

In this chapter, we're going to cover the following main topics:

- Understanding the concept of blue/green deployments
- AWS services that you can use for blue/green deployments
- Benefits of blue/green with AWS
- Updating Auto Scaling group launch configurations
- Using best practices in your data tier with blue/green deployments

Understanding the concept of blue/green deployments

When you use the blue/green deployment technique, you employ a technique that reduces downtime and risk. You do this by running a duplicated environment where one is taking the active traffic and the other is receiving the changes. Once the changes have been completed and any testing has been commenced, the traffic that was previously directed at the initial environment, the blue environment, can then be switched over to the green environment. This switch can happen all at once or in gradual phases, depending on both your deployment plan and the services that you use in your deployment. If there is any issue with the deployment, then you can quickly reroute the traffic back to the previously known good state environment, the blue environment, while you remediate the new environment.

The foundation of blue/green deployments is two distinct environments. One environment, the *blue environment*, refers to the current environment where your application or workload is currently running. The second environment, or the *green environment*, refers to a duplicated environment where you can deploy your updated changes to your application code or workload. Using both of these separate environments in symphony allows for near zero-downtime release and rollback proficiencies.

When you perform a blue/green deployment, you are working with an immutable infrastructure. This means that you are not upgrading or updating the current infrastructure in place. Instead, you are creating a new set of resources for each deployment process.

Deployments are not easy

Traditional deployments favor in-place upgrades. However, with in-place upgrades, there are many risk factors to consider:

- Resource constraints
- Potential downtime
- Dependencies from other systems
- Difficulties in rolling back unsuccessful deployments

With the cost and complexity involved in the deployment process, teams will sometimes choose to deploy to an existing infrastructure. This is a viable deployment strategy, but it contains inherent risks, especially when performing a deployment such as an all-at-once deployment where all of the instances or applications are updated simultaneously. If there was an issue or failure during the deployment, this often leads to downtime, loss of revenue, loss of trust in the product brand, and loss of customer confidence, depending on how long the outage lasts.

> **Question**
>
> When starting to map out a deployment, try and ask yourself, *What solutions reduce downtime, handle dependencies, and can coordinate workloads in an improved manner?*

There are multiple risks to navigate when performing deployments, including the following:

- Application failure
- Infrastructure failure

Using blue/green deployment strategies helps mitigate these risks and business impacts by allowing an almost seamless cutover from one environment to the next.

The term *environment* is used a great deal when talking about deployments and especially blue/green deployments. Having an understanding of the definition of environment is critical for understanding what resources are subject to change. Knowing what is being defined as the environment is also crucial when navigating some of the questions on the AWS DevOps professional test.

An environment is a boundary where things change and where things need to be deployed. This can be as small as one component of your application or it can be as large as a full tier of the application, such as the web tier.

AWS services that you can use for blue/green deployments

There are a number of tools available natively from AWS that allow you to perform blue/green deployments. These tools allow for a range of deployment options, from full control of all aspects of the environment using a service such as CloudFormation to granular changes in services such as Route 53 or Auto Scaling. These more granular options would only allow you to perform modifications to a specific section of your application but can be just as effective in implementing blue/green deployments.

Let's look at the different AWS services that are available to assist us in blue/green deployments.

AWS CloudFormation

Using **AWS CloudFormation**, you have the ability to use the service's templating capabilities to both describe the AWS resources you are deploying along with quickly creating a copy of the environment with any updates that are needed. This is all done with one of the two languages that CloudFormation supports: JSON or YAML.

The templates can be smaller pieces of a more extensive infrastructure. You have the ability with CloudFormation templates to create groups of dependent items, such as an autoscaling group, and then either use the output of this created group to update a previously created template that holds the load balancer to update where it is pointing or make the switch manually with the newly created Auto Scaling group.

AWS Elastic Beanstalk

Elastic Beanstalk is a service that helps developers concentrate on their code by managing the underlying infrastructure. This includes items such as ELB, EC2 instances, storage for the EC2 instances with EBS volumes, Elastic IPs, Auto Scaling groups, security groups, and even monitoring via CloudWatch metrics.

When you are performing blue/green deployments in Elastic Beanstalk, you can easily clone your environments. These clones can be exact replicas of your current application code base, or they can even be the latest version of the code if you have pushed features and changes to Beanstalk since you last made your deployment. Elastic Beanstalk makes it easy for those using the service to switch from one environment to the next using the Swap Environment URL feature. This feature takes and then performs the DNS switch in the background and reroutes the traffic from the previous (blue) environment to the new (green) environment.

AWS CodeDeploy

AWS CodeDeploy is a managed deployment service that helps automate the deployment of your software to on-premise servers, EC2 instances, AWS Lambda functions, and AWS Fargate containers. When creating a deployment with AWS CodeDeploy, you have the option to choose either an in-place deployment or a blue/green deployment.

With the blue/green deployment option, you have the ability in your environment configuration to have CodeDeploy automatically copy your EC2 Auto Scaling group or manually provision instances for a blue/green deployment. There are also options to enable load balancing. There is also an option to reroute the traffic automatically or allow for manual rerouting of the traffic after the deployment to the new instances or Auto Scaling group is complete.

There is even the ability to integrate CloudFormation templates using the CodeDeploy service to perform blue/green ECS deployments.

AWS ELB

AWS ELB (**ELB**) is a compute service that allows you to route and distribute your traffic to multiple instances, IP addresses, Lambda functions, containers, and even virtual containers. As ELB is a managed service, it can also perform health checks to determine which instances are healthy and which ones need to stop having traffic being directed to them:

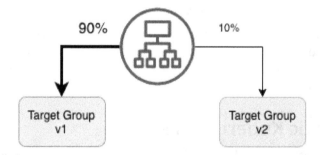

Figure 13.1 – An Application Load Balancer deploying traffic to two versions of an application at once using target groups

Using target groups with Application Load Balancers, you can roll out a new version of your application to a subset of your users with a canary deployment, which is a variety of the blue/green deployment. Using multiple target groups that are connected to the same Application Load Balancer, you can determine how much traffic will be directed to each version of the application.

Amazon ECS

Amazon **ECS** (**ECS**) uses tasks to make groups of Docker containers easy to run, stop, and manage on AWS EC2 instances. With ECS, you can schedule when you want your containers to be placed on an instance using the service scheduler. ECS also lets you stretch multiple containers running the same task across multiple ECS-compatible instances in the same Region but in different Availability Zones:

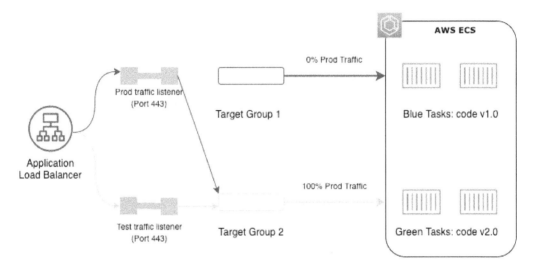

Figure 13.2 – Switching from Target Group 1 to Target Group 2 in ECS using an ALB

The use of containers can make deployments, and especially blue/green deployments, more straightforward to perform. Containers are not as complex as full instances, and multiple containers can run in a single EC2 instance, even those that are running different versions of an application.

Amazon Elastic Kubernetes Service

Amazon **Elastic Kubernetes Service** (**EKS**) allows you to run Kubernetes applications and scale Kubernetes applications both in the AWS cloud and on-premises. It helps the management of the clusters by providing secure clusters and allows the clusters to be highly available.

EKS can be run on specialized EC2 instances or can be run on **AWS Fargate**, which provides on-demand compute capacity for containers. The use of Fargate removes the need for provisioning instances, choosing server types, or managing virtual machines.

You can perform blue/green deployments with EKS on AWS Fargate with the help of the blue/green deployment feature of the CodeDeploy service. When you create a new blue/green deployment for EKS, you specify your Application Load Balancer name fronting your Kubernetes task, and the CodeDeploy service handles deploying the new green service and then phasing out the older blue task.

AWS OpsWorks

AWS OpsWorks is a configuration management service that allows you to configure stacks based on the Chef or Puppet frameworks.

Blue/green deployments are simplified with OpsWorks by simply cloning the entire stack.

Amazon CloudWatch

Amazon CloudWatch is the metrics and monitoring service that allows users to track and observe their deployed resources. The CloudWatch service also has the ability to set alarms and send notifications via other services, such as Amazon SNS or Amazon SES.

When metrics have been set up on the resources in both the blue and green environments, then you have the ability to assess the new environment as you start to direct traffic to it. Keeping an eye on the metrics that you have set in CloudWatch and making sure that all services stay in a steady state while making the switch to the new environment can help ease the anxiety of switching to a new environment.

Amazon Route 53

Amazon Route 53 is the DNS service that can be used in blue/green deployments by allowing the pointing of DNS records to the new green environment. This allows DevOps professionals and network administrators to quickly and easily redirect traffic by updating the DNS records. Route 53 also includes advanced capabilities, such as the ability to adjust the **Time To Live** (TTL) for resource records and using advanced techniques such as weighted policies so that traffic can be shifted gradually to the new environment instead of redirected all at once.

Now that we have looked at the different services that can be used to implement blue/green deployments, let's look at the benefits of using blue/green deployments.

Benefits of blue/green deployments with AWS

The use of blue/green deployment strategies provides multiple benefits over in-place deployments. It is important to note that although these benefits are substantial, there are extra costs for the additional environments that are created during the blue/green process. The secondary environments may be taken down after the deployment has been validated or, in the case of a failed deployment, after a rollback has been completed.

Techniques for performing blue/green deployments in AWS

Implementing blue/green deployments in AWS can be done in a variety of ways. Recognized patterns have emerged for successful ways to implement these deployments. As we look at each technique described, specific services used will be featured. Different applications lean toward different patterns.

Updating DNS routing with Route 53

The Route 53 service allows you to use hosted zones once you have brought up your new environment. Adding the additional record to the record set will then create a seamless transition to the new application deployment for your end users:

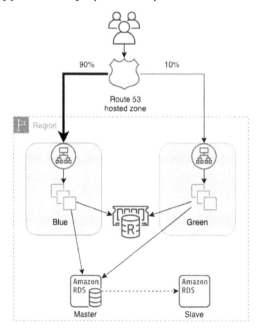

Figure 13.3 – Showing blue/green deployment using Route 53-hosted zones

This switch can be done all at once to force all the traffic to the new green environment. You can also use weighted records to send a portion of the traffic to the green environment, initially as a canary test. These canary users would then generate logs and metrics that could be evaluated over a period of time. If there are no errors being reported in your new environment, then you can change the weight of your policy so that 100% of the traffic is now being directed to the green environment.

This technique is not limited to creating a new green environment in the same set of Availability Zones or even the same Region for that matter. You can create your new green environment in a totally separate Region. If you are planning to switch from Region to Region in your environment, be sure that you have taken into account your data tier and how that would be affected during a regional switch.

This is also not limited to instances or services listening for requests behind a load balancer. Using a DNS routing switch with Route 53 could be used in all of the following scenarios:

- Groups or clusters of EC2 instances fronted by ELB
- Instances in an Auto Scaling group that are fronted by ELB
- Single instances that have either a public address or an Elastic IP address
- Elastic Beanstalk web applications in specified environments
- Services running in ECS or EKS

Process for deployment – DNS routing

The process for implementing a blue/green deployment using Route 53 to switch hosted zones is detailed as follows:

1. Start with 100% of your traffic being directed to the **blue** environment, with the current version of the application deployed.
2. Deploy the new version of the application into the **green** environment.
3. Test that the deployment of the green stack was successful either by running a series of manual or scripted tests.
4. Update the weighted record in the Route 53-hosted zone to direct a portion of the traffic over to the new green environment.
5. Monitor the new environment for errors or failures.

6. Update the weighted record in the Route 53-hosted zone to shift the remaining traffic to the **green** environment.

7. If there was an issue with the deployment, update the Route 53 record to redirect all the traffic back to the **blue** environment.

Now that we have seen how to do blue/green deployments using DNS and Route 53, let's look next at how we can do blue/green deployments without changing the DNS settings.

Swapping the Auto Scaling group behind ELB

In our second blue/green deployment option, we will take the DNS option out of the picture. In many organizations, the team that does the application deployment is not the same team that deals with the networking configurations, including the DNS records hosted by the Route 53 service. Hence, we need to be prepared for those scenarios:

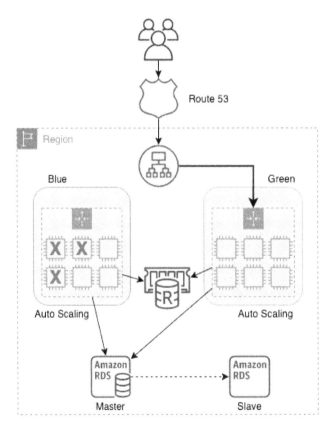

Figure 13.4 – Deploying blue/green environments by swapping Auto Scaling groups

Once the new Auto Scaling group has been launched, you perform a set of tests on the new green stack before registering the new Auto Scaling group, the green Auto Scaling group, with ELB.

> **Important Note**
>
> We go through the hands-on exercise of creating and deploying Auto Scaling launch templates in *Chapter 18, Autoscaling and Lifecycle Hooks*.

An important item of note here is that ELB is only part of the deployment process as far as registration and deregistering are concerned. As you are deploying new versions of the software, you are not deploying a new load balancer.

Process for deployment – swapping the Auto Scaling group

The process for carrying out a blue/green deployment by swapping the Auto Scaling group is as follows:

1. Before you start, make sure that your ELB instance is not part of your deployment environment.
2. Begin with the ELB instance pointing to the **blue** Auto Scaling group.
3. Deploy the new **green** Auto Scaling group.
4. Test the **green** Auto Scaling group.
5. Register the **green** Auto Scaling group with the ELB instance.
6. Deregister the **blue** Auto Scaling group from the ELB instance.

Now that we have seen how to carry out a blue/green deployment by swapping Auto Scaling groups, let's next look at another technique for blue/green deployments that also uses Auto Scaling groups.

Updating Auto Scaling group launch configurations

Each Auto Scaling group is associated with a launch configuration. The launch configuration contains the information needed to launch new instances whenever a scaling event occurs.

Create a new launch template (or launch configuration). This new launch configuration contains either an updated AMI, updated user data, or both:

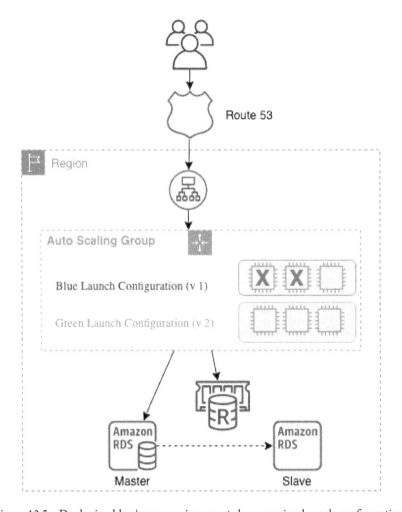

Figure 13.5 – Deploying blue/green environments by swapping launch configurations

Process for deployment – updating the Auto Scaling group launch configurations

The process for performing a blue/green deployment by updating Auto Scaling group launch configurations is as follows:

1.   Ensure that your ELB instance is not part of the deployment process.

2. Start with your traffic being directed to the current Auto Scaling group using the blue launch configuration.

3. Create a new launch configuration (the green launch configuration) and attach it to the Auto Scaling group.

4. Scale out the Auto Scaling group to **twice** its original size.

5. Once the instances have been launched and have become healthy from the green Auto Scaling group, scale down the Auto Scaling group back to its original size.

Updating ECS

Packaging your application in a container makes it easy to deploy.

Use target groups, which allow you to run multiple services behind a single load balancer. One can be the blue service and one can be the green service.

One of the key pieces to executing a blue/green deployment with ECS is the Application Load Balancer. It is at the Application Load Balancer where the ECS tasks are registered:

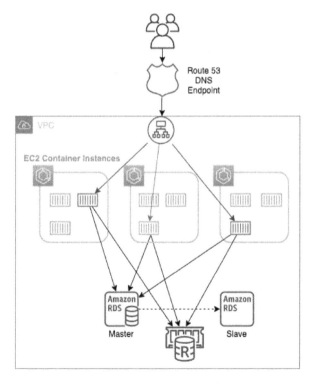

Figure 13.6 – Using an ECS service update for a blue/green deployment

When using this approach, you need to take the following considerations into mind:

- Your code needs to be completely stateless.

- Canary deployments aren't possible.

- Long-running connections will be abruptly terminated during the task switch.

Process for deployment – updating ECS

The process for conducting a blue/green deployment by updating ECS is as follows:

1. Start with a blue service that has a task definition defined that is pointing to the Application Load Balancer.

2. Next, create a new task definition that is established on the new application version, which has been created in a new container; this is your green version.

3. Scale up the green service with the green task definition and map this green service to the Application Load Balancer.

4. Scale down the blue service by setting the number of tasks to zero.

Now that we have seen how to perform a blue/green deployment using containers and ECS, let's look at using the Elastic Beanstalk service to quickly swap application versions in a blue/green deployment.

Swapping the environment of an Elastic Beanstalk application

If you are going to use a blue/green deployment strategy with Elastic Beanstalk, then you must make sure that your application environments are separated from your database:

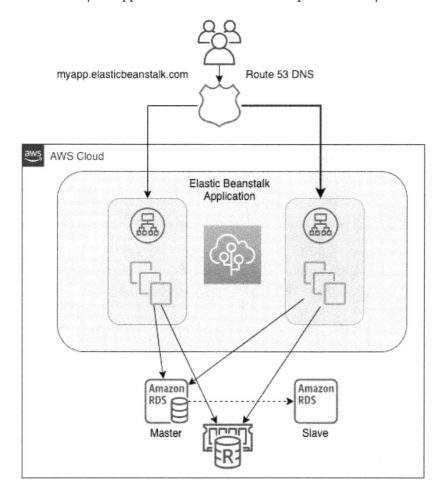

Figure 13.7 – A blue/green deployment by swapping an Elastic Beanstalk application

Process for deployment – swapping an Elastic Beanstalk environment

The process for executing a swap of an Elastic Beanstalk environment to perform a blue/green deployment is as follows:

1. Open up your AWS Management Console to the Elastic Beanstalk service and ensure that you are in the correct Region for your Beanstalk application.

2. Clone your environment. You can do this without any changes to the current platform or choose **Clone with latest platform** to use the newest version of the platform's Git branch. This new platform becomes your green environment.

3. If you only cloned the environment, you will need to deploy the new version of the application to the green environment.

4. You can now test the new environment using the unique DNS name given to the green environment's ELB.

5. On the environment overview page, choose **Environment Actions | Swap URLs**.

6. The traffic coming from Route 53 will now be directed to the green environment.

We have looked at how to easily do a blue/green deployment with the Elastic Beanstalk service using the Swap URLs feature. Next, let's look at how we can perform blue/green deployments in the OpsWorks service by cloning OpsWorks stacks.

Cloning an OpsWorks stack and then updating the DNS record

When you create an application in AWS OpsWorks, you start by creating a stack. An OpsWorks stack consists of one or more layers. Once the stack has been created, it can easily be cloned to an exact copy of itself to create an entirely new environment. Within this new environment, you can update your recipes and deploy a new version of your application. You can even use the local DNS name on ELB to test the new version of the application before routing any traffic to the new stack.

Combining the OpsWorks stack hosting your application and the Route 53 service to direct the traffic for your CNAME, you can quickly and easily switch between stacks as you deploy new versions of your application.

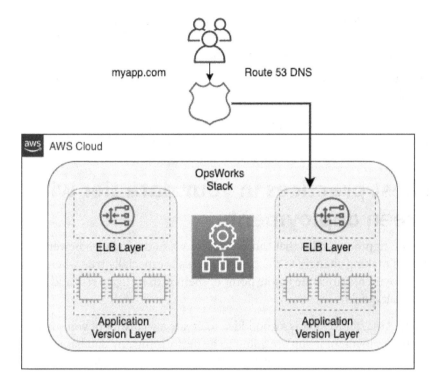

Figure 13.8 – Cloning an OpsWorks stack

Process for deployment – cloning the OpsWorks stack

The process for undertaking a blue/green deployment by cloning an OpsWorks stack is as follows:

1. Start with your current stack in OpsWorks; this is your blue stack that contains the current version of your application.

2. Next, create the new stack by cloning the stack, which is now your green environment. You can do this by clicking on the clone link in the console or by using the CLI.

3. Deploy the new version of the application to the application layer of the green environment. There should not be any traffic being directed to the green environment at this point.

4. If necessary, pre-warm your ELB so that it can handle the traffic for your customer base.

5. When you are ready to promote your green stack to the one being used as the production stack, update the DNS records to point to ELB using Route 53. This can be done all at once or in a gradual process.

6. When you are satisfied with your deployment, then you can decommission the blue stack.

Now that we have examined a multitude of different blue/green deployment techniques using a variety of different AWS services, we will next move on to the data tier.

Using best practices in your data tier with blue/green deployments

One of the more significant risks that can be present when deploying a newer version of an application is making changes to the database. This is especially true when performing blue/green deployments since the whole point of them is to mitigate risk and create the ability to roll back quickly.

If you are using Amazon RDS, it is a good idea to create a snapshot of your database prior to starting your deployment if you are going to be performing any database changes. This will allow you to restore from that snapshot if the data deployment doesn't go as planned and have as little downtime as possible.

Separating schema changes from code changes

When performing deployments, it is vitally important to separate any database changes, such as schema changes, from the application deployment. The order in which you perform some of the database updates may depend on what type of schema changes you are performing.

There are generally two approaches that you can take when schema changes are necessary, yet when to use either of the two approaches depends on whether the schema approaches are backward compatible and will work with the current version of the application:

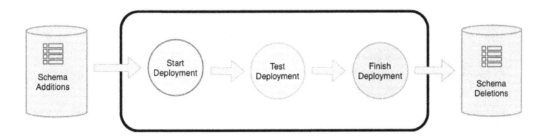

Figure 13.9 – The process for making database schema changes during blue/green deployments

The second approach is to compete for the schema changes after you have conducted your deployment. This is best for any changes that would not be compatible with the current version of the application and, if done prior to deployment, would cause errors to the current application version.

There are cases when you should perform two separate schema changes. This would be when you have both a set of non-breaking changes as well as a set of breaking changes. In splitting up the changes into two separate deployments, you have made smaller incremental changes to your database and therefore are mitigating the risk by breaking the steps into smaller, more manageable pieces.

Summary

In this chapter, we covered blue/green deployments, what they are, and how to deploy them successfully. We also talked about the services in AWS that can be used to successfully perform blue/green deployments, along with the processes for performing the deployment with each of the services. Finally, we looked at how to deal with data updates when implementing deployments. We concentrated on which schema changes should be performed first.

In the next chapter, we will begin to look at monitoring and logging for our environments and workloads. This starts with a look at the roles that the CloudWatch and X-Ray services play in DevOps.

Review questions

1. You have been contracted by a company to help architect their application on AWS. The company has a requirement to have a hardened AMI that can be auto scaled as part of the application. The application is also listening on both HTTP and TCP ports, so you have decided to use a Classic Load Balancer that can handle both protocols. There is a vanity CNAME, which is being hosted on Route 53. Blue/green deployments must be a part of this architecture. Which routing policy can you use in Route 53 to achieve blue/green deployments?

    a. Multi-answer

    b. Latency

    c. Weighted

    d. Simple

2. You are performing a blue/green deployment, updating an application environment within Elastic Beanstalk. Once you have created an identical *green* environment to the existing *blue* environment and deployed the new version of the application to the green environment, what should be done next to switch over to the new *green* environment?

    a. Update the DNS records to point to the green environment.

    b. Redirect traffic to the new green environment using Amazon Route 53.

    c. Replace the Auto Scaling launch configuration that is currently pointed at the environment's load balancer.

    d. Select the Swap Environment URLs option.

3. A company has hired you because they need help implementing their blue/green deployment process on AWS. After deployment of the new environment, they want to be able to gradually shift the traffic from the blue environment over to the new green environment. The application has been deployed on EC2 instances, which are in an Auto Scaling group placed behind the Application Load Balancer. Route 53 is routing the consumer traffic to the load balancer. Finally, the application's data tier consists of PostgreSQL RDS Multi-AZ database instances.

    Which three steps will successfully implement the blue/green deployment process?

    a. Create a new Application Load Balancer and a new Auto Scaling group.

    b. Create a new Auto Scaling group behind the current load balancer.

    c. Create a new alias record in Route 53 that points to the green environment and has a failover policy for the two records.

d. Create a new alias record in Route 53 that points to the green environment and uses weighted routing between the two records.

e. In your new Auto Scaling group, set the EC2 instances to use the same RDS instance.

f. In your new Auto Scaling group, set the EC2 instances to use the failover node of RDS.

Review answers

1. C
2. D
3. A, D, and E

Section 3: Monitoring and Logging Your Environment and Workloads

In this section, you will learn about the native monitoring options available in AWS, along with how to receive feedback from your environment and applications via the different accessible logs.

This part of the book comprises the following chapters:

- *Chapter 14, CloudWatch and X-Ray's Role in DevOps*

- *Chapter 15, CloudWatch Metrics and Amazon EventBridge*

- *Chapter 16, Various Logs Generated (VPC Flow Logs, Load Balancer Logs, CloudTrail Logs)*

- *Chapter 17, Advanced and Enterprise Logging Scenarios*

14
CloudWatch and X-Ray's Role in DevOps

Once you have your application running in the cloud, you need a way to keep an eye on it to make sure that it stays healthy and is performing effectively. **CloudWatch** can aggregate the logs of your services and applications, but combining this with **X-Ray** allows you to trace the performance of your application to see where you can enhance performance even further.

In this chapter, we're going to cover the following main topics:

- CloudWatch overview
- Using CloudWatch to aggregate your logs
- CloudWatch alarms
- Adding X-Ray for application tracing

CloudWatch overview

Monitoring is important for several reasons that range from both the operational side to the business side. First, it gives you visibility into not only what is running in your environment but how the resources in your environment are performing. Second, monitoring matters from an operational level when you are trying to troubleshoot issues in real time. From a business perspective, monitoring lets you know things such as whether your deployments have been completed successfully, and whether your customers don't see any adverse effects.

In today's world, where information can be dispersed quickly throughout the internet and especially social media channels, anything that affects customer experience can and usually does spread quickly. This can affect your brand and your business bottom line. Monitoring your resources with CloudWatch allows you to stay ahead of this and be proactive rather than reactive:

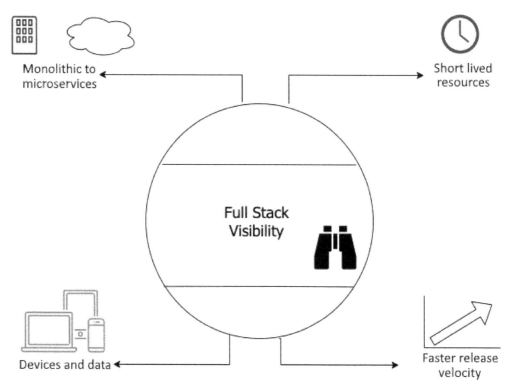

Figure 14.1 – How monitoring continues to evolve

How you monitor your systems and resources today may be different than how you monitored things 5 to 10 years ago. Things such as infrastructure changes and waterfall deployment philosophies drove what was going to be monitored.

Gone are the days where you would be provisioning a server, and that same server would be running for up to a decade before being sunsetted. With services in the cloud such as auto-scaling, which constantly adds and removes instances based on demand, the metrics that you monitor need to not only be relevant to the health of your system but the performance of your customers:

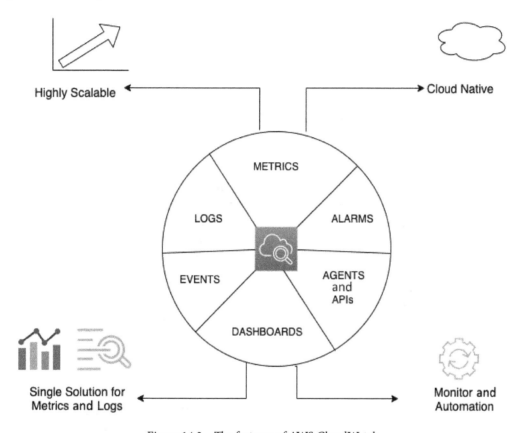

Figure 14.2 – The features of AWS CloudWatch

Amazon CloudWatch is an AWS native service that helps you monitor your services and resources. It is part of the management tools offered by AWS. The primary function of CloudWatch is to help you track and monitor the performance of your resources and applications. While monitoring, the CloudWatch service can use alarms to notify either personnel via the SNS service or use events to trigger automation in response to those alarms. CloudWatch can also be used to collect and monitor log files. CloudWatch consists of three main components: *metrics*, *alarms*, and *events*.

Understanding and using the CloudWatch unified agent

AWS provides a unified CloudWatch agent that can help you do several things both for on-premises servers and on EC2 instances. Let's review a few common scenarios where using the CloudWatch unified agent comes in handy.

Running as a different user

When running the CloudWatch agent on Linux servers, the CloudWatch agent will run as the root user by default. If your company does not allow agents and programs to be run as the root user, then you can create a custom user for the CloudWatch agent to use and then, in the configuration file, tell the agent to `run_as_user`.

Having multiple CloudWatch agent configuration files

Organizations that base their build on an approved **Amazon Machine Image** (**AMI**) that must be used for any development or production builds can have a CloudWatch configuration file pre-baked into the instance. This configuration file would allow standard configurations to be collected across all instances. Development teams can then add another configuration file that the CloudWatch agent can read and process at startup, which would include any specific application metrics or logs.

An example of this could be if an application team is running an NGINX server to front their web application; otherwise, they may be using the proxy powers of NGINX to redirect traffic. Their custom configuration file can be designated to look for 4XX and 5XX errors from NGINX, and they can also be configured to consume the logs and ship them back to CloudWatch Logs.

Sending metrics and logs to a different account than the one the instance is running in

The CloudWatch agent has the flexibility to send either metrics, logs, or both to another AWS account for monitoring purposes by specifying `role_arn` in the configuration.

Adding custom dimensions to metrics collected by the CloudWatch agent

CloudWatch will create its own rollups of the metrics that it collects. These might not always be the optimal grouping for you and your team; hence, there are ways to customize how things are grouped by using the `append_dimensions` field.

We just saw how the unified CloudWatch agent, whether used on an AWS EC2 instance or an on-premises server, allows you to collect logs and metrics in a multitude of scenarios. Now, let's go through the process of installing the CloudWatch agent on an EC2 instance.

Installing the CloudWatch agent on an EC2 instance

The best way to get a good understanding of the CloudWatch unified agent is to go through the process of installing the agent on an EC2 instance. The following tutorial will take you through the steps of standing up an EC2 instance and then installing and configuring the agent. Finally, we will send some traffic to the EC2 instance so that we can look at the metrics and logs that have been generated.

> **Note**
>
> We will go much deeper into CloudWatch metrics in the next chapter. *Chapter 15, CloudWatch Metrics and Amazon EventBridge*, will discuss both generic metrics and custom metrics.

So far, we have been using Amazon Linux for any EC2 instances in our examples. Amazon Linux is an excellent operating system and comes pre-installed with many packages for use on the AWS cloud computing environment. This is the precise reason that for this example, we are going to use a different operating system – *Ubuntu*. The Ubuntu OS will not have some of the packages, such as the CloudWatch unified agent, installed by default. This allows us to go through the process of installing it on this EC2 instance, but the process would be the same if you had an EC2 instance running in your data center, except for the fact that you would need to authenticate your instance in your data center with a keypair, which would be rotated regularly. EC2 instances, on the other hand, may assume roles and would not need the access key and secret key to be stored on the instance, and in fact, assuming the role is a more secure practice. Let's get started:

1. First, let's create the IAM role for our instance; we are going to need our instance to have permissions for CloudWatch and AWS Systems Manager. To create the role, we will need to save an initial JSON policy locally, and then we can attach the two managed policies that we need. Copy the JSON policy to a file named STS.json:

```
{
    "Version": "2012-10-17",
    "Statement": {
        "Effect": "Allow",
        "Principal": {"Service": "ec2.amazonaws.com"},
        "Action": "sts:AssumeRole"
```

```
        }
    }
```

2. With our initial policy saved, we can now create the role and attach the two
    managed policies:

    ```
    aws iam create-role --role-name CW-EC2 --assume-role-
    policy-document file://STS.json

    aws iam attach-role-policy --policy-arn
    arn:aws:iam::aws:policy/AmazonSSMManagedInstanceCore
    --role-name CW-EC2

    aws iam attach-role-policy --policy-arn
    arn:aws:iam::aws:policy/CloudWatchAgentServerPolicy
    --role-name CW-EC2
    ```

3. Since we will be launching our instance with the CLI, we will need to create an
    instance profile. Once our instance profile has been created, we can attach our new
    role to `instance profile`:

    ```
    aws iam create-instance-profile --instance-profile-name
    CW_SSM

    aws iam add-role-to-instance-profile --role-name CW-EC2
    --instance-profile-name CW_SSM
    ```

4. Then, we will need to query the `SSM` parameter store for the AMI, which we want to
    use to spin up our image:

    ```
    AMI='aws ssm get-parameters --names \        /aws/service/
    canonical/ubuntu/server/16.04/stable/current/amd64/hvm/
    ebs-gp2/ami-id \
        --query 'Parameters[0].[Value]' --output text
    --region us-east-2'
    ```

    If you would like to see the value of the AMI that is returned, then you can use the
    following command:

    ```
    echo $AMI
    ```

5. We are going to want to get into our instance later, but rather than creating a key
    pair, we will add the following script, which will install both SSM Agent and the
    unified CloudWatch agent for Debian. We will create this script and save it to a file
    named `agents.sh` so that we can use it when launching our EC2 instance later in
    the `user-data` parameter.

The script to install SSM Agent and the unified CloudWatch agent is as follows:

```bash
#!/bin/bash
mkdir /tmp/ssm
mkdir /tmp/cw-agent
# Download and Install the SSM Agent
cd /tmp/ssm
wget https://s3.amazonaws.com/ec2-downloads-windows/
SSMAgent/latest/debian_amd64/amazon-ssm-agent.deb
sudo dpkg -i amazon-ssm-agent.deb
sudo systemctl enable amazon-ssm-agent
# Install CollectD
sudo apt-get update -y
sudo apt-get install -y collectd
# Download and Install the Unified CloudWatch agent
cd /tmp/cw-agent
wget https://s3.amazonaws.com/amazoncloudwatch-agent/
ubuntu/amd64/latest/amazon-cloudwatch-agent.deb
sudo dpkg -i -E ./amazon-cloudwatch-agent.deb
sudo systemctl enable amazon-cloudwatch-agent
```

> **Note**
>
> This script can also be downloaded from the Chapter-14 folder of this book's GitHub repository under the same name – agents.sh.

6. Once we have the AMI value stored in our variable, we can launch our image with the following command:

```
aws ec2 run-instances \
--image-id $AMI \
--instance-type t2.micro \
--user-data file://agents.sh \
--tag-specifications
'ResourceType=instance,Tags=[{Key=Name,Value=CW_Agent}]'
\
--region us-east-2
```

You will notice that we have added a few more parameters in our instance for the startup, including the `user-data` script, which we just created, as well as a `Name` tag for our instance. This name will help us identify our instance when we try to find it later.

7. While our instance is initiating and running the startup scripts that we have given to it, we can log into the AWS Console and go directly to the SSM session manager using the following URL: `https://console.aws.amazon.com/systems-manager/session-manager`. Once logged in, at the top right-hand side, check that you are in the correct region. We specified for our instance to be launched in the Ohio region (`us-east-2`), and if another region is listed, then switch to the region where you have spun up your instance. As an alternative, you could also do this from the **EC2** screen by selecting the instance and then choosing **Connect** from the **Actions** dropdown. On the **Connect to instance** screen, choose **Session Manager** and then the orange **connect** button. A new window containing your session should appear as you connect to your instance.

8. Now that we are connected to our instance securely without having to use keys, we can configure the agent. We have already installed the CloudWatch agent using some of the commands in our `user-data` script. To get some of the logs flowing to CloudWatch Logs, we will need to tell the agent which logs specifically to push. We will need to run the configuration script as a `sudo` user, which is located at `/opt/aws/amazon-cloudwatch-agent/bin/amazon-cloudwatch-agent-config-wizard`. To simplify the commands, we will jump into the root user using the `sudo su` command:

    ```
    sudo su
    /opt/aws/amazon-cloudwatch-agent/bin/amazon-cloudwatch-agent-config-wizard
    ```

9. Running the CloudWatch agent configuration manager will prompt you for quite a few questions, but the setup should take less than 5 minutes. The questions shown here are the ones that do not follow the defaults:

*   Which default metrics config do you want? Standard (2) Log file path: `/var/log/amazon/ssm/amazon-ssm-agent.log`.

*   Do you want to specify any additional log files to monitor? No (2).

*   Do you want to store the config in the SSM parameter store? No (2).

10. When the config wizard has completed, since we are using the Ubuntu operating system, we will copy the newly generated `config` file to where the agent is expecting it to be and give ownership to the `cwagent` user:

```
cp /opt/aws/amazon-cloudwatch-agent/bin/config.json /opt/
aws/amazon-cloudwatch-agent/etc/amazon-cloudwatch-agent.
json
chown cwagent /opt/aws/amazon-cloudwatch-agent/etc/
amazon-cloudwatch-agent.json
```

11. With our CloudWatch agent file now configured, we will have to restart the agent for the changes to take effect. We can do this with the following command:

```
/opt/aws/amazon-cloudwatch-agent/bin/start-amazon-
cloudwatch-agent &
```

The & symbol at the end of the command will put the script in the background so that you can do other things, and the script will still be running.

12. One of the first ways we can generate a log is by clicking the orange **Terminate** button at the top right-hand side of our **SSM session** window. Confirm that you want to terminate the session. This tab should close automatically once terminated. Find one of your other **AWS Management Console** tabs, which should still be open, and using the top search bar, search for the `CloudWatch` service. Once the service name appears, click on **CloudWatch** to be taken to the service.

13. Once in **CloudWatch**, from the left menu, find the main item named **Logs**. Expand the **Logs** item so that you can see **Log groups**. You should see a group named amazon-ssm-agent.log. Click on this log group's name to be taken to the log streams.

14. There should only be one log stream in the log group. The log stream's name will be the identifier for your AWS EC2 instance. Click on this **Log stream** so that we can view the logs:

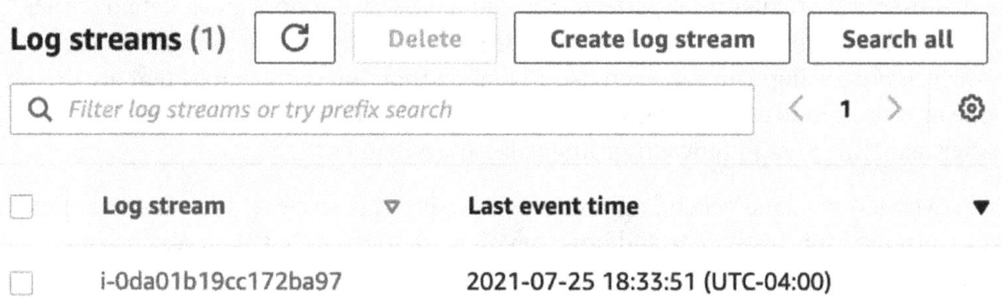

Figure 14.3 – A single log stream in our CloudWatch log group

15. If you did log out of your **Session Manager** session, then you can use the search box to search for the term `Session worker closed`. Any time that you have logged out of **Session Manager** while the CloudWatch agent was running should appear in the list of logs. You can click the small triangle on the left-hand side of the log to expand the contents and see the full log entry.

> **Note**
>
> We are going to continue to use this instance for further exercises in this chapter. If you are going to continue at a later point in time, I would suggest that you put this instance into hibernation so that you don't get charged for it. Then, you don't have to reconfigure everything when you are ready to go again.

Now that we have added the unified CloudWatch agent to a non-Amazon instance, we will look at some other features of CloudWatch Logs. If you plan on doing the next exercise with the CloudWatch alarms in a relatively short manner, I would suggest that you either leave this instance running or simply stop the instance so that you have access to the metrics that the CloudWatch agent aggregates and pushes for alarming.

Using CloudWatch to aggregate your logs

Amazon's CloudWatch service is not only a powerful monitoring tool, but it also allows you to route multiple types of logs such as operating systems, applications, custom log files, and even CloudTrail logs to the dependable storage of CloudWatch Logs.

CloudWatch Logs allow you to group logs that come from the same source (log streams) and then search through those groups using filter patterns. Filter patterns are like the CloudWatch version of regular expressions and allow you to search through the different fields of the logs in your log streams and groups.

Using subscriptions, you can push either all the logs from a particular log stream or only those that meet a particular filter pattern. You can have subscriptions push data to either an Amazon Kinesis stream for real-time data processing or to a Lambda function for event-driven processing. You can even use a Lambda function to push logs that are driven into one or more CloudWatch log groups directly into the managed Elasticsearch service for easier searching and graphical trending using the Kibana interface.

With an overview of CloudWatch Logs out of the way, let's take a closer look at some of the terms we just used for this service and how they all work together in CloudWatch Logs.

CloudWatch Logs terminology

As we start to get deeper into CloudWatch Logs, there are some terms that we need to be familiar with, especially in the context of both the following exercise and the professional exam:

Filter pattern: The filtering expressions that restrict which logs get forwarded to the AWS destination resource.

Log events: A record of some activity that's been recorded in CloudWatch Logs is known as a log event. Event messages must be in UTF-8 format.

Log streams and **Log groups**: A group of log streams that share the same source is grouped in the CloudWatch Logs console as log groups. There is no limit on how many log streams can be a part of one log group.

Metric filters: Using metric filters, you can extract data from ingested events. You can then convert them into data points on a CloudWatch metric. Metric filters are assigned to log groups. All log streams in a particular log group get the particular metric filters assigned.

Retention settings: How long your log files are kept in CloudWatch Logs is determined by the retention settings. By default, logs are kept indefinitely and never expire. This can lead to extra costs if you do not need your logs over a specified time. Instead, you can choose a retention period between 1 day and 10 years for each log group, and once the retention period is met, then the logs will be automatically deleted:

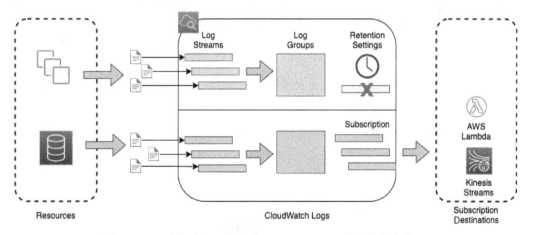

Figure 14.4 – The flow of logs from resources to CloudWatch Logs

Using the terms we just learned, in the preceding diagram, we can see how a log that has been generated from an AWS resource becomes a log stream. One or more streams are combined to form a log group. Log streams are then retained or deleted based on the retention settings.

Next, we will learn how to use one of the features of CloudWatch logs, Insights, to analyze the data that is being captured by the CloudWatch Logs service.

CloudWatch alarms

In addition to shipping your logs for storage and searching, another aspect of monitoring your system is being alerted when something goes awry. These could be simple notifications that let you know that a service or server is not being responsive. It could also be proactive alerts, letting you know that the platform that you are running your application on is running out of CPU or memory and needs to be scaled before larger issues arise.

You can use the **CloudWatch** service to monitor either a single metric or multiple conditions to create alarms. These alarms can be raised when the metrics of the underlying resources meet a certain criterion. There are two types of alarms that you can create in CloudWatch: metric alarms and composite alarms.

A **metric alarm** monitors a specific metric of CloudWatch. It has a threshold for monitoring that is set when it's initially created, along with the number of periods that can break the threshold before going into an alarm state. Once that alarm state has been triggered, then an action can be configured. Available actions include sending a notification to an SNS topic, performing an EC2 action, performing an AutoScaling action, or creating an OpsItem or incident in Systems Manager.

A **composite alarm** uses multiple alarm states that you have created to allow you to create specific conditions when the alarm goes off.

Once the alarm goes off, you can have the alarm perform a variety of actions, such as the following:

- Stop, terminate, or reboot an EC2 instance.
- Have an autoscaling group scale up or down.
- Send a notification message to an AWS SNS topic.

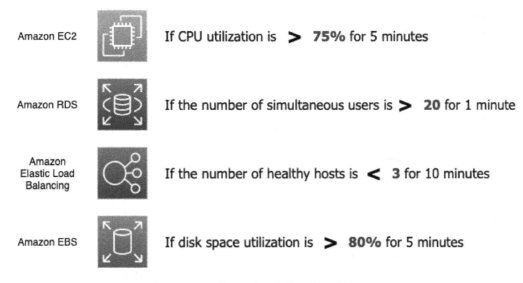

Amazon EC2		If CPU utilization is > 75% for 5 minutes
Amazon RDS		If the number of simultaneous users is > 20 for 1 minute
Amazon Elastic Load Balancing		If the number of healthy hosts is < 3 for 10 minutes
Amazon EBS		If disk space utilization is > 80% for 5 minutes

Figure 14.5 – Examples of CloudWatch alarms

There are certain facts about CloudWatch alarms that you should know about when studying for the DevOps Professional exam. While the Professional exam will not test you on these facts verbatim, they can be incorporated into a larger scenario:

- Alarm names can only be comprised of ASCII characters.
- You can create up to 5,000 alarms per region, per account.
- You can add alarms to CloudWatch dashboards.
- You can test alarms by using the `SetAlarmState` setting (either to engage or disengage the alarm).
- The CloudWatch service saves alarm history for 2 weeks.

Next, we will create a CloudWatch alarm and create some events to trigger the alarm. To receive the notifications for the alarm, we will also need to create an SNS topic. However, you can skip this first step if you already have an SNS topic and would like to use that previously created topic. Just make sure that if you are going to skip this step, you are subscribed to the topic.

Creating a CloudWatch alarm

You should have an EC2 instance that is creating metrics that have been stood up from our previous exercise. We will use the metrics from this instance to monitor for an alarm and then stress the instance so that an alarm is created.

For us to receive notifications for the alarm, we will need an SNS topic that we can subscribe our email address to:

1. Open your terminal and type the following commands to create `topic`:

    ```
    $aws sns create-topic --name cwatch
    ```

    If `topic` was created successfully, then it should return something like this:

    ```
    {
        "TopicArn": "arn:aws:sns:us-east-
    2:470066103307:cwatch"
    }
    ```

2. Now that we have our `topic`, we need to use `subscribe` with our email address:

    ```
    $ aws sns subscribe \
        --topic-arn arn:aws:sns:us-east-2:470066103307:cwatch \
        --protocol email \
        --notification-endpoint devopsproandbeyond@gmail.com
    ```

    This should return a JSON statement telling you that the subscription is pending:

    ```
    {
        "SubscriptionArn": "pending confirmation"
    }
    ```

3. Now, we need to go to our email account and find the email that the SNS service has just sent. Then, we must click on the link that says **Confirm Subscription**.

4. With our SNS topic created, we can now back to the AWS Management Console and navigate to the **EC2** service. Click on the **EC2** dashboard and click on the **Instances (running)** link under the resources section. Find the instance we created named CW_Agent and make a note of the instance ID. Once you have this marked down, go to the CloudWatch service under the **Services** dropdown at the top left of the navigation bar. Open the CloudWatch service in a new tab; we will need to come back to this instance to test our alarm later.

5. Once you're in the CloudWatch service, find the **Alarms** menu setting from the left-hand menu. Click on this to expand the menu item and view the sub-menu items. Once expanded, click on **All alarms** so that you can be taken to the **Alarms** screen.

6. On the **Alarms** page, click on the orange button labeled **Create Alarm** to create a new alarm. This will bring up the prompts for creating the alarm.

7. Click on the **Select metric** button. This will bring up a dialog where you can choose the metric. Find **Custom Namespaces** and click on **CWAgent**. Once you're inside the **CWAgent** namespace, choose the grouping labeled **ImageId, InstanceId, InstanceType** and click on this link.

8. Find the metric named `mem_used_percent`. However, before clicking on the box to select it, make sure that the **InstanceId** matches the instance ID that you found earlier. If everything matches, then select the checkbox to the left of the metric to use this metric in the alarm. When you select this metric or any metric, a graph will appear at the top, showing the recorded values for the metric you selected. Click on the orange **Select metric** button at the bottom to continue:

Figure 14.6 – Selecting the single metric to monitor when creating a CloudWatch alarm

9. Scroll down to the **Conditions** section of the next page. Under **Threshold type**, choose **Static** and **Greater**, and set the defined value to 20. Once you have changed these values, click the orange **Next** button.

10. Now, on the **Notifications** screen, use the following selections. Once you have filled in all the sections, scroll to the bottom of the page and click the orange **Next** button:

    a. **Alarm state trigger**: In Alarm.

    b. **Select an SNS Topic**: Select an existing SNS topic.

    c. **Send notification to**: Choose the cwatch notification topic you just created.

11. Next, on the **Add name and description** page, use `chapter14` as the alarm name. Click the orange **Next** button.

12. Finally, on the **Preview and create** page, scroll down to the bottom of the page, checking the values of the items on the way down. If everything looks right, then click on the orange **Create Alarm** button at the bottom of the page.

13. Now that we have created our alarm, we need to test it. We could test it by changing `SetAlarmState` using the CLI. However, we are going to test our alarm by performing a stress test on the EC2 instance itself. Find the tab that contains the instance ID that we opened earlier. Select the checkbox next to the instance, and then at the top of the **Instances** screen in the AWS Management Console, find the dropdown labeled **Actions**. Click on the **Actions** menu to expose the sub-menu items. Choose **Connect**.

14. On the **Connect** screen, use **Session Manager** to connect to the instance. Once on the **Session Manager** tab, click on the orange **Connect** button at the bottom to start our session inside of our EC2 instance.

15. Now that we are inside the instance, we will need to install a package to help us stress test the instance and set off the alarm. Run the following command to install the package:

```
sudo apt-get install -y stress
```

16. Once the `stress` package has been installed, we can run it with the following options to stresstest the memory of the instance:

```
stress -vm 2 -vm-bytes 126M
```

17. After the **Memory used** percent on the instance goes above 20% for 5 minutes, you should receive an email notification to whatever email you subscribed with in the SNS topic.

Now that we have seen how the CloudWatch service can collect and aggregate our logs and help us with our monitoring duties with sending alerts, we will take a look at another tool in our monitoring and debugging toolbelt: **AWS X-Ray**.

Adding application tracing with X-Ray

X-Ray is a service that's used to monitor modern web applications. Modern applications are essentially service-oriented applications. These can be serverless architecture applications or applications that run inside containers. With these modern applications, the app itself is broken into multiple pieces. While this presents many advantages, including ease of scaling in a horizontal nature and taking full advantage of cloud-native services, it can also present some challenges. Understanding where errors are ultimately impacting your service (or business) becomes more problematic.

Tracing allows you to connect the dots in a modern application by letting you do the following:

- Discover multiple services.

- Get insights into individual operations.

- See issues isolated within a segment.

- Perform root cause analysis for a specific issue.

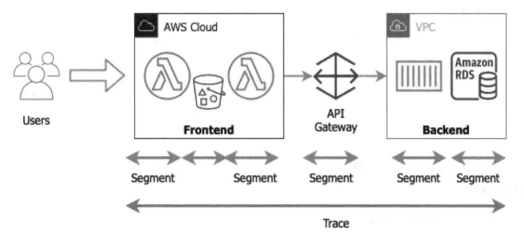

Figure 14.7 – How an X-Ray trace breaks down pieces into segments

Tracing allows you to quickly look at and easily examine what happened for a particular API call or a particular user.

A **trace** is a holistic view, and it encapsulates the end-to-end transactions from a customer standpoint where the customer created the transaction.

At that point, X-Ray breaks the trace down into segments. Segments are chunks that come from individual servers.

Now that we understand traces, which are one of the main concepts in X-Ray, let's look at how the X-Ray service itself works.

The trace is then broken up into various segments. A **segment** furnishes the name of the resource, the specifics about the request, and the work being done. Segments can also show issues that occur in the segment, such as errors, faults, and exceptions.

How does the X-Ray service work?

You start by integrating the X-Ray SDK into your applications. This can be customized for different languages such as Python, Java, and Node.js. Next, an on-instance daemon starts to collect the data. The daemon then ships this data to the X-Ray backend. At the X-Ray backend, the traces are then recorded. There may be feeds coming for different services at multiple points in time, yet the X-Ray service can piece all of this information together using a **trace id**. Once the trace has been collected, the X-Ray service creates an aggregated view called a **Service Map**. Finally, X-Ray presents a set of analytics capabilities that allow you to dive into and answer three valuable questions.

Another item to note is that the X-Ray service is cloud-agnostic. This means that the code you have written doesn't have to only run in the AWS cloud. The code can run in other places to take advantage of the X-Ray service's tracing capabilities, such as on a developer's laptop or in a corporate data center. This assumes that the environment that it is running in has connectivity back to the AWS X-Ray service and a set of credentials that will enable it to run.

X-Ray helps you answer three questions

The X-Ray service helps developers answer three specific questions:

1. **How** is my application doing?

2. **Why** is my application performing the way it is?

3. **Who** is impacted by the issues?

The graphical nature of the X-Ray interface, which is depicted as a service graph, lets you and your development team see where the users of your application are consuming resources. It also gives you the times of responses for the different resources so that you can see if a particular resource is the cause of latency or trouble issues.

Now that we understand how X-Ray can help us with developing and troubleshooting our applications that are running on the AWS cloud, let's take a look a how the X-Ray service integrates with serverless services.

X-Ray and serverless services

When using X-Ray in conjunction with serverless services and Lambda, in particular, the X-Ray service brings a few unique benefits to the table that you would not get with CloudWatch monitoring. X-Ray lets you obtain information on the timing of AWS Lambda cold starts.

When a Lambda service first receives a request to run a function, it needs to prepare an execution environment. This means that it needs to retrieve the code from an S3 bucket where the code is stored and then allocate the runtime environment, which includes the memory and CPU. After performing any last initialization steps, Lambda can then run the handler.

Implementing X-Ray on a Lambda function

In *Chapter 12*, *Lambda Deployments and Versioning*, we went over AWS Lambda. This is the Function-as-a-Service offering from AWS. We can take the functions we build and then enable the X-Ray service on one of them to see the traces and segments from AWS X-Ray:

> **Note**
>
> If you have either not done the *Creating a Lambda function* exercise in Chapter 12, *Lambda Deployments and Versioning*, where we went through the process of creating the word count function, or you have deleted this function from your account, please go back and do/redeploy this exercise. We will be using this function to continue our AWS X-Ray exercise.

1. Log into **AWS Management Console** and navigate to the **Lambda** service.

2. Once at the **Lambda** service, find the function we created previously in *Chapter 12*, *Lambda Deployments and Versioning*, named my_word_count_python. Click on the name of this function to be taken to this function. If you have not created this function or had deleted it after performing the exercise, you have two options. You can go back and recreate the function or you can try and follow the steps for another function that you have created in your account to implement the X-Ray tracing service.

3. Now that you are in the **Lambda** service, scroll down the page until you find the horizontal menu bar and click on the menu item named **Configuration**:

Figure 14.8 – Configuration highlighted on the Lambda horizontal menu

4. The **Configuration** menu choice will bring up a vertical menu on the left-hand side of the screen. Locate the option named **Monitoring and operations tools** and click on it. This will bring **Monitoring and operations view** into focus. At the top right of this view, choose **Edit**.

5. On the **Edit monitoring tools** page, find the section labeled **AWS X-Ray**. Click on the **Slider** button to turn X-Ray tracing on for this Lambda function. Then, click **Save** at the bottom of the page:

AWS X-Ray Info

When you enable and choose Save, the Lambda console verifies your execution role's permissions. If your execution role does not have the required permissions, the Lambda console will attempt to add them to the role.

🔵 Active tracing

> ⓘ Couldn't find the required permissions. The Lambda console will attempt to add them to the execution role. Be sure that your execution role does not deny the required permissions.

Figure 14.9 – Adding AWS X-Ray to our Lambda function

6. Clicking **Save** will take you back to the **Monitoring and operations tools** page under the **Configuration** header. You should see that **Active tracing** is now enabled. With tracing enabled, let's run a test to see X-Ray in action. From the horizontal navigation bar, click the menu item named **Test**.

7. If you need to create another test event, then you can use the default data and save it as XRtest. If you still have the Lambda function up from the previous exercise, then you should have a test event named Test1 that we previously created. With your test event created, click the orange **Test** button in the top-right corner of the **Test event** section:

Figure 14.10 – The orange test button to the right of the Test event section

8. Once your test event has run, we can click on the **Monitor** menu item on the horizontal menu to be taken to the **Monitoring** page. By default, this page will be on the metrics section. However, we will click on the item named **Traces** so that we can see the X-Ray trace that was produced.

9. On the trace screen, we can now see a **Service Map**, which was produced along with the trace ID of the function, which was traced in the following table:

Service Map

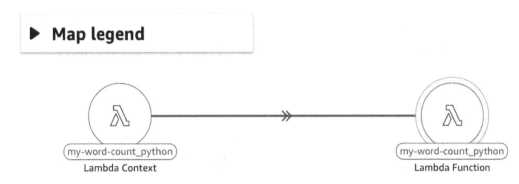

Figure 14.11 – The generated X-Ray Service Map

10. Go back to the **Test** section and click **Test event** about two to three more times. Once the Lambda invocations have been completed, you can go back to the **Monitor** tab and look at the trace table once again:

ID	Trace status	Timestamp	Response code	Response Time	HTTP Method	URL Address
1-60fdaa70-1abb925f2 6a8847f0cb4e669	⊘ OK	48.7s (2021-07-25 14: 16:16)	200	1.012s	-	-
1-60fdaa69-52fb84640 5113a9d065ce1db	⊘ OK	55.7s (2021-07-25 14: 16:09)	200	2.016s	-	-
1-60fdaa6d-36c73bd4 573c19884f98ae87	⊘ OK	51.7s (2021-07-25 14: 16:13)	200	1.024s	-	-
1-60fda764-4fa3b5ee6 ebd132b05bca152	⊘ OK	13.8min (2021-07-25 14:03:16)	200	1.948s	-	-

Figure 14.12 – Table of traces from the same Lambda function

If you clicked the test event multiple times in a row, then in the **Response Time** column in the table, you should see that there will be some functions that have a much faster response time than others. This is because there is no cold start needed for these functions.

We just went through an exercise that took a previous Lambda function that we had running, and then allowed us to gain more information on it by adding X-Ray traces. We could see the path of the calls, starting from the customer, as well as the variations in the time that the function takes to respond initially.

Next, let's wrap up what we learned in this chapter regarding AWS CloudWatch and X-Ray.

Summary

In this chapter, we looked at the importance of monitoring for our different environments and applications, especially the CloudWatch service.

In the next chapter, we will continue to look at the CloudWatch service by looking deeper at the monitoring and metrics capabilities of the service. We will also learn how to automate reactions from the CloudWatch service using AWS EventHub.

Review questions

Answer the following questions to test your knowledge of this chapter:

1. You have been hired by a company to help the DevOps team. The team currently needs help with their monitoring. The company wants to use all native AWS services rather than third-party services. There is currently a production RDS PostgreSQL database that needs to be monitored closely as it is the main data store for the customer orders. Which services can be used to monitor and alert in real time if the IOPs metrics exceed the normal levels and allow the DevOps teams to provision more IOPs? (Choose two)

    a. Amazon CloudWatch

    b. Amazon CloudTrail

    c. Amazon Simple Notification Service

    d. Amazon Route 53

2. You have developed a modern application using AWS Lambda functions for a company that is invoked when someone places a file in an S3 bucket. The company would like more visibility into the application and has requested that you integrate AWS X-Ray into the Lambda function so that they can see the traces. How should you do this to ensure that you make sure that all non-instrumented services that invoke your Lambda function are traced?

    a. Under the AWS Lambda function configuration, enable Start tracing.

    b. Under the AWS Lambda function configuration, enable Active tracing.

    c. Lambda functions do not support tracing when invoked by non-instrumented services.

    d. Lambda functions do not need any additional configuration to record traces when invoked by non-instrumented services.

3. You have been brought on to a team that is using a mainly serverless architecture consisting of Lambda functions. They want to be able to analyze the invocations of the functions during the testing phase of their SDLC. Which of the two (2) following tools will help them achieve this?

    a. Amazon CloudTrail

    b. Amazon CloudWatch

    c. Amazon Inspector

    d. Amazon X-Ray

Review answers

1. a, c
2. b
3. b, d

15
CloudWatch Metrics and Amazon EventBridge

Metrics are one of the central tenants of AWS CloudWatch. They record the performance of your services and can be used to trigger alarms and actions based on the data they provide. There is a multitude of out-of-the-box metrics available, but there is also the ability to create custom metrics. This can expand the capabilities of CloudWatch even further.

Having the ability to harness the metrics captured when they reach certain thresholds is a capability of Amazon EventBridge. As a DevOps engineer, this is a powerful tool that can help you automate your systems so that they are self-healing, as well as constantly scaling up and down to meet your customers' generated capacity needs. These automated responses can be triggered by both native AWS services and third-party systems. This is especially useful in routine monitoring and creating routines when certain events occur.

In this chapter, we're going to cover the following main topics:

- A closer look at CloudWatch metrics
- Basic metrics in CloudWatch for AWS services
- Using CloudWatch metrics to create dashboards
- Amazon EventBridge overview

A closer look at CloudWatch metrics

In the previous chapter, we looked at the CloudWatch service and examined a few of the features that it offered. We even touched on the topic of metrics when creating our alarm.

Metrics, when it comes to applications and monitoring, is data. Many times, this is lots of data streaming constantly. This data is used not only from a technical perspective but also from a business perspective to see how the company is performing.

Metrics are fundamental in CloudWatch. When recorded, a **metric** represents a time-ordered set of data points that are then published to the CloudWatch service.

Metrics endure in a single region. This means that if you have a multi-region environment, then the metrics for the different resources will be gathered and stored in the same region where the resources have been created and have been running at a point in time. Although you cannot delete metrics, they do expire automatically after 15 months.

A **namespace** is a container for CloudWatch metrics. The namespace where you will find the metrics will be the same as the AWS service names for many services:

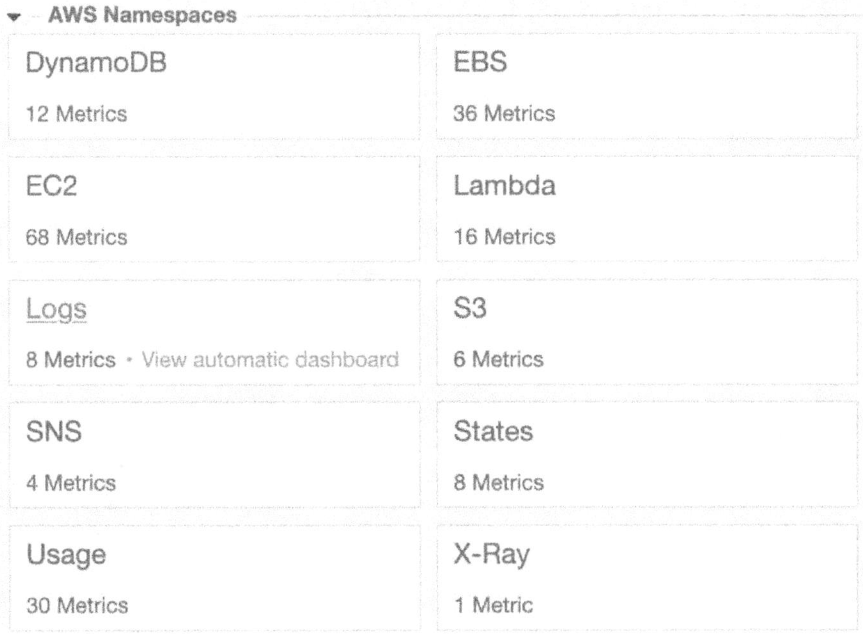

AWS Namespaces

DynamoDB 12 Metrics	**EBS** 36 Metrics
EC2 68 Metrics	**Lambda** 16 Metrics
Logs 8 Metrics · View automatic dashboard	**S3** 6 Metrics
SNS 4 Metrics	**States** 8 Metrics
Usage 30 Metrics	**X-Ray** 1 Metric

Figure 15.1 – Namespaces inside the CloudWatch console

Dimensions are name/value pairs that help you identify unique metrics. You can define the dimension when you use the `PutMetricData` command with the CloudWatch agent.

The metric data is aggregated over specific periods into **statistics**. These aggregations are correlated using the namespaces, metric name, dimensions, and data point of measure. You can measure the statistics in one of five ways: average, minimum, maximum, sum, or sample count. When choosing sample count, CloudWatch will count the number of data points.

One of the good things about CloudWatch is that the resource does not have to be running to access the metrics. You can even access metrics for resources that you have terminated, such as terminated EC2 instances, deleted Elastic Load Balancers, Fargate containers, and deleted EBS volumes.

Viewing your metrics in CloudWatch

You can view graphs of your metrics by logging into the AWS Management Console, going to the **CloudWatch** service, and choosing **Metrics** from the left-hand menu:

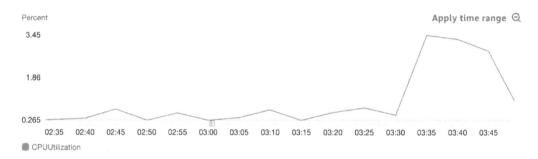

Figure 15.2 – An example of a graph from a CloudWatch metric

In the **All metrics** view, you can search by things such as `InstanceID`, `functionName`, `Invocations`, `Namespace`, and others. One thing that doesn't work is searching by **Amazon Resource Name** (**ARN**).

Streaming metrics with CloudWatch metric streams

As we stated earlier, **CloudWatch** keeps your metric data for 15 months and then deletes the data. If you would like to push that data for long-term storage in a data store such as an S3 bucket or a data lake via **Amazon Kinesis Data Firehose**, this is possible via a metric stream. You also have the option to push your metric data using metric streams to a third-party provider.

Data from CloudWatch metrics can be pushed to metric streams in either JSON or OpenTelemetry format.

Now that we have an understanding of what CloudWatch metric streams are, let's take a look at why we would use metric streams.

Why would you push your metrics to a third party via metric streams?

In the past, partners who specialized in services such as monitoring and dashboarding relied on API calls via the CloudWatch service to get the data from your account into their service. As your account grew, this could add additional expenses on your side. The `GetMetricData` API call is $0.01 per 1,000 requests. Metric streams reduce this cost substantially by only charging $0.003 per 1,000 requests.

Now that we have learned how to store or share our metrics using metric streams, let's examine the different types of metrics available in CloudWatch metrics.

Basic metrics in CloudWatch for AWS services

CloudWatch automatically monitors a basic set of metrics at 5-minute intervals for free. Most AWS services send metrics automatically for free to CloudWatch metrics. These include foundational services such as EC2, S3, EBS, Kinesis, Lambda, and many others.

Basic monitoring for the EC2 service

When an instance is created, seven metrics are pushed out to CloudWatch at a frequency of every 5 minutes. You can change this frequency to 1-minute intervals for an additional charge. CloudWatch also makes a binary status check available as part of their Free Tier. Using this check is an essential measurement to see if your instance is up and running. It is not a good check to ensure that your application is performing correctly. The status check can be a canary in the coalmine for things such as AMI issues, accidentally (or purposefully) terminating an instance, or even Availability Zone or regional failures.

Besides the status check, there are three standard categories that EC2 metrics fall into:

- CPU
- Disk I/O
- Network

CPU metrics contain metric data points for CPU usage. This can be one of the main metrics that you can use for AutoScaling events while gauging to see if your instances are starting to breach their compute capacity. For burstable instances, such as the T family of EC2 instances, you will also get metric data on `CPUCreditUsage` and `CPUCreditBalance`.

> **Remember**
> With a burstable performance EC2 type instance, you earn a set rate of CPU credits every hour that can accrue until needed. When tasks that require more CPU than the baseline are processed by the instance, then that instance will spend its CPU credit balance.

Almost every service from AWS integrates with CloudWatch metrics. As you build and deploy your applications, think of some of the most critical metrics that need to be monitored so that you can ensure the basic health of your application and your environment. These can include the following:

- `CPUUtilization` for EC2 instances
- Number of *errors* for Lambda functions
- *Duration* for Lambda functions
- `DatabaseConnections` for RDS instances
- `DiskQueueDepth` for RDS instances
- `NumberOfObjects` for S3 buckets
- `ActiveConnectionCount` for Elastic Load Balancers
- `HealthyHostCount` for Elastic Load Balancers
- `TargetResponseTime` for Elastic Load Balancers

Now that we have looked at some of the primary metrics that CloudWatch provides for us with a few of the common services, let's look at how we can use custom metrics in cloudwatch.

Using custom metrics in CloudWatch

The AWS CloudWatch service not only allows you to watch and monitor the metrics coming from the resources themselves, such as CPU, memory, and network usage. It also allows you to create custom metrics that can be correlated to the number of errors in an application or tied directly to key performance indicators for business measurement.

High-resolution metrics in CloudWatch

When monitoring your custom metrics, sometimes, 1-minute intervals just don't provide enough granular detail. Using the put-metric-data API, either through the CLI or from one of the SDKs, you can publish custom metrics in up to 1-second intervals.

If you have an application that can have short-lived spikes whose behavior would not be captured by the default 1-minute intervals of CloudWatch metrics, then enabling high-resolution metrics allows you this visibility. Also, if you desire real-time monitoring, then once again, high-resolution metrics can fulfill this need.

After looking at the high-resolution metrics that are available to us, let's look at how we can create custom metrics that would be even more useful in our own scenarios.

Creating custom metrics in CloudWatch

CloudWatch metrics allow you to create metrics and namespaces for the items that matter to you. These can be incorporated into your scripts by using the SDKs available from AWS for your specific language or by using the AWS CLI and the put-metric-data command.

You can define metrics for things such as the **Key Performance Indicators** (**KPIs**) that you want to track. These custom metrics don't always tie into a specific technology metric; they can be used to count the number of ERRORS instances in a log file or track items such as the number of items in a cart at checkout for an e-commerce application.

When you create and publish a custom metric, you can either define it as standard resolution, and it would be measured in 1-minute intervals, or you can define it as a high-resolution metric, and it can be measured in 1-second intervals.

Now, let's use a Lambda function to create some custom metrics.

Publishing a custom metric

Custom metrics can be published from several services, including **AWS Lambda**, **Elastic Beanstalk**, **Amazon EC2**, or even container services such as ECS, EKS, or Fargate.

For our hands-on example, we will use a Lambda function to create our custom metrics and then send them to the AWS CloudWatch service. Our example scenario contains some example code where we are trying to track signups from a particular marketing campaign. This way, both the marketing department and the executive team can instantly know how effective the dollars being spent on this particular campaign have been, almost in real time. We are pushing these metrics to a CloudWatch metric named custom_metric.

The Lambda code that we will use for our example is located in this book's GitHub repository, in the `Chapter-15` folder, under the filename `cw_events.py`. We have also included an abbreviated version of the function that doesn't include the section for CloudWatch events:

```
cw_events.py:
import boto3
import random

# Resources
cw = boto3.client('cloudwatch')

# The Lambda handler
def lambda_handler(event, context):
    put_metric = custom_metric()
    return put_metric
###################################
# Create CW Custom Metric
###################################

def custom_metric():
    create_metric = cw.put_metric_data(
        Namespace='custom_metric',
        MetricData = [
            {
                'MetricName': 'Signups',
                'Dimensions': [
                    {
                        'Name': 'EMAIL_CAMPAIGN',
                        'Value': 'cableTV_spot2'
                    },
                ],
                'Unit': 'None',
                'Value': random.randint(1,100)
            },
```

```
        ],
    )
    return create_metric
```

The steps for creating a custom metric and then sending them to the AWS CloudWatch service are as follows:

1. Log into **Amazon Management Console** and navigate to the **Lambda** service.

2. Once you're on the **Lambda** service, click on the orange **Create function** button.

3. Once you're on the **Create function** screen, make sure that the option that's been selected to create the function is **Author from scratch**. Under the **Basic** information, use the following options when creating the function:

   - **Function name**: `custom_metric`

   - **Runtime**: Python 3.8

   - **Permissions**: Create a new role with basic Lambda permissions:

Basic information

Function name
Enter a name that describes the purpose of your function.

```
custom_metric
```

Use only letters, numbers, hyphens, or underscores with no spaces.

Runtime Info
Choose the language to use to write your function. Note that the console code editor supports only Node.js, Python, and Ruby.

```
Python 3.8                                                          ▼
```

Permissions Info

By default, Lambda will create an execution role with permissions to upload logs to Amazon CloudWatch Logs. You can customize this default role later when adding triggers.

▼ **Change default execution role**

Execution role
Choose a role that defines the permissions of your function. To create a custom role, go to the **IAM console**.

○ Create a new role with basic Lambda permissions

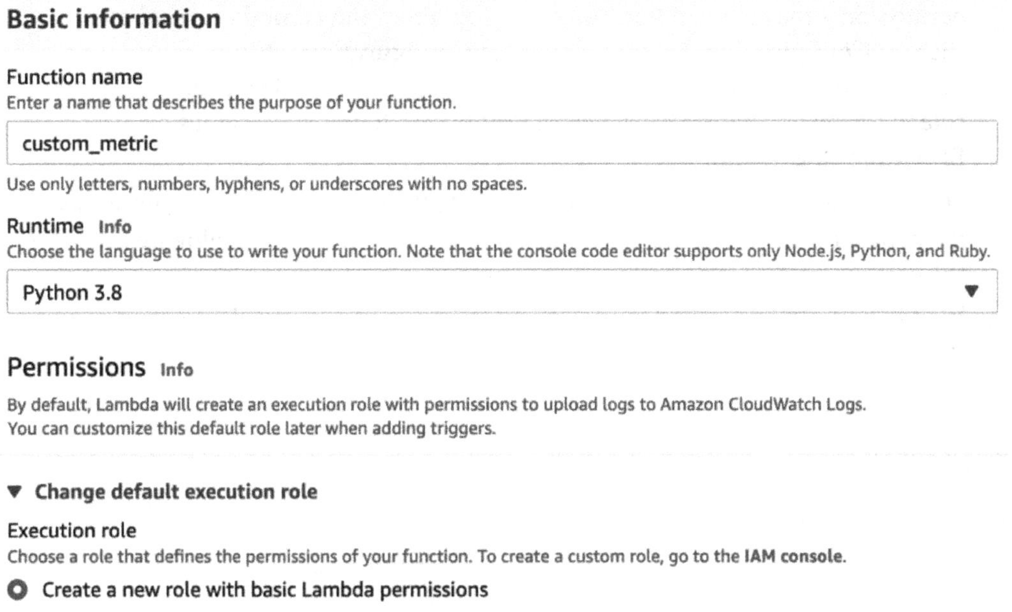

Figure 15.3 – Basic information for creating the Lambda function

4. Once you have filled out all these options, press the orange **Create function** button.

5. Once the function has been created, go into the **Code** section and click on the `lambda_function.py` file in the **Environment** tab. We can now either type the code mentioned previously or cut and paste the code from this book's GitHub repository, in the `chapter-15` directory, replacing what is currently in the `lambda_ function` tab. Once you have replaced the code, click on the **Deploy** button at the top of the code window. Doing so will ensure that the **Changes not deployed** message disappears and is replaced with a green **Changes deployed** message:

Figure 15.4 – Lambda function showing the changes that have been deployed

6. With our function now created, we can start creating a test event so that we can both test the function and see the custom metric appear in the CloudWatch metrics. Click on the orange **Test** button to configure the test event. No special test data is needed, so just set the event name to **Test** and then click the orange **Create** button at the bottom of the dialog window.

7. Before we test our function, we will need to give our function one more permission – the ability to `PutMetricData`. From the **Lambda** vertical menu, click on **Configuration**. Once you're in the **Configuration** settings, click on the **Permissions** menu item on the left-hand menu. This should bring up **Execution role** in the main window. Click on the **Edit** button to the right-hand side of the **Execution role** heading.

8. This will bring you to a **Basic Settings** page. At the bottom of the page, underneath the name of your existing role, there should be a link in blue that allows you to **View the custom_metric_role** on the **IAM console**. Upon clicking this link, a new tab will open for the IAM service.

9. When you're on your role in the **IAM console**, click on the **Add inline policy** link. Use the following values:

   - **Service: Choose | CloudWatch**

   - **Actions: Filter | PutMetricData:**

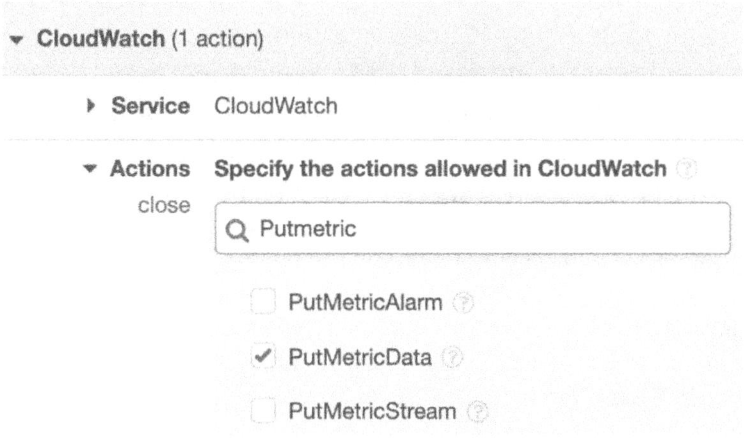

Figure 15.5 – Adding the inline policy from IAM to our Lambda role

10. Once you have added the extra permission, click the blue **Review policy** button at the bottom of the page.

11. Name the policy `PutMetricData` and then click the blue **Create policy** button at the bottom of the page.

12. Now that we have modified our IAM role so that it has the `PutMetricData` permission, go back to the tab that contains the Lambda function. You should still be on the **Basic Settings** page. Click on the orange **Save** button at the bottom of the screen. This will take you back to the main Lambda screen's **Configuration** menu. Click on the **Code** tab at the top of the horizontal menu. Now, press the orange **Test** button to send the test event to our Lambda function.

13. Go to the **Services** dropdown at the top left-hand side of the **Navigation** bar. You should see **CloudWatch** under your **Recently visited** services. Right-click on **CloudWatch** to open it in a new tab. If you don't see **CloudWatch** in your recently visited services, then just search for `CloudWatch` in the top search bar and right-click on it to open it in a new tab.

14. From the left-hand menu of **CloudWatch**, find **Metrics** and then click on the **All metrics** sub-menu item. We should see a value of `custom_metric` in **Custom Namespaces**:

Figure 15.6 – Our custom metric in Custom Namespaces

15. Click on the **custom_metric** namespace. At this point, we will see our secondary namespace; that is, **EMAIL_CAMPAIGN**. Click on this value to be taken to the metric data. Check the box next to the **cableTV_spot2** value to see the data plotted on the graph. The number will vary since we used a random value in our function.

With that, we have created a Lambda function that creates and publishes a custom metric to CloudWatch metrics. Next, we will look at how to incorporate this custom metric, along with others, in CloudWatch dashboards to provide a quick view of our environment to ourselves, our team, as well as others.

Using CloudWatch metrics to create dashboards

Looking at individual metrics in CloudWatch can provide many insightful details. However, sometimes, it can be more useful to glance at a single pane of glass where the most relevant metrics are displayed all at once for a quick view. CloudWatch dashboards allow us to quickly and easily create these views – not only of the metrics created by the Amazon resources in our account but also custom metrics, along with text and hyperlinks to runbooks for documentation purposes in case of an emergency or other helpful documentation.

The CloudWatch service even comes with automatic dashboards for many of the most used AWS services, such as DynamoDB, EC2, Lambda, S3, EBS, and others. Each of these pre-configured dashboards is interactive and can be viewed based on custom date ranges.

You can even share the dashboards that you create with people that don't have direct access to your AWS account. This can be done in a few ways. The first is by projecting the dashboard on a large screen so that a team of users, or anyone who comes into the room where the screen or projection is displayed, can view the metrics and graphs that are displayed on the dashboard.

The second method is the built-in ability to share a dashboard with a specified email address using both a username and password.

This second way of sharing access to a dashboard can be beneficial when you're trying to provide real-time metrics to a stakeholder of a particular project. This user may not be incredibly technical but is looking for the correct business information to help make decisions. Adding some of the custom metrics that we discussed earlier in this chapter allows the business stakeholder to review the particular KPIs at their leisure, without requesting a special report to be generated for them.

When you create a dashboard in CloudWatch Dashboards, it becomes available globally. This is because dashboards are not region-specific.

With that, we've looked at how dashboards can allow us, as DevOps engineers, our development team, and even stakeholders of a project, to quickly view what is going on with our environment or project. Now, let's go through the hands-on process of creating a dashboard.

Creating a base dashboard to monitor our resources

Let's use some of the metrics that we created previously and incorporate them into a custom dashboard:

1. Open **Amazon Management Console** and go to the **CloudWatch** service. You may need to log into your account if you have lost your session. Also, make sure that you are in the Ohio region (or whatever region you have been using to create your resources): `https://console.aws.amazon.com/cloudwatch/`.

2. Once you're in the CloudWatch service, find and click on the **Dashboards** menu item at the top of the left-hand menu.

3. This should bring you to the **Custom Dashboards** screen. We will start the process of creating our dashboard by pressing the orange **Create dashboard** button:

Create dashboard

Figure 15.7 – The Create dashboard button

4. Pressing the **Create dashboard** button will bring up a dialog where you can name the dashboard. Name this new dashboard `Chapter15`. Then, press the orange **Create dashboard** button to close the dialog box and start building the dashboard.

5. A new dialog should appear, asking us to add a widget to our dashboard. We will begin with the **Explorer** widget:

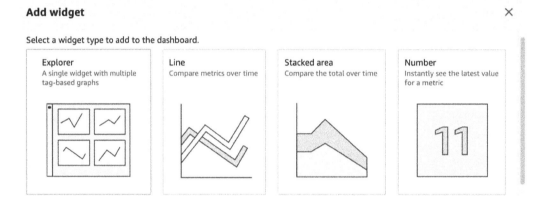

Figure 15.8 – Adding a widget to a CloudWatch dashboard

6. Start your dashboard off by clicking on the **Number** widget. Scroll down to **Custom namespaces** and click on the **custom_metric** namespace. Change **Period** from **5** minutes to **1** day so that your dashboard will keep the data. Click through the **EMAIL_CAMPAIGN** secondary namespace and check the **cableTV_spot2** box. Once this value has been checked, select the orange box labeled **Create widget** at the bottom of the dialog window:

Figure 15.9 – Adding the values for the Number widget

7. Once our number widget has been added, click on the orange **Add widget** button to add another widget to our dashboard.

8. Choose the **Line** widget and then select **Metrics**. In the search box, use the `custom_metric` search term so that we can find the metrics for the Lambda function that we created in the previous exercise. Scroll past **Custom namespaces** and go to **AWS Namespaces**. You can click on either the **Lambda | By Resource** or **Lambda | By Function** name as they should both have the same set of metrics available. Find the **Invocations** metric and select the checkbox to the left of the function name. Once you have selected the checkbox, click on the orange box and then click on the orange **Create widget** button.

9. Let's add one more widget to our dashboard by clicking the **Add widget** button one last time. This time, scroll down and find the **Logs table** widget. From the dropdown box that contains the **select log group(s)** option, find the log group for the `custom_metric lambda` function. This should be named `/aws/lambda/custom_metric`. Use the checkbox to select this log group. Once you've selected the log group, click the orange **Add to dashboard above** button.

10. You will back on your dashboard, which should now contain three widgets. Click the blue **Save dashboard** button at the top of the dashboard. You now have a working dashboard that you can view and share:

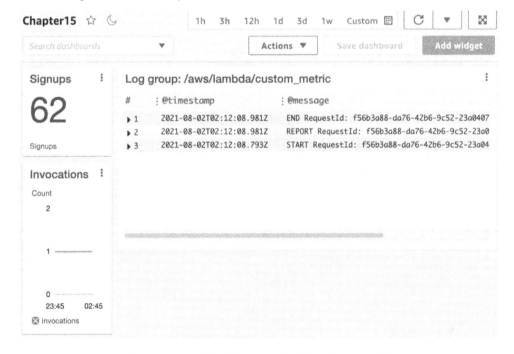

Figure 15.10 – The Chapter15 dashboard we created

Now that we have learned how to incorporate our metrics into a dashboard so that we can quickly and easily monitor our systems, along with any custom metrics that we need to view at a glance, let's look at how we can use CloudWatch to kick off event-driven architectures using the **Amazon EventBridge** service.

Amazon EventBridge overview

Amazon EventBridge is a serverless event-driven bus that makes it easy to ingest and process data from a variety of sources. These sources include AWS services, your applications, and third-party SaaS providers. It removes the discord of writing point-to-point integrations between services. EventBridge is a managed service from AWS. This means that you don't need to worry about having to provision more or less of the service as your needs fall and rise. The EventBridge service takes care of this for you:

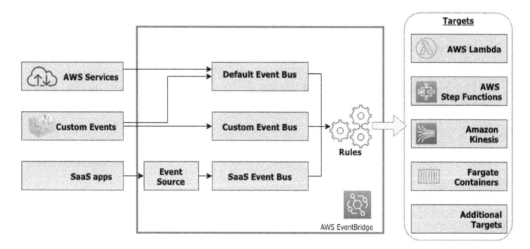

Figure 15.11 – AWS EventBridge flow from events to targets

An **Event Source** can be almost any of the AWS services, custom applications, or SaaS applications.

For SaaS applications, there is special support for partner applications called an **Event Source**. This Event Source provides a logical connection between the third-party SaaS provider and your AWS account, without the need to provision any cross-account IAM roles or credentials.

Event Buses are the core of the **Event Bridge Service**. There is a default event bus to handle AWS service events. Event Buses can be custom-created for your application.

Once you have set up an event bus, you can create **rules**. With the use of rules, you can match values in the metadata or payload of the event that has been inspected by the event bus. The rule then determines which events should get routed to which destination:

EVENT   **RULE**

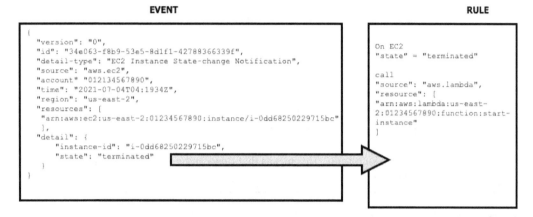

```
{
    "version": "0",
    "id": "34e063-f8b9-53e5-8d1f1-42788366339f",
    "detail-type": "EC2 Instance State-change Notification",
    "source": "aws.ec2",
    "account" "012134567890",
    "time": "2021-07-04T04:1934Z",
    "region": "us-east-2",
    "resources": [
     "arn:aws:ec2:us-east-2:01234567890:instance/i-0dd68250229715bc"
    ],
    "detail": {
        "instance-id": "i-0dd68250229715bc",
        "state": "terminated"
    }
}
```

```
On EC2
"state" = "terminated"

call
"source": "aws.lambda",
"resource": [
"arn:aws:lambda:us-east-
2:01234567890:function:start-
instance"
]
```

Figure 15.12 – An example event and the rule it triggers

Once a rule has been triggered, you can associate one or more **Targets** with that rule. Targets are various AWS services such as Lambda functions, Step Functions, Kinesis Streams, and ECS or Fargate clusters.

> **Note**
> The CloudWatch Events service is now known as Amazon EventBridge. If you had used CloudWatch events in the past, then that capability is still available through the default event bus in EventBridge.

Now that we have looked at the Amazon EventBridge service, let's take a look at some of the service limits that are automatically imposed on EventBridge.

EventBridge service limits

As you start to build event-driven services with EventBridge, it is a good idea to keep the service limits that are initially imposed on the EventBridge service in mind. This can help you in cases where you are sending too many events to your event bus at the same time. It can also help you as you are building out your applications since you know how many event buses and rules are allowed by default in a single region:

Invocations	750 per second. After 750 per second, invocations start to get throttled.
Rules	100 per region, per account.
Put Events	400 requests per second, with up to 10 events per request.
Event Buses	100 per region, per account.

Table 15.1 – AWS EventBridge service limits

> **Note**
> All of these limits are soft limits. This means that they can be raised by opening a service request ticket with AWS.

Now that we understand what service limits we are working with, let's look at how to build event-driven architectures using AWS EventBridge.

Event-driven architectures with EventBridge

Modern cloud applications are based on decoupled services.

There are three critical components to event-driven architectures: event producers, event consumers, and event routers. The producer is the service or trigger that produces the event and then sends it to the router. The router or event bus then filters the specific events and sends specific events to event consumers.

Multiple benefits of event-driven architectures

When you're using architectures that are decoupled, meaning that each component performs a specific task, you are gaining multiple benefits:

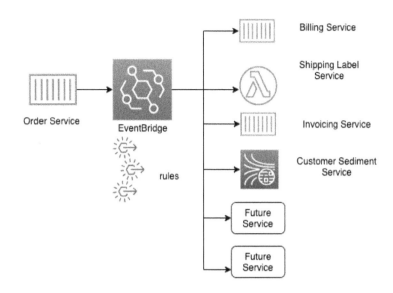

Figure 15.13 – A single service using EventBridge and custom rules to push to multiple targets

Using EventBridge to capture AWS service events

We can use the EventBridge service to automatically trigger events through the use of rules and the default event bus. Let's start by capturing any time an EC2 instance has an instance state change and send that to a log file so that we can view the event:

1. Log into **AWS Management Console** and navigate to the **CloudWatch** service. From the left-hand menu, find and expand the menu for **Events**. In the **Events** sub-menu, click on the **Rules** link.

2. Once the **Rules** screen appears in the main window, click on the blue **Create rule** button:

Figure 15.14 – The EventBridge Rules page with the Create rule button

3. We should now be on a screen called **Step 1: Create rule**. Under the **Event Source** heading, make sure that the radio button next to **Event Pattern** is selected so that we can start building out our pattern.

4. For our event pattern, use the **Service Name** selection dropdown to choose EC2 service. Then, from the **Event Type** selection dropdown, select **EC2 Instance State-change Notifications**.

5. We don't want every event from EC2; we only want to know when instances are being spun up or terminated. Choose **Specific state(s)** and then select the **terminated** and **pending** states for the states of the rule. You will have to use the dropdown menu twice to populate both selections. Leave the radio checkbox next to **Any instance** in the box:

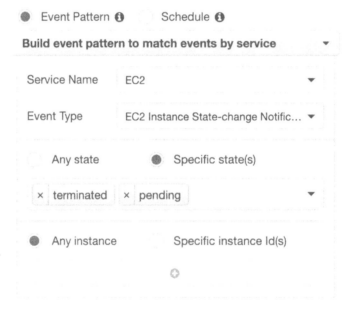

Figure 15.15 – The EventBridge rule's Event Pattern populated

6. Next, we can move on to the right-hand side of *Step 1*, where you can find the **Targets** heading. Click the **Add target*** button.

7. Under this first target, find **CloudWatch log group** via the drop-down box. Keep the radio button selection as the top choice. Then, in the text field, insert EC2_STATE:

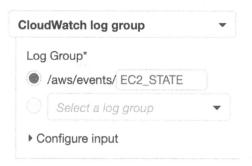

Figure 15.16 – Creating the custom log group as our EventBridge target

8. Scroll down to the bottom of the page and click the blue button labeled **Configure details**.

9. For the name of the rule, use chapt15-ec2. You can insert a description if you like. Once you have filled in the name and the description, click the blue **Create rule** button.

10. Our rule should appear on the **Rules** page with the status of a green dot, meaning that it is in good health. Now, let's quickly switch to our terminal so that we can quickly start some ec2 instances and trigger the rule. Don't close your browser window – we're going to want to go back and check on the CloudWatch log group afterward.

11. With your terminal window open, use the create-instance command to spin up an instance quickly (hopefully, these commands look a bit familiar to you if you completed the exercises in *Chapter 14, CloudWatch and X-Ray's Role in DevOps*):

```
# Capturing the AMI to a variable
AMI='aws ssm get-parameters --names \
        /aws/service/canonical/ubuntu/server/16.04/
stable/current/amd64/hvm/ebs-gp2/ami-id \
    --query 'Parameters[0].[Value]' --output text
--region us-east-2'
# Creating a new instance with the AMI variable
aws ec2 run-instances \
--image-id $AMI \
--instance-type t2.micro \
--iam-instance-profile 'Name=CW_SSM' \
```

```
--tag-specifications
'ResourceType=instance,Tags=[{Key=Name,Value=
EVENTBRIDGE}]' \
--region us-east-2
```

Once your instance starts, it will be a good idea to take note of `InstanceId` so that you can quickly reference it when terminating your instance in the next step.

12. After about 2 to 5 minutes, we are going to terminate our instance to create another event for our rule:

```
aws ec2 terminate-instances \
--instance-ids {YOUR INSTANCE ID} \
--region us-east-2
```

13. Now, we should wait about another minute or two so that our instance can terminate completely. As it starts to terminate, go back to the browser that had **Amazon Management Console** open.

14. In **Amazon Management Console**, we should already be on the **CloudWatch** service, so we simply need to find the **Log groups** sub-menu under the **Logs** heading and click on it.

15. Now, you should be able to find the custom log group we quickly created, called `EC2_STATE`. If you have too many log groups in your region, simply search for the term `EC2_STATE`, and it should appear. Click on the log group's name so that we can see what EventBridge has generated for us.

16. You should now have two log entries in your log group. One will correspond to the **pending** event, while the other will correspond to the **termination** event.

With that, we've learned how to use events that occur in our AWS account and the AWS EventBridge service to build out an event-driven architecture. Although we only used a simple example in our hands-on exercise, this could be expanded to perform actions such as sending out SNS notifications in concert with recording a log entry, or even creating a new resource if this was a critical piece of infrastructure. Now, let's recap everything that we have learned in this chapter.

Summary

In this chapter, we took a deeper look at the **AWS CloudWatch** service. We focused on the metrics and what makes up a metric. We looked at the different types of metrics available from AWS, starting with basic metrics on the Free Tier, then moving on to detailed metrics, and finally learned how to create custom metrics. We also learned how to use these metrics to create custom dashboards in CloudWatch and discovered how the dashboards could be shared with not only team members who had IAM access, but also how they can be shared with others outside of our AWS account.

We also looked at **EventBridge**, the service that has taken over CloudWatch Events. We learned how using event buses for AWS services, custom application events, and even SaaS providers can help drive event-driven architectures.

In the next chapter, we are going to look at the various types of logs that can be generated from the different Amazon services. This includes VPC Flow Logs, Elastic Load Balancer logs, CloudTrail logs, and how these logs can help us troubleshoot issues with our application or security incidents.

Questions

1. You have been hired by a company to help develop an e-commerce application on AWS. The stakeholders of the company want to know how many orders have been placed via this application with second-level granularity. To get this information, you will need to create a custom CloudWatch metric using the AWS CLI. You know that by default, custom metrics have a 1-minute granularity. How can you get the application to send the custom metric in sub-minute intervals?

    a. Use the AWS CLI `put-metric-data` command to publish the data and set the `StorageResolution` option to 1 second to specify the metric as a high-resolution metric.

    b. Update the CloudWatch agent config file and then add `line high-resolution: true`.

    c. Go to the graph in the CloudWatch service on the Amazon Management Console and set the resolution to 1-second intervals.

    d. Add the `flag -dimensions=1` to the AWS CLI `put-metric-data` command to specify a high-resolution metric.

2. You are currently working for a mid-sized e-commerce company that has built a serverless shopping cart system using AWS Lambda and DynamoDB. An executive for the company has asked you to create and share a dashboard with some of the board of directors, showing the number of purchases per cart and the number of abandoned purchases per cart. The board members do not currently have IAM accounts. How can you give the board members real-time access to the data as simply and cost-effectively possible?

    a. Use social logins with Amazon Cognito. Have Cognito assume a role that has access to the specific dashboard so that the board members have access.

    b. Create IAM users for each of the board members. Create an IAM group that has access to CloudWatch dashboards but has a condition to show only the ARN of the specific dashboard that they need to see.

    c. Gather the emails of the board members. Share access to the CloudWatch dashboard via a username and password using the email access feature.

    d. Gather the emails of the board members. Incorporate SAML for the CloudWatch dashboards. Allow the board members to use single sign-on to access the specific dashboard.

Review answers

1. a

2. c

16

Various Logs Generated (VPC Flow Logs, Load Balancer Logs, CloudTrail Logs)

Logs are a river of information, and they flow from various sources. The logs that come from the Load Balancer can be a valuable source of data or a resource for troubleshooting. Knowing how to enable these resources can be vital when setting up or running your environment. Any action taken in the AWS environment, either via the AWS Management Console, the CLI, or an SDK, is recorded via the underlying API call to CloudTrail. As a DevOps engineer, it's essential to know who and what is making changes to your environment and be able to retrieve that data, especially when requested.

In this chapter, we're going to cover the following main topics:

- The power of AWS CloudTrail
- Enabling Elastic Load Balancer logs
- Using VPC Flow Logs
- Cleaning up the resources

Previous logs discussed

So far, we have mainly been discussing logs that are generated from the application itself. Also included in some of those earlier exercises with CloudWatch Logs were some logs that AWS gives us as wrappers around those logs; however, these are still, for the most part, just application and AWS service logs.

When we want to understand how users are interacting with our environment, be it our network environment or how they are adding and removing resources within our account, then we would not be able to find that information in the application logs. Instead, we must look at some of the other logs available in AWS.

Knowing which logs to use for which purpose can also help us when it comes to other services to protect our environment, such as **GuardDuty**.

> **Note**
> We will discuss GuardDuty in *Chapter 22, Other Policy and Standards Services to Know About*.

Now that we've looked at where we have been and where we are going, let's start with our first set of logs – CloudTrail logs.

The power of AWS CloudTrail

CloudTrail enables governance, compliance, risk auditing, and operational auditing with either your AWS account or multiple accounts using AWS organizations.

In AWS, every action is performed by an API call. This is true if you are using the AWS Management Console, the Amazon CLI, or any of the available SDKs. All of these use API calls to perform the necessary actions, then those API actions are recorded by the CloudTrail service if it has been turned on:

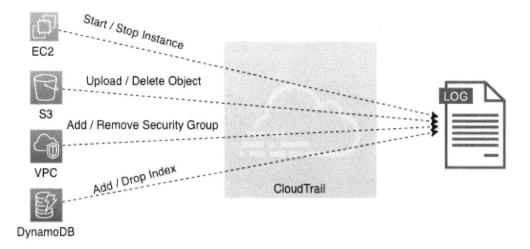

Figure 16.1 – The flow of actions to logs via the CloudTrail service

These include recording calls to start and stop EC2 instances, uploading and deleting objects from S3, adding or removing security groups from a VPC, adding or dropping indexing from a DynamoDB table, and many more. When an activity occurs within your account, CloudTrail will capture and record that activity as a CloudTrail event. This CloudTrail event contains the following details:

- Who performed the request
- The date and time that the request was made
- The source IP of the request
- How the request was made
- What action is being performed
- The region the action is being conducted in
- The response to the request

It's also important to note that CloudTrail logs are not pushed in real-time to the S3 bucket that they are stored in. Instead, the CloudTrail service publishes updated log files every 5 minutes with a batch of events it has collected.

When it comes to securing the CloudTrail logs themselves, by default, the service encrypts the files, which it stores in Amazon S3 with **S3 server-side encryption** (**SSE**). You also have the option to create an encryption key using the **KMS** service and encrypt the CloudTrail logs with that key.

The CloudTrail service offers several benefits. The first is that it records user and resource activity. Using these recordings, you can identify who did what and when they performed an action on a resource in your AWS account. Secondly, since event logs are automatically stored and recorded, compliance reporting becomes much more manageable. Third, you gain the ability to monitor, alarm, and react to the events that are happening by sending the CloudTrail events to CloudWatch Logs. The fourth and final benefit that we will mention here is the ability to search through the logs using the CloudWatch service using a SQL-like syntax. This enables you to perform powerful queries on the large amounts of data that CloudTrail produces.

Now that we know about the CloudTrail service, we will set up CloudTrail in our account.

Setting up CloudTrail

We are going to set up CloudTrail before we look at the logs that CloudTrail has recorded. We're doing this so that as we perform other exercises in this chapter, we are sure that the CloudTrail service is on and recording our actions. This will also ensure that we have a comprehensive set of recordings to search through when we perform the CloudTrail exercise later.

Amazon has updated the default way that it creates CloudTrail trails so that all the regions are included when it's initialized. In this section, we want to create a trail that will be specific to the region that we are working in. This is still possible, but only if we use the AWS CLI:

1. Open your terminal so that you have access to the AWS CLI. First, we will need to create an S3 bucket so that our CloudTrail logs can be captured and placed. Use the following *example* command, remembering that each S3 bucket name is unique and that you will need to create your own S3 bucket:

    ```
    aws s3 mb s3://devopsproandbeyond-trail --region
    us-east-2
    ```

2. For the CloudTrail service to be able to put logs into the S3 bucket, we need to attach a bucket policy to our bucket. Cut and paste the following bucket policy into a local file (where you are performing your terminal commands) called `cloudtrail_s3.json`. Look for the two instances of the word `BucketName`; you will need to replace those with the name of the bucket that you created in the previous step. A copy of this file can be downloaded from the `Chapter-16` folder of this book's GitHub repository:

    ```
    {
        "Version": "2012-10-17",
    ```

```
    "Statement": [
        {
            "Sid": "AWSCloudTrailAclCheck20150319",
            "Effect": "Allow",
            "Principal": {"Service": "cloudtrail.
amazonaws.com"},
            "Action": "s3:GetBucketAcl",
            "Resource": "arn:aws:s3:::BucketName"
        },
        {
            "Sid": "AWSCloudTrailWrite20150319",
            "Effect": "Allow",
            "Principal": {"Service": "cloudtrail.
amazonaws.com"},
            "Action": "s3:PutObject",
            "Resource": "arn:aws:s3:::BucketName/*",
            "Condition": {"StringEquals": {"s3:x-amz-
acl": "bucket-owner-full-control"}}
        }
    ]
}
```

3. Once you have created the policy file, you can attach it to your bucket using the following command. Be sure to change your bucket name so that the policy attaches to your bucket:

```
aws s3api put-bucket-policy \
   --bucket devopsproandbeyond-trail \
   --policy file://cloudtrail_s3.json
```

4. With our S3 bucket attached, we can create our single-region trail. Use the following command to create your trail, remembering to switch out the name of the S3 bucket in the command for the bucket that you created in *step 1* of this exercise. Note how we named our trail sixteen. We will be referencing this trail name later in this chapter:

```
aws cloudtrail create-trail \
   --name sixteen \
   --s3-bucket-name devopsproandbeyond-trail \
   --region us-east-2
```

If the trail was created successfully, then you should see JSON returned, similar to the following:

```
{
    "Name": "sixteen",
    "S3BucketName": "devopsproandbeyond-trail",
    "IncludeGlobalServiceEvents": true,
    "IsMultiRegionTrail": false,
    "TrailARN": "arn:aws:cloudtrail:us-east-
2:470066103307:trail/sixteen",
    "LogFileValidationEnabled": false,
    "IsOrganizationTrail": false
}
```

1. Now, it's time to start the trail. Just because we created the trail doesn't make it automatically start recording events. Use the following command to start the trail so that it captures all the API calls in the region:

```
aws cloudtrail start-logging --name sixteen --region
us-east-2
```

2. Next, to stream our logs to CloudWatch Logs (so that we can search for them later), we will need to log into the AWS Management Console and make a quick edit to our trail. After logging into the console, navigate to the CloudTrail service. When you are brought to the CloudTrail dashboard, you should see the trail that we created named **sixteen**. Click on this trail's name:

Figure 16.2 – The sixteen CloudTrail trail on the dashboard

3. When you are in the sixteen cloud trail, you should see a section named **CloudTrail logs**. Click the button to the right of the section labeled **Edit**.

4. On the **CloudWatch Logs – Optional** screen, check the **Enabled** box. Keep the setting for creating a new log group as-is but rename the log group to *CloudWatch*. This will make it easier to find later. Allow the CloudTrail service to create a new IAM role for you and name it `CloudTrailRole-sixteen`. Then, click the orange **Save changes** button at the bottom of the screen.

Now that we have set up the CloudTrail service to record the API actions that we will perform going forward, we will look at Elastic Load Balancer logs.

Enabling Elastic Load Balancer logs

The **Elastic Load Balancing** service lets you capture more data about your environment. This can help with troubleshooting, especially regarding latency. Elastic Load Balancer access logs can also let you see the path that a user or service took from an originating address to the destination service. Sometimes, this information is not captured on application logs since the originating address that is captured is the Elastic Load Balancer address. The Elastic Load Balancer access logs include the following information:

- The client's IP address
- Request paths taken
- The time and date that the request was received
- Server responses (in numerical format)

We looked at how load balancing helps spread the load between both instances and services when we examined in-depth services such as Elastic Beanstalk and OpsWorks. At this point, we should also understand that Elastic Load Balancing can be used to attach multiple instances and even instances that are part of an auto-scaling group:

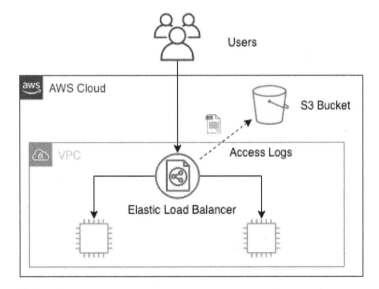

Figure 16.3 – The flow of the access logs from an Elastic Load Balancer to an S3 bucket

Once you have enabled access logs for your Elastic Load Balancer, there is no additional charge for logging. There is a charge, however, for storing the logs in S3.

Setting up an Elastic Load Balancer and enabling logging

In our first hands-on example for this chapter, we will use two cross-referenced CloudFormation templates to stand up a VPC; the second template will stand up an Elastic Load Balancer with two EC2 instances serving a simple website. The child template will also create an S3 bucket for capturing our access logs. Once we have stood up our test environment, we will need to go into the AWS Management Console and turn on logging for our Elastic Load Balancer. Once logging has been turned on, we can try to access the website a few times. Doing this should put some records in our S3 bucket that we can access and analyze. The templates referenced in this exercise are located in the `Chapter-16` directory of this book's GitHub repository. You should download all the templates before starting. Let's get started:

1. Log into **AWS Management Console** and navigate to the **CloudFormation** service.

2. Once you're on the **CloudFormation** page, if you are brought to the main CloudFormation service page, then click on the orange **Create stack** button. Otherwise, if you are brought to a listing of your current stacks, then click on the **Create stack** button at the top right of the main window and choose the **With new resources (standard)** option:

Figure 16.4 – The Create stack button from the stacks listing page in CloudFormation

3. Whichever way you got here, we should now be on the **Create stack** page in CloudFormation. On the second heading of the page, labeled **Specify template**, click on **Upload a template from a file**. When the **Choose file** button appears, select the `vpc.yaml` template to upload it. Then, click the **Open** button in the dialog box. Once the file has been uploaded, you can click the orange **Next** button.

4. This will bring you to the **Specify stack details** page. Use `Chapter16-VPC` when naming your stack. This name will be important as this stack will create some of the resources that will be used by the next stack, and it will reference those resources by the name of the stack. Once you have entered the name, click the orange **Next** button at the bottom of the page.

5. On the **Configure stack options** page, scroll down to the bottom and hit the orange **Next** button.

6. At this point, we will be on the **Review** page for the first `Chapter16-VPC` stack. Scroll down to the bottom of the page and fill in the box under the **Capabilities** section, acknowledging that this template will create an IAM role. After this, you can click the orange **Create stack** button so that our initial stack can be initialized and created.

7. Clicking on the **Create stack** button should take you to the `Chapter16-VPC` stacks page of the CloudFormation service. Once the stack has been created, go to **Outputs** via the horizontal menu. At this point, you should see the six outputs that the VPC stack has created, including the VPCid, two private subnets, and two public subnets:

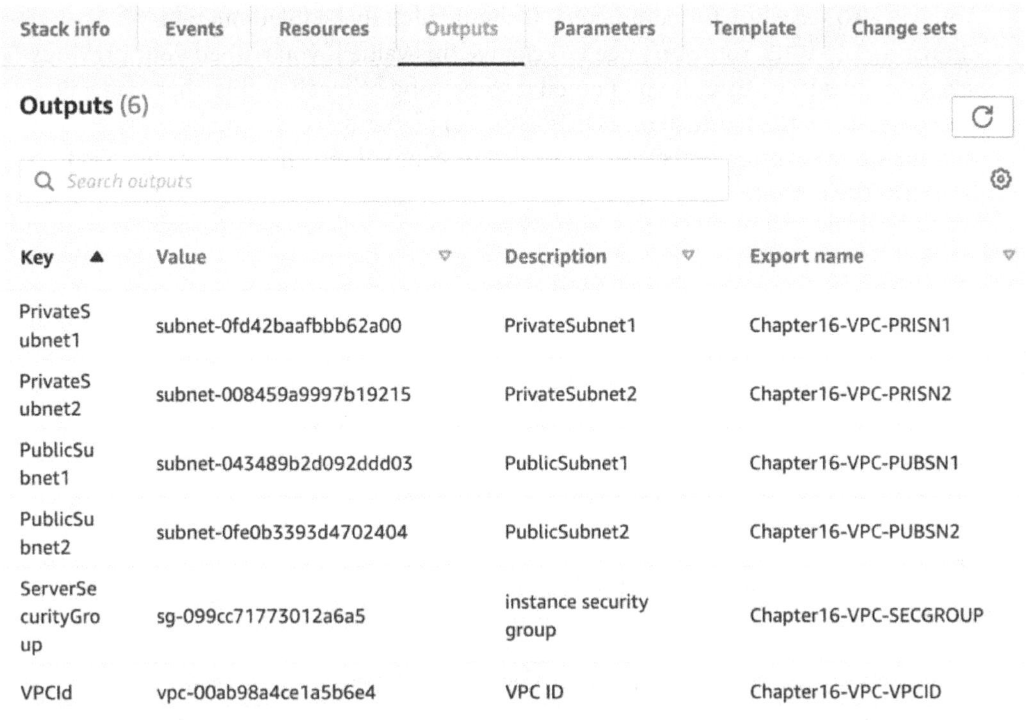

Stack info	Events	Resources	Outputs	Parameters	Template	Change sets

Outputs (6)

Key ▲	Value	Description ▽	Export name ▽
PrivateSubnet1	subnet-0fd42baafbbb62a00	PrivateSubnet1	Chapter16-VPC-PRISN1
PrivateSubnet2	subnet-008459a9997b19215	PrivateSubnet2	Chapter16-VPC-PRISN2
PublicSubnet1	subnet-043489b2d092ddd03	PublicSubnet1	Chapter16-VPC-PUBSN1
PublicSubnet2	subnet-0fe0b3393d4702404	PublicSubnet2	Chapter16-VPC-PUBSN2
ServerSecurityGroup	sg-099cc71773012a6a5	instance security group	Chapter16-VPC-SECGROUP
VPCId	vpc-00ab98a4ce1a5b6e4	VPC ID	Chapter16-VPC-VPCID

Figure 16.5 – The outputs from the initial stack creation

8. With our VPC template created and showing the outputs, we can move on to our next template. This next template will set up the Load Balancer, along with two EC2 instances running the Apache web server. Each server is running a static web page, but you will know the difference if you are being directed to instance number one or instance number two based on the page displayed.

9. Since we are already on the **CloudFormation** page of the **Outputs** section, we can simply go to the top right-hand corner and click on the white **Create stack** button. When the drop-down list appears, choose the **With new resources (standard)** option.

10. Now, back on the **Create Stack** page, under the **Specify Template** section, choose **Upload a template file**. Then, click on the **Choose File** button that appears and select the `cross-stack-website.yaml` template to upload it. Once the template has been uploaded, click on the orange **Next** button at the bottom of the page.

11. You should now be on the **Specify stack details** page. It's time to name our new stack. The suggested name for this stack is `Chapter16-Elastic Load Balancer`. Enter this value in the **Stack name** box.

12. You will also see a box for parameters. It should already be filled out with `Chapter16-VPC`. You won't need to change this value if you named your previous stack `Chapter16-VPC`. If you named it something other than this, then you will need to provide the name here, as this is the value that drives all intake for the outputs we saw earlier. Once you have filled in your values, click the orange **Next** button at the bottom of the page:

Stack name

Stack name

> Chapter16-ELB

Stack name can include letters (A-Z and a-z), numbers (0-9), and dashes (-).

Parameters

Parameters are defined in your template and allow you to input custom values when you create or update a stack.

NetworkStackName

Name of an active CloudFormation stack that contains the networking resources, such as the subnet and security group, that will be used in this stack

> Chapter16-VPC

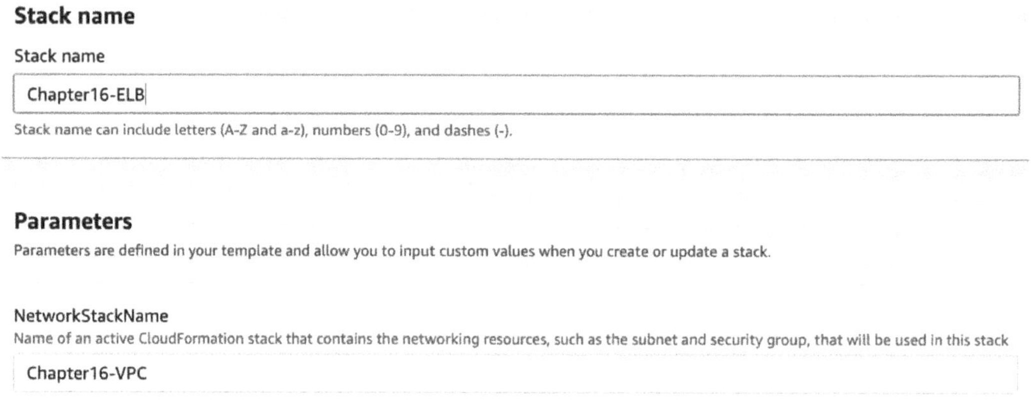

Figure 16.6 – The Stack name and Parameters fields on the Specify stack details page

13. At this point, you will be taken to the **Configure stack options** page. There is nothing to configure on this page, so we will scroll down to the bottom of the page and click the orange **Next** button.

14. Finally, we will be on the **Review stack** page. Briefly look over the options you have chosen for your stack. If you don't see any errors, then click on the orange **Create stack** button at the bottom of the page.

15. It will take a few minutes for our stack to be created as it is creating the two EC2 instances and installing the software and web page. It is also creating a classic Load Balancer, registering those instances with that Load Balancer, and performing the initial health checks.

16. Once the stack has completed, click on **Outputs** from the horizontal menu, as we did with the previous stack. This time, you will find the URL of a key there and the value of the public URL for the Elastic Load Balancer. Right-click on the URL to open it in a new tab.

17. Now, click on the **Resources** menu item in the same horizontal menu where you found **Outputs**. Click on the **Physical ID** link for the Elastic Load Balancer so that you are taken directly to the Elastic Load Balancer's details page:

Logical ID ▲	Physical ID ▽	Type ▽	Status ▽
ElasticLoadBalancer	Chapter16-ElasticL-10E9H11HC6Z4I ⬀	AWS::ElasticLoadBalancing::LoadBalancer	⊘ CREATE_COMPLETE

Figure 16.7 – The resource listing of the Elastic Load Balancer shown in CloudFormation

18. Now, scroll down to the bottom half of the screen until you find the **Attributes** heading. Under this heading, you will find the **Access logs** section. It should have a value next to it that is currently set to **Disabled**. Click the gray button labeled **Configure Access logs**.

19. When the dialog box appears, you will need to fill in the following values:

   • Check the **Enable Access logs** box.

   • Change the interval in which the logs are being pushed to **5 minutes**.

- Choose a name for a **NEW** S3 bucket where the Elastic Load Balancer access logs can be stored.

- Check the box to have the S3 bucket created for you:

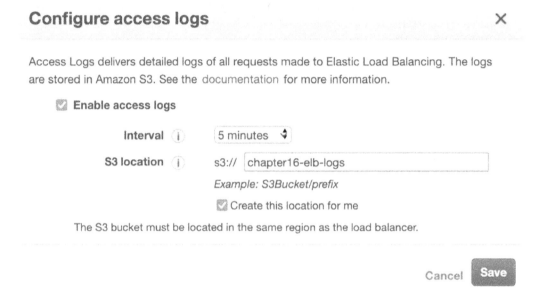

Figure 16.8 – Configure access logs

Once you have filled out all the values, click on the blue **Save** button.

1. Now, with the logs turned on, it is time to go back to the browser tab that you opened previously that contained the URL for the Elastic Load Balancer. Refresh the page multiple times; if both servers came up healthy for the Elastic Load Balancer, then you should see a mix of server one and server two appear on the screen.

2. Once you have generated some traffic, we can start to navigate to the S3 bucket that we created for our Elastic Load Balancer access log storage. However, remember that the logs are only pushed once every 5 minutes, so you might need to be a bit patient as you wait for the logs to appear.

3. At this point, it's time to navigate to the S3 service and find the name of the bucket you created to store your Elastic Load Balancer access logs. Click the name of the bucket. You should see a *folder* named `AWSLogs` and then a *subfolder* underneath that is numerically named based on an account number. In that account folder, there should be further subfolders named `elasticloadbalancing` and then subfolders based on region, year, month, and day. Finally, you will get to the log files. Click on one of the log files, download it, and open it in Textpad or Notepad.

With that, we've learned how to enable logs and see the traffic patterns coming from our Elastic Load Balancers. Don't take down this set of CloudFormation templates just yet, however. We are going to use the VPC that we created here in the next section to examine VPC Flow Logs. But first, let's look at the use cases for Elastic Load Balancer logs.

Use cases for Elastic Load Balancer logs

You may be wondering why you would be interested in turning on Elastic Load Balancer logs when you could get information such as a client's address from your application's log files? Let's look at a few use cases in detail:

- Understanding the latency from requests – how long it takes to respond to a request.

- Monitoring access requests – where the requests are coming from and going to.

- Measuring how efficient the operation is between the client and the resources – are there any bottlenecks in the process that can be detected easily?

Now that we understand when we would use Elastic Load Balancer logs, let's look at VPC Flow Logs.

Using VPC Flow Logs

Flow logs help you capture information regarding the IP traffic going in and out of the network interfaces of your **Virtual Private Cloud** (**VPC**). Once this data has been captured, it can be written to either an S3 bucket or pushed out to a CloudWatch log group.

Once a flog log group has been created and has started writing logs, the logs do not appear immediately. It can take up to 5 minutes for the logs to appear in either the S3 bucket or the log group:

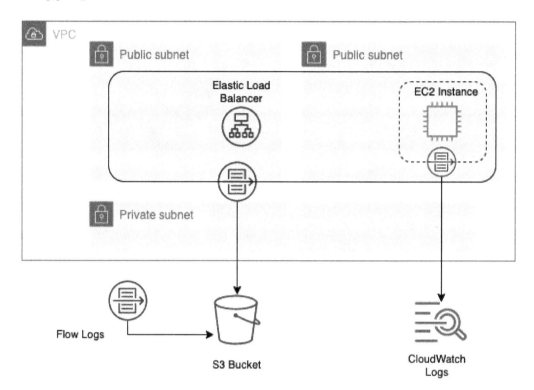

Figure 16.9 – VPC Flow Logs traveling to and from different sources

Flow logs can be created for network interfaces. These include the network interface of a VPC itself or even other services that contain network interfaces, such as the following:

- Elastic Load Balancers
- Amazon RDS databases
- Amazon ElastiCache caches
- Amazon Redshift databases
- Amazon WorkSpaces
- Transit Gateway
- NAT Gateway

Now that we understand what VPC Flow Logs are and what they can be attached to, let's see what kind of limitations they have before we set up the flow logs on the VPC we stood up earlier.

Limitations regarding VPC Flow Logs

Although VPC Flow Logs allow you to capture most traffic either coming from or going to the different network interfaces in your **VPC**, there are cases where this isn't possible. Let's take a quick look at those. In this case, if one of these scenarios comes up either in the DevOps professional exam or in real life, we know that flow logs would not be the optimal solution. An example of this is when flow logs cannot be enabled for peered VPCs unless the VPCs have been peered within the same account. Another thing to note is that Flow Logs do not capture all IP traffic. There is specific traffic that is *dropped* and specifically not captured by the flow log files:

- Any traffic from instances trying to contact the Amazon DNS server.

- Any traffic from a Windows instance for Amazon Windows license registration.

- All traffic to and from 169.254.169.254 for metadata information

- All traffic to and from 169.254.169.123 for the Amazon Time Sync service

- Any DHCP traffic

- Any traffic to the VPC router that travels on a reserved IP address

- All traffic between an endpoint network interface and a Network Load Balancer interface

Enabling VPC Flow Logs

As we mentioned previously in our previous hands-on exercise, we are going to use the same **VPC** that we stood up in this hands-on exercise to capture the VPC Flow Logs.

When you create a flow log, you need to specify the following items:

- The AWS resource that you want to have

- If you wish to capture accepted traffic, rejected traffic, or all the traffic in the flow log

- Where you want the data to be published to (an S3 bucket or CloudWatch Logs)

Let's get started:

1. Start by going to the **CloudWatch** service. On the left-hand menu, find **Logs** and click on **Log groups**.

2. At the top right, click on the orange **Create log group** button.

3. Create a log group called `FlowLogs`. You can change the retention settings to allow the logs to expire after 7 days (1 week). Once you have updated these settings, scroll to the bottom of the page and click the orange **Create** button.

4. Navigate back to the **CloudFormation Service** page. From your stacks listing, find the stack that we created called Chapter16-VPC. Click on this stack name to get to the details of this stack.

5. Then, from the top horizontal menu of the stack, click on the **Resources** menu item. Once the table of resources appears, scroll down until you find the listing for the VPC (it might be near the bottom since it was one of the first resources we created). Click on the blue **Physical ID** link for the VPC to be taken to the VPC. (Make a note of the physical ID before you click it; remember the last 4 alphanumeric characters of the VPC).

6.   Once you are on the **Your VPCs** page, *check* the box to the right of the name of the VPC. Once the box has been checked, click on the white button named **Actions** at the top right of the screen. Once you've done this, a drop-down menu should appear. Select the **Create flow log** menu item.

7. At this point, you will be brought to the **Create flow log** page. Use the following settings to fill out the flow log options:

    *   **Filter**: `All`
    *   **Maximum aggregation interval**: `1 minute`
    *   **Destination**: Send to CloudWatch Logs
    *   **Destination Log Group**: `FlowLogs`
    *   **IAM Role**: (search for `VpcFlowLogRole`)

8. After you have filled out all these settings, click the orange **Create flow log** button.

9. Now, it's time to go back to your Elastic Load Balancer URL and send some traffic to the websites that have been loaded on the EC2 instances. Doing this will generate logs for the VPC Flow Logs. Once you have generated some traffic, go back to the CloudWatch Log group called FlowLogs and look at the records that it generated.

We just learned how to create a custom CloudWatch log group and turn on VPC Flow Logs so that we could capture traffic from our VPC. Next, we will look at some use cases when it makes sense to turn VPC Flow Logs on.

Use cases for VPC Flow Logs

With so many types of logs available, let's look at some clear use cases for using VPC Flow Logs.

Using VPC Flow Logs to monitor remote logins

Remotely logging in to instances on your cloud infrastructure should only be done through authorized personnel, as well as trusted addresses. You can use VPC Flow Logs to see what users and IP addresses are gaining access to or trying to gain access to via protocols such as **Remote Desktop Protocol** (**RDP**) or **Secure Shell Access** (**SSH**).

> Note
>
> You can connect to AWS Mac instances using a **Virtual Network Connection** (**VNC**) client; however, this type of connection is insecure. To make the connection secure, you can wrap the connection in an SSH tunnel, as mentioned here: `https://aws.amazon.com/premiumsupport/knowledge-center/ec2-mac-instance-gui-access/`.

Detecting threats with VPC Flow Logs

If you choose to capture all events, both ingress and egress, with VPC Flow Logs, then you can detect threats to your environment such as port scans being performed on your network, network scans, someone trying to look for weakness or entry points into your system, or data being pushed out to unauthorized sources.

If you find that your system has been impacted by an event, then you can use VPC Flow Logs to trace the path that the offender took through the network.

Diagnosing and troubleshooting network issues

With layered security, there will be times where you are trying to figure out why access to a particular instance or service is not being allowed, which is troublesome.

Gaining an understanding of your network traffic

You can use VPC Flow Logs to analyze how the users of your network and AWS account are behaving and generate reports of any unsafe behavior that might be occurring. This can include using unprotected ports or allowing access from the world, rather than restricting access to origin assets to specific IP addresses and leaving world access to content delivery servers such as CloudFront.

Now that we've looked at the various use cases for VPC Flow Logs, we will turn our attention to the initial logs that we looked at – CloudTrail logs.

Going back to our CloudTrail logs

Now that we have created some resources through the exercises we have performed in the chapter, as well as several different API calls, we can go back to our CloudTrail logs and see how they have been populated with events.

Searching through CloudTrail logs

Since we have performed enough activities to generate CloudTrail logs at this point, we will start to search through our logs to see what we have captured:

1. Navigate back to the **CloudWatch** service. From the left-hand menu, find **Logs**, but this time click on the submenu named **Log Insights**.

2. From the drop-down menu for selecting log groups, choose the **CloudTrail** log group.

3. Place the following query in the query box and then press the orange **Run query** button:

```
filter eventSource="ec2.amazonaws.com"
  | stats count(*) as eventCount by eventName, awsRegion
  | sort eventCount desc
```

4. You should see a graphic similar to the following:

Figure 16.10 – The CloudWatch Logs insights visualization graph

5. You can try and see what other things you can find in your CloudTrail log if you like.

With that, we've learned how to query CloudTrail logs using CloudWatch Log Insights. Now, let's clean up the resources we created before we summarize this chapter.

Cleaning up the resources

We created a lot of resources in this chapter that, if left up and running, may leave you with a higher AWS bill than you may expect. After completing this chapter, be sure to delete the CloudFormation templates that were used to create the instances and Elastic Load Balancer.

Summary

In this chapter, we looked at three different sources of logs that can provide information for your AWS accounts that are not CloudTrail or application logs. Initially, we learned how to set up a CloudTrail trail to record all the API calls that happen inside of an account. Next, we looked at Elastic Load Balancer access logs and how they record the IP addresses, time, and responses coming into an Elastic Load Balancer. Finally, we looked at how VPC Flow Logs can capture network traffic from a variety of network interfaces.

In the next chapter, we will wrap up our discussion on logging by going over how enterprises complement the AWS services that capture logs. Then, we'll learn how log storage and advance searching are handled.

Review questions

1. You are working as a DevOps engineer at a company that has implemented multiple CI/CD pipelines. One pipeline is used to push out the application code and its features. Another pipeline is used to update the underlying infrastructure and security settings of the account. After the last set of security group updates for the application, all the users at one of the company's remote offices can no longer access the instances in the autoscaling group. These users can still access the application from the web protocol via the Elastic Load Balancer. These users contain members of multiple IAM groups, including developers, power users, and even an administrator. Where can you go for information to try and find out where the issue is occurring?

   a. Gather the IAM usernames that have been denied access. Use these usernames to search through the IAM log group in CloudWatch.

   b. Make sure that VPC Flow Logs have been turned on. Search the VPC Flow Logs for both the internal and external IP addresses for the remote office.

   c. Gather the IAM usernames that have been denied access. Go to the CloudTrail service and search for the usernames to find denial.

   d. Turn on logging to the application's Elastic Load Balancer. Check the Elastic Load Balancer logs for both the internal and external IP addresses for the remote office.

2. You have been brought into a company that is about to make a major production push and release of their application. As part of this release, they are implementing a new governing model that requires all activity on the AWS account to be monitored. How can you quickly and effectively help the company achieve this goal?

   a. Set up the Amazon Inspector service to constantly inspect all of the activity happening on the account.

   b. Turn on VPC Flow Logs for all the VPCs, making sure that you are capturing both inbound and outbound traffic.

   c. Set up the AWS CloudTrail service to monitor and record all the activity in all regions.

   d. Create a designated CloudWatch Logs log group so that any creation or termination events can be filtered specifically to this log group.

Review answers

1. b
2. c

17
Advanced and Enterprise Logging Scenarios

As we look to wrap up our section on logging, we will discuss how to implement enterprise-grade logging systems. While CloudWatch has the ability to search through logs and even present some visualizations, we will look at other native solutions offered by AWS that are more in tune with capturing, processing, storing, and visualizing massive amounts of logs streaming in constantly.

In this chapter, we're going to cover the following main topics:

- Using QuickSight to visualize data
- Streaming logs to Amazon Elasticsearch
- Using Amazon Kinesis to process logs

Using QuickSight to visualize data

Although there are multiple third-party visualization tools available to analyze your data and create graphical representations of your logs, there is a native service that Amazon has created for its customers, **QuickSight**. QuickSight was developed to be a cloud-scale **Business Intelligence** (**BI**) service that is easy to use and can ingest data from multiple sources.

Amazon QuickSight uses a proprietary SPICE engine to calculate and serve data. **SPICE** stands for **Super-fast, Parallel, In-memory Calculation Engine**. This technology has been built to achieve blazing-fast performance at an enterprise scale. The SPICE engine can do this by automatically replicating data, allowing thousands of users to perform queries and analysis on that underlying data at immensely fast speeds.

Another key feature about Amazon QuickSight is that you can share created dashboards with members of your IAM organization. Still, it also has the ability to share access via email to those that do not have an IAM or federated account to your AWS organization. QuickSight also has an iPhone and Android app that is available for access:

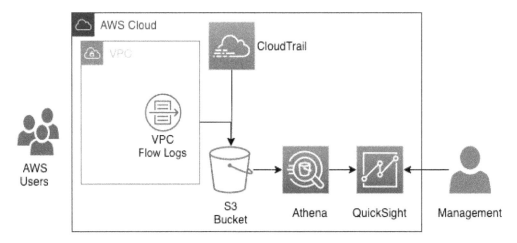

Figure 17.1 – A flow of logs to Athena and AWS QuickSight to create visualizations

In the previous figure, we show a flow of AWS users creating events from actions that they are taking.

When you set up QuickSight in your AWS account, you create a namespace, which is a logical container that is used to organize your teams, clients, and others that you will invite to your QuickSight visualizations. You can create multiple namespaces and can isolate the data viewed by the users to that namespace. Namespaces can also span over multiple regions. Once you set up the namespace, there is no further administration needed from that point forward.

With an understanding of the value that the QuickSight service brings when creating visualizations to not only those in our Amazon account but others in our organization, let's now look at how the Athena service can expand on these capabilities, using the files that we already have stored in our S3 buckets.

Querying data with Amazon Athena

AWS has created a service that allows you to query data stored in S3 buckets. The service is serverless, and hence there is no need to provision servers, and the service only charges you for the queries that you run.

The Presto query engine backs **Amazon Athena**. This is an open source SQL engine that allows users to query large datasets with low latency. The Presto engine also fully supports joins, arrays, and window functions.

The key features of Amazon Athena are as follows:

- Since it's serverless, there's no administration or infrastructure to manage.
- It uses standard SQL to query the underlying data.
- It has extremely fast performance without the need for tuning.
- It allows for federated queries across multiple data sources.
- It is secure, allowing you to take advantage of both IAM and S3 bucket policies to control access to data.
- It is highly available with S3 as the data source.

Now that we have seen how to use Amazon QuickSight to create more powerful visualizations using Amazon Athena, let's look at a few Amazon QuickSight use cases.

Amazon QuickSight use cases

The next section will explore some of the use cases where Amazon QuickSight can be used in conjunction with other AWS services to create enterprise-grade systems to create dashboards and analysis systems to monitor logs and analytics.

Using QuickSight to visualize logs and usage analytics with the help of Athena and Glue

Amazon has built an interactive query service that allows you to use standard SQL statements to query your data, named Athena. One of the best parts about Athena, besides the fact that it uses standard SQL and hence there's no new special language to learn to use it, is that it's serverless. This means that there are no servers to provision, and you are only charged for the queries that you run on the system and the data that it scans.

The ability to add machine learning insights to your dashboards

Amazon QuickSight extends the normal proficiencies of just displaying your data in a dashboard format by adding natural language capabilities and machine learning insights to help you gain a much grander understanding of your data. These features help users discover patterns and hidden trends arising from the underlying data without having specialized technical expertise or machine learning skills.

Connect user dashboards to your data warehouse or data lake

If you have your data stored in a data warehouse, such as using the **Amazon Redshift** service, then you can create a connection in Amazon QuickSight to connect to your Redshift cluster. This Redshift cluster becomes an auto-discovered dataset and secures the connection between the Redshift cluster and QuickSight using SSL automatically without extra configuration. You can then either choose the tables you want to use in your QuickSight visualizations or create a custom SQL statement to import your data into the SPICE engine to analyze and visualize your data.

If you are storing data in a data lake, especially using **Lake Formation** from AWS, your data resides in Amazon S3 buckets. You can then use **AWS Glue** to crawl the data and create a data catalog. Once the data catalog has been created, you can query the data with Amazon Athena and create a table and database. These tables and databases serve as containers for the schema of data held in the S3 buckets. Amazon QuickSight can then connect to Athena databases and create visualizations of the data or even perform further SQL queries:

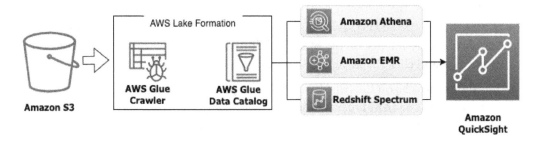

Figure 17.2 – Connecting Amazon QuickSight to a data lake in AWS

Now that we have gone over some of the use cases where enterprises would find value in Amazon QuickSight, let's go through a hands-on exercise using QuickSight to cement the concepts of this service. This way, if a question comes up regarding visualizations on the DevOps professional exam, we have a solid basis of when to choose Amazon QuickSight versus CloudWatch dashboards.

Creating a dashboard with Amazon QuickSight

When you create a data dashboard in Amazon QuickSight, you are publishing a collection of interactive graphs and charts for your users to explore, using the underlying data that not only shows them insights but gives them the tools to explore further should they feel the need.

Let's go through the process of creating a dashboard in Amazon QuickSight. We will need the assistance of the Amazon Athena service in order to get our data to a temporary database so that we can connect to it from QuickSight:

1. Log on to the Amazon Management Console, and in the top search box, search for QuickSight. Once on the **QuickSight** screen, verify your AWS account number and then press the blue button labeled **Sign up for QuickSight**. Be sure to change the default of **Enterprise** to **Standard**:

Figure 17.3 – The QuickSight icon from the search menu

You can also use the following URL: `https://aws.amazon.com/quicksight/pricing/`. Click on the link in the middle of the page that says **Standard Edition**. Once on the **Standard Edition** page, scroll down to the bottom of the page and click the large yellow button labeled **Start Your Free Trial**.

You should now be on the page labeled **Create your QuickSight account**. Keep the first selection, **Use IAM federated identities & QuickSight-managed users**, for **Authentication method**. Next, for **QuickSight region**, change the region to **US East (Ohio)**:

Standard

Authentication method

◉ Use IAM federated identities & QuickSight-managed users
   Authenticate with single sign-on (SAML or OpenID Connect), AWS IAM credentials, or QuickSight credentials

○ Use IAM federated identities only
   Authenticate with single sign-on (SAML or OpenID Connect) or AWS IAM credentials

QuickSight region

Select a region   ⓘ

US East (Ohio)   ⌄

Figure 17.4 – Setting the Authentication method and region to create the QuickSight account

2. For your QuickSight account name, which is where you declare your namespace, choose something unique that you can remember. You will also need to enter an email address.

3. Now, we can specify which data will be available to QuickSight in our current setup. Mark the boxes next to the following items:

    a. Enable auto discovery of data and users in your Amazon Redshift, Amazon RDS, and AWS IAM services.

    b. Amazon Redshift

    c. Amazon RDS

    d. IAM

e. Amazon Athena.

f. Amazon S3 (select the bucket shown – **chapter16-elb-logs**):

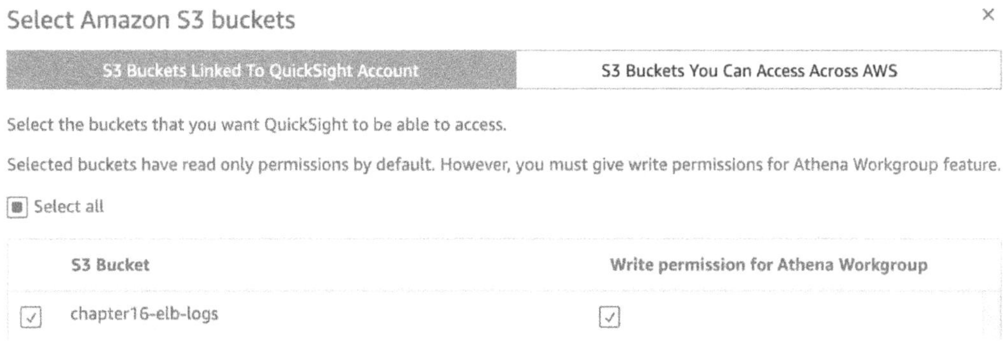

Select Amazon S3 buckets   ✕

S3 Buckets Linked To QuickSight Account	S3 Buckets You Can Access Across AWS

Select the buckets that you want QuickSight to be able to access.

Selected buckets have read only permissions by default. However, you must give write permissions for Athena Workgroup feature.

☑ Select all

S3 Bucket	Write permission for Athena Workgroup
☑ chapter16-elb-logs	☑

Figure 17.5 – Selecting the S3 bucket for QuickSight

We have added a bucket from our previous hands-on exercise, which should have data to query with Amazon QuickSight.

4.   You will now be on a screen with a dancing graph as AWS creates your QuickSight account. Once it is created, then you will be presented a blue button to click on labeled **Go to Amazon QuickSight**. Click on this button to proceed.

5.   Once you click the button, it will start to create a few samples for you and display a pop-up window welcoming you to QuickSight. Click the blue **Next** button on the popup to close them, or click **X** in the top right-hand corner.

6.   We will need to create a dataset from the data in our S3 bucket. Find the datasets icon in the left-hand vertical menu and click on it. Once on the datasets page, click on the dark blue **New dataset** button in the top right-hand corner.

7. Choose **Upload a file** as the data source for your dataset. Download the MOCK_
    DATA.csv file from the Chapter-17 folder of the GitHub repository if you have
    not already done so and upload it to QuickSight. Click the blue **Next** button on the
    Confirm file upload settings popup. Once it is done, click the blue **Visualize** button:

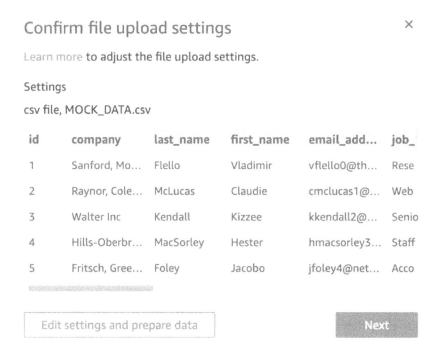

Figure 17.6 – Confirming the data has been uploaded correctly to QuickSight

8. Once the data has been loaded, you should be taken to the **Visualize** section of
    Amazon QuickSight. It's now time for us to create a visualization from the data we
    just imported. We need to initially select some fields to show on the graph.

9. Choose the *filled map* type of graphic, which is in the visual types to the right-hand side as the last value. Then, with that type of visualization selected, drag the `state_province` value to the **Location** field well and the `zip_postal_code` value to the **Color** field well. Click on some of the other visualization types to see how QuickSight can change how your data is presented:

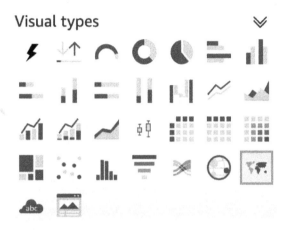

Figure 17.7 – The filled map visualization type in Amazon QuickSight

> **Important Note**
>
> Amazon QuickSight Standard edition has a price of $12 per month when paying month to month at the time of publishing. There is a noted free 30-day trial for authors; however, if you have already used the service, you will be charged for it if you go through this tutorial.

Now that we have seen how to create interactive dashboards and visualizations using the QuickSight service, let's look at our next service used for managing logs at an enterprise level – the Amazon Elasticsearch service.

Searching and grouping logs with managed Elasticsearch

Many people associate Elasticsearch with ELK; however, the two have differences. **ELK** stands for **Elasticsearch, Logstash, and Kibana**. In this configuration, Elasticsearch serves as the storage, Logstash serves as the log parser, and Kibana serves as the visualization frontend of the system where users interact with the system:

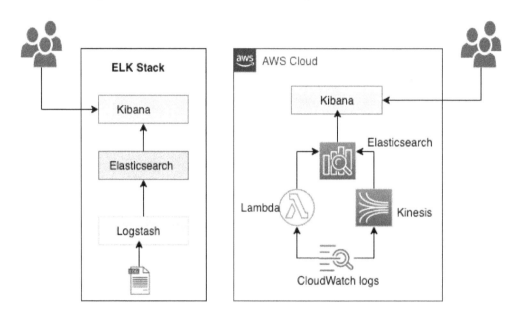

Figure 17.8 – A comparison of the ELK stack versus Amazon's managed Elasticsearch service

With Amazon's managed Elasticsearch service, there is no Logstash installed by default; however, there are other options to get the logs that you generate into your Elasticsearch cluster.

Use cases for managed Elasticsearch

There are several use cases for using the managed Elasticsearch product from AWS. Let's examine them next.

Store and search logs for application monitoring

You can stream logs that have been placed into AWS CloudWatch Logs into Amazon's managed Elasticsearch service. Once the logs have been put into the Elasticsearch cluster, they can be searched by the Lucene-backed Elasticsearch search engine.

Security Information and Event Management (SIEM)

Storing logs from multiple events and applications across your network in a central system allows you to detect and alert on security events in near real time, using the capabilities of Amazon's managed Elasticsearch service.

An enterprise-grade search engine

Although the ELK stack is most synonymous with collecting and displaying logs, Elasticsearch is a powerful search engine, built using the Lucene library, which allows for performing searching in near real time. You can connect to Elasticsearch using a RESTful API to send results back to your application or deliver new data to store in your search engine.

Monitoring your infrastructure

As you collect logs from the different pieces of infrastructure you manage, whether they reside in the cloud or on-premises, you can collect them in a single solution using managed Elasticsearch from AWS. This can help you quickly research any issues that arise from all angles, helping to cut down your **Mean Time To Resolution (MTTR)**.

Please note that Amazon's Elasticsearch service is being renamed to Amazon OpenSearch.

Now that we understand the use cases for the Elasticsearch service, let's see how we can get our CloudWatch logs into an Elasticsearch cluster with a hands-on example.

Streaming logs from CloudWatch Logs to the Elasticsearch service

For the hands-on exercise with the managed Elasticsearch service, we will deploy a simple Lambda function so that we can generate some logs. We can then stand up a single-node Elasticsearch cluster to receive the logs. Use the Lambda function's logs that go into the CloudWatch Logs log group. We will then subscribe that log group to the Elasticsearch cluster we just created. The CloudFormation script also includes an AWS Events rule to fire off the Lambda function every five minutes so that logs will be periodically sent to our CloudWatch Logs group. Finally, we will go to the Kibana visualization interface and look for our logs. Let's get started:

1. We will first need to download the CloudFormation template named `lambda_stack.yml` from the `Chapter-17` folder in the GitHub repository for the book.

2. Start by logging into the AWS Management Console. Once inside the console, navigate to the **CloudFormation** service so that we can quickly get our Lambda function up and running. Create a stack with new resources by either pressing the orange **Create stack** button from the main service page or, if you are already on the **Stacks** page with previous stacks you have created, the white **Create Stack** button on the top right-hand side.

3. Once on the **Create stack** page, keep the top setting as **Template is ready**. Under the **Specify template** heading, select the radio button next to the option of **Upload a template file**. Use the **Choose a file** button to select the `lambda_stack.yml` file that you had downloaded from the `Chapter-17` folder in the GitHub repository. Once the file has been uploaded, click the orange **Next** button at the bottom of the screen.

4. Now that you're on the **Specify stack details** page, enter `Logging-Lambda` as the name for this CloudFormation stack in the text box for **Stack name**. After this is entered, click the orange **Next** button at the bottom of the screen:

Stack name

Stack name

> Logging-Lambda

Stack name can include letters (A-Z and a-z), numbers (0-9), and dashes (-).

Figure 17.9 – Entering the name of the CloudFormation stack

5. There is nothing to do on the **Configure stack options** page. Scroll all the way down to the bottom of the page and click the orange **Next** button.

6. On the **Review** page, scroll down to the bottom of the page and click the checkbox, acknowledging that this template needs to create an IAM role. After you have done this, you can click the orange **Create stack** button.

7. It should take 1 to 5 minutes for our stack to completely create our resources; after it has, we can click on the **Resources** tab on the horizontal menu and find the logical ID named `LambdaLogGenerator`. This will be our actual Lambda function. Click on the blue link next to this name, which will be in the column named **Physical ID**. This will open up a new window directly to our Lambda function:

Logical ID ▲	Physical ID ▽	Type ▽	Status ▽	Status reason ▽	Module
LambdaLogGenerator	Logging-Lambda-LambdaLogGenerator-XsEJ09CvzOho 🔗	AWS::Lambda::Function	⊘ CREATE_COMPLETE	-	-

Figure 17.10 – The Logical ID and Physical ID of the created Lambda function

8. We need to wait at least 5 minutes for our Lambda function to be invoked, so in the meantime, we will create our Elasticsearch cluster so that we can stream the logs when they are ready. In the top search box of the AWS Management Console, search for the term `Elasticsearch`. When you see the **Elasticsearch Service** icon, right-click on the icon to open up the service in a new tab.

9. When you come to the **Amazon Elasticsearch Service** page, click the blue button that is labeled **Create a new domain**.

10. For **Deployment type**, choose **Development and testing**, as we only need a single availability zone. Under the heading of **Version**, choose the latest version. At the time of writing, the latest version available was 7.10. After you have made your selections, click the blue **Next** button at the bottom of the screen.

11. We should now be on the **Configure domain** page. There are only a few settings on this page that need to be set. The first is **Elasticsearch domain name**. Use `chapter-17` for the domain name. The second setting will be under the heading of **Data nodes**. It is suggested for this exercise that you use the `T3.medium.elasticsearch` instances, since we are only performing a short test and not storing much data. Once you have made this change, scroll down to the bottom of the page and click the blue **Next** button.

12. On the **Configure access and security**, to simplify the access to the Kibana interface, use the following settings:

    a. Choose **Public Access**.

    b. Check the box for **Enable fine-grained access control**.

    c. Select the radio button to create a master user:

    * **Username**: `devops`

    * **Password**: `Chapter17*`

    d. Under **Access policy**, select the following:

    *

    Keep all other settings the same, and then click the blue **Next** button at the bottom of the page. Click the blue **Next** button on the **Tags** page, and finally scroll down to the bottom of the **Review** page and click the blue **Confirm** button.

13. It will take a few minutes for our Elasticsearch cluster to spin up, and this will give us time to go back to our Lambda function. Go back to the other tab in our browser window where we previously had our Lambda function.

14. Once on the `Logging-Lambda-LambdaLogGenerator` function, click the **Monitor** item in the horizontal menu. This will not only allow us to see a line where our Lambda function has been invocated but there will also be a white button near that horizontal menu, now labeled **View logs in CloudWatch**. Click on this button to be taken directly to the logs:

Figure 17.11 – The View logs in CloudWatch button directly under the horizontal menu

15. We should now be in the CloudWatch Logs group for our Lambda function. Above the heading of **Log streams** will be a horizontal menu. Click on the menu item labeled **Subscription filters**.

16. Once the heading of **Subscription filters** appears, click on the white button on the right-hand side labeled **Create**. A menu of choices will appear. Choose the item named **Create Elasticsearch subscription filter**. This will bring you to a destination page. Choose the Elasticsearch cluster named `chapter-17` that we had just created from the drop-down list. For **Lambda IAM Execution Role**, start typing `Logging-Lambda` and select the role we created for our logging Lambda function. Under **Log Format**, select **Common Log Format** from the drop-down list. For **Subscription filter name**, type in `log-test`. Finally, scroll to the bottom of the page and click the orange **Start streaming** button.

17. Once you click the **Start streaming** button, a Lambda function will be created. Now our logs should start streaming into our Elasticsearch cluster.

18. Let's now go back to the tab where we had created our Elasticsearch cluster and click on the **Dashboard** menu item in the left-hand vertical menu. We should now see **chapter-17** as a linking domain. Click on **chapter-17** to be taken to the details of the domain. There will be a link to Kibana on the **Overview** page. Click on this link. Enter in the username (`devops`) and password that was created when we provisioned our Elasticsearch cluster.

19. We are now in the Kibana interface for our Elasticsearch cluster. Click on the text that says **Explore on my own**. When the **Select your tenant** dialog box appears, just click the blue **Confirm** button.

Now that we have looked at how to capture logs and store them on the managed Elasticsearch cluster, let's now look at how we can incorporate the Amazon Kinesis service with Elasticsearch.

Understanding the Amazon Kinesis service

There is a service in AWS that has been created especially for the real-time processing of streaming data. That service is **Amazon Kinesis**. As more and more items in the world produce data, and more and more applications want to consume that data, there need to be services that can quickly consume and do some pre-processing on that data. The Kinesis service also provides a bit of redundancy in case your main application goes down, storing the records on its shards. By default, the records can be accessed for 24 hours from when they were written. This can also be extended in the settings to save the data for up to 7 days. Data that is sent to Kinesis can be a maximum of 1 MB in size.

The key features of Amazon Kinesis to understand for the test are the following:

- It allows for real-time ingesting, processing, and streaming of data.

- It is a fully managed service, and hence there is no infrastructure for you to manage.

- It integrates with a number of other AWS services, such as Amazon Redshift.

Knowing that the service comes with these built-in benefits, it also has evolved since its initial introduction to meet the needs of the customers who have been using it. Although it is useful for processing immense amounts of incoming data, such as logs from multiple sources, this is only one of its many uses.

The Kinesis service comes to us from AWS with four separate capabilities:

- **Kinesis Video Streams** – This capability allows you to easily process incoming video data securely and allows for analytics, machine learning, or other processing.

- **Kinesis Data Streams** – This capability allows you to stream data from applications to the managed Kinesis service.

- **Kinesis Data Firehose** – This capability gives users a simple way to capture, transform, and load data streams into AWS data stores, such as S3, ElasticSearch, and Redshift.

- **Kinesis Data Analytics** – This capability allows users to easily process data streams in real time, using SQL or Apache Flink.

There are scenarios when you would use more than one of the capabilities at the same time.

Tip for the Test

While you are required to know the Amazon Kinesis service in much greater detail when taking the Data Analytics – Specialty certification, there are cases where scenarios are presented in the DevOps professional exam where knowing when to use the Kinesis service would be the correct solution (or incorrect solution).

Now that we have a basic understanding of the Amazon Kinesis service, let's look at when it would be appropriate to use the Amazon Kinesis service to process our logs in an enterprise-type scenario.

Using Amazon Kinesis to process logs

Kinesis Firehose allows us to add an automatic call to an Amazon Lambda function that can transform our records before insertion into the Elasticsearch cluster. This works very similarly to the way a Logstash instance in an ELK stack would take incoming logs and transform them before sending them off to an Elasticsearch instance:

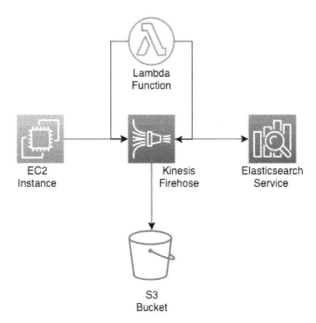

Figure 17.12 – Kinesis Firehose inserting logs into S3 and the Elasticsearch service simultaneously

One of the reasons that there is an S3 bucket in the diagram is the fact that any failed records can be retried for a specified period by Kinesis Firehose to be resent to the Elasticsearch service if there is a failure during delivery.

What types of things could cause failure? Well, running out of space in your Elasticsearch cluster is one that comes to mind. If you are constantly streaming logs and do not have the correct way to phase out older data or have not added enough nodes to your cluster, there will come a point where your cluster will run out of space and can no longer accept any new data, or logs in our case. Rather than lose that information or have to manually try and insert it into the cluster, Kinesis can queue the misfires to an S3 bucket and then try to resend the data at a later time.

Now that we have an understanding of how Amazon Kinesis can be used to enhance our logging setup, let's take a look at the importance of properly tagging and categorizing our logs.

Using tagging and metadata to properly categorize logs

When categorizing your cloud resources and subsequent logs for ingestion, tags can help classify where the resources are coming from, especially in the case where you are pushing all of the logs to a larger enterprise-logging solution.

As you create your assets in AWS as a seasoned DevOps engineer, you should have a number of tags that you place on your resources as they come out of your CI/CD pipeline so that you know how to manage them effectively:

Technical Tags	Automation Tags	Business Tags	Security Tags
Name	Backup	Team	PII
Application	DR	Owner	Lifecycle
Environment	Expire	Country	Compliance
OS	Maintence_Window	Business_Unit	
Patch		Customer	
		COTS	

Figure 17.13 – Examples of tag categories and keys for use within an enterprise system

When we talked about processing in the previous example with the Lambda function using Kinesis, we could look at the tags contained in the metadata.

Cleaning up resources

Once again, we have created a number of resources in our AWS account during the course of the hands-on exercises of this chapter. If not taken down, items such as the managed ElasticSearch cluster could bring about a larger bill than you might like. The Lambda function, which fires every 5 minutes, could eat into your Free Tier allowance, as there are on average 43,800 minutes in a month, so this function would be invoked 8,760 times. Simply delete the resources that you are no longer testing with and delete any CloudFormation stacks that you are no longer using to be sure that your AWS bill stays as low as possible.

Also, remember to cancel your QuickSight subscription, whether it was a free trial or a single-month subscription, so that you do not get a recurring charge for the QuickSight service.

Summary

In this chapter, we have discussed how to implement enterprise-grade logging systems. We had a look at other native solutions offered by AWS that are more in tune with capturing, processing, storing, and visualizing massive amounts of logs streaming in constantly.

In the next chapter, we will begin looking at how to ensure that our workloads are highly available, starting with Auto Scaling groups and their life cycles.

Review questions

1. Your company has asked you to create a visual dashboard that shows logs in a visual format that can be accessed by the management team. The management team does not log on to the AWS account via IAM users, nor do they have AWS Management Console access. They also want data enrichment from one of the AWS Redshift databases and a CSV file that is created from an AWS Batch process and stored in an S3 bucket. How can you create an easy-to-access, secure, and dynamic dashboard for the management team to use?

    a. Create a CloudWatch dashboard using all of the sources. Gather the management team's email addresses and send out a link for access to the dashboard.

    b. Stream all the data into a managed ElasticSearch cluster via CloudWatch Logs. Create a Kibana dashboard that will display the required visualizations. Share the link to the Kibana dashboard with the executive team.

c. Use Kinesis Firehose to stream the data from Redshift to a managed Elasticsearch cluster. Create a Kibana dashboard that will display the required visualizations. Share the link to the Kibana dashboard with the executive team.

d. Use Amazon Athena to create a temporary table of all the logs that are required to be used in the QuickSight dashboard. Import the CSV file into QuickSight for visualization purposes. Import the Redshift database from QuickSight as a data source. Gather the management team's email addresses and send out a link for access to the dashboard.

2. You have recently joined a company that has been storing multiple different log files in S3 buckets. These logs have come from a variety of sources, including being pushed from on-premises servers using S3 sync, Application Load Balancer logs, VPC Flow Logs, and others. The company would like you to quickly do an analysis of the logs and find out how old the logs are from each category. How could you perform this task quickly and cost-effectively?

a. Set the S3 bucket, which contains the logs as a data source in Amazon QuickSight. Use the SPICE engine to analyze the logs. Create two quick visualizations, one showing the age of the logs and the other the types of logs.

b. Use the S3 inventory function to search through the files in the S3 bucket. Import the report into Excel and then sort the files by age and the types of logs.

c. Create an AWS Glue job catalog of all the items in the S3 buckets. Use Amazon Athena to be able to query the types of logs. Create a query to group the types of logs. Sort by date descending to show the oldest logs first in each group.

d. Import all the logs into a managed Elasticsearch cluster. Using the Kibana interface, run a query on the types of logs and then group the logs by age to get a count of the number of logs.

Review answers

1. D
2. B

Section 4: Enabling Highly Available Workloads, Fault Tolerance, and Implementing Standards and Policies

Failures happen within any environment, however, with the regional and global capabilities of the AWS cloud, the barriers to creating highly available and fault-tolerant workloads have been taken away. Couple this with the enforcement of standards and policies both on a schedule and in real time and your CI/CD journey becomes smoother.

This part of the book comprises the following chapters:

18
Autoscaling and Lifecycle Hooks

One of the critical features of automation is taking advantage of autoscaling. Although many use this **Amazon Web Services** (**AWS**) service in its essential capacity, the educated **Development Operations** (**DevOps**) professional understands and takes advantage of some of the more advanced capabilities that Auto Scaling can provide. Insight into these components will not only help you pass the certification exam but will also allow you to manage your AWS environment more effectively.

In this chapter, we're going to cover the following main topics:

- Understanding AWS Auto Scaling
- Deploying **Elastic Compute Cloud** (**EC2**) instances with Auto Scaling
- The Auto Scaling lifecycle
- Using Auto Scaling lifecycle hooks

Understanding AWS Auto Scaling

Auto Scaling, which is a subset of the **Amazon EC2 service**, revolves around automatically provisioning and managing your EC2 instances without the need for any manual intervention. The Auto Scaling service can be used to constantly maintain a fixed number of servers at any one time across one or more **Availability Zones** (**AZs**) within an Amazon Region. The service also provides you elasticity in that it can scale up to meet spikes of demand that your customers or applications present without constantly monitoring the system.

It does this by tapping into the power of a complementary service, Amazon CloudWatch. It watches a metric such as **Central Processing Unit** (**CPU**) utilization on an instance and makes sure that if it rises above 80 percent, meaning that 80 percent of the available CPU of the instance is being used for a specific time, such as 5 minutes, a scale-out event will then occur. This helps ease the load on this particular instance and should bring the total percent of CPU utilization back under that 80 percent threshold once the new instance comes online.

The Auto Scaling service is constantly performing periodic health checks on the instances in its **Auto Scaling Group** (**ASG**). The time between health checks can be configured, but the default setting is 300 seconds (5 minutes). You also configure how many health checks an instance can fail before it is taken out of the ASG for being unhealthy. On a similar note, whichever number of health checks that an instance must fail in order to be removed from the ASG is the same number of health checks that it must pass before being added as a healthy instance to the ASG.

Auto Scaling also plays a key component in infrastructure event planning. If you know that you are about to have an increase in customer traffic, such as the marketing department has bought a television spot on a popular television show or there is a special sale about to happen, then you can simply increase both the desired capacity and the maximum capacity of your ASG to ensure that your customers not only have a good experience but also that your servers do not get overloaded.

Now that we have a basic understanding of the Auto Scaling service, let's look at the key components that make up this service.

Understanding the differences between vertical and horizontal scaling

As we discuss the Auto Scaling service throughout this chapter, we will be talking about how to horizontally scale your instances in and out to meet the demands of your workload. This is usually a more cost-efficient way to scale and gain resources as you can benchmark your workload and know how much memory and CPU that your workload actually needs. This leads to a number of benefits, including your workload having a much higher availability since there are multiple instances or containers running the application at the same time. This prevents the instance (or container) that was scaled up vertically from becoming a **Single Point of Failure (SPoF)**. You also don't have a limit on hardware. With vertical scaling, there comes a point in time when you will hit a resource limit. Although there are instances in AWS that have massive amounts of **Random-Access Memory (RAM)** and large numbers of CPUs, this is not a way to correctly construct and deploy your application.

The processes of horizontal and vertical scaling are depicted in the following diagram:

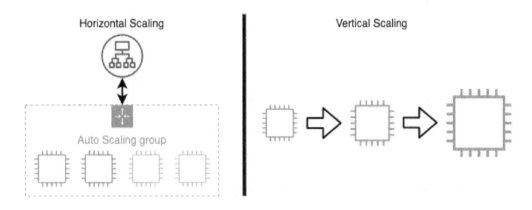

Figure 18.1 – Horizontal scaling versus vertical scaling

Now that we know the differences between horizontal and vertical scaling, let's take a look at the key components of AWS Auto Scaling.

The key components of Auto Scaling

One of the main components of Auto Scaling is the ASG. This ASG is a logical group of instances for your application or service. You set three main variables for the ASG: the minimum number of instances, the maximum number of instances, and the desired number of instances. The minimum number of instances tells the ASG what is the lowest number of instances that can be running at any one time in a combination of all AZs specified. The maximum number of instances tells the ASG what is the greatest number of instances that it can allocate in that region. These maximums must stay within the service limits. Finally, the desired capacity is the number of running instances at any given moment in time if there are no scheduled actions or scaling events driven by scaling policies.

You can see a depiction of an ASG in the following diagram:

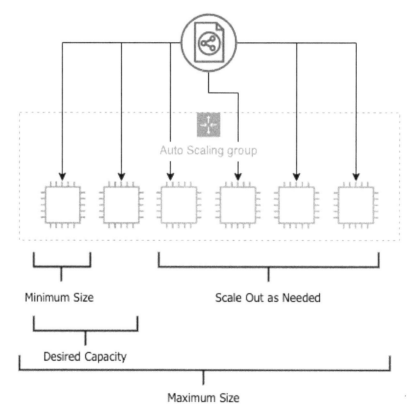

Figure 18.2 – ASG visualization

Another key component of Auto Scaling is the **launch template**. The launch template determines which properties the instances inside the ASG will launch with. These properties are provided inside of the launch template. You can determine items such as the instance size, image **Identifier (ID)**, **Virtual Private Cloud (VPC)** ID, along with others, when crafting your launch template.

If you have used ASGs in the past, then you may be familiar with launch configurations. Launch configurations are very similar to launch templates, but launch templates replaced launch configurations in a few respects. Firstly, launch configurations are immutable. Each time that you want to make a change to your launch configuration, you need to clone your launch configuration, make the change, and then attach it to the ASG. This is in contrast to launch templates. Launch templates allow for versioning. Launch templates also allow for the latest EC2 features, such as the use of unlimited T2 instances and **Elastic Block Store (EBS)** tagging.

ASGs are fully integrated with **Elastic Load Balancing (ELB)**. It doesn't matter which one of the three types of ELBs that you use; ASGs fully integrate without issues. This means that if an ASG is associated with a load balancer, whenever an instance is provisioned, it will automatically get registered with the load balancer. Conversely, when an instance is either de-provisioned or terminated, the load balancer will drain the traffic from the instance and deregister the instance from the load balancer before terminating the instance.

> **Note**
> If you have old launch configurations but would like to take advantage of the latest EC2 technology, then AWS has a documentation page on how to convert a launch configuration to a launch template. It can be found here: `https://docs.aws.amazon.com/autoscaling/ec2/userguide/copy-launch-config.html`.

The final key component that we are going to discuss is going to be the **scaling plan**. A scaling plan can be set to use either predictive scaling or dynamic scaling. If you put your scaling plan to use predictive scaling, the ASG will use **Machine Learning (ML)** to look at how taxed your workload is, especially in respect to certain hours and days of the week, and then generates scheduled scaling actions so that your application has the capacity to meet those needs, as illustrated in the following diagram:

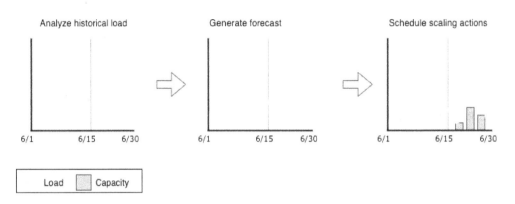

Figure 18.3 – How predictive scaling works to schedule resources in Auto Scaling

Dynamic scaling policies are the component that you configure when using dynamic scaling with a scaling plan.

It's important to note that even if your scaling calculation states that you should go beyond the maximum number of instances that you have set, that maximum is a hard limit and will not be surpassed by the scaling policy. You can raise the maximum number of instances if you hit the ceiling while scaling, or you may want to create a new version of your launch configuration to account for a different type or family of instances that may handle the type of traffic your application is receiving in a better fashion.

There is an old saying when it comes to Auto Scaling that you should *scale up like a rocket and down like a feather*. This is due to the fact that scaling up instances is expensive from a time perspective. If you are starting one or three instances based on a scaling event, they should all take the same amount of time to come online. Bringing in three instances will allow you to have a bit more capacity rather than scaling up a single instance at a time. If that capacity is unjustified, then you can scale down a single instance at a time in order to keep your spending at an appropriate level using Auto Scaling.

When you plan out your ASG, you want to take into consideration the following factors:

- How long does it take to launch and configure a server that will be part of the ASG?
- Which metrics would be most applicable when monitoring your workload's performance?
- Do you want the ASG to span multiple AZs? If so, how many?
- What type of role should Auto Scaling play in your application?

With the key components of Auto Scaling under our belt, let's look at the main use cases for AWS Auto Scaling.

Understanding the different types of Auto Scaling

Amazon's EC2 Auto Scaling gives you a number of options on how you can scale your instances to meet both the demand of your business and the goals of your budget.

If you wanted total control of how to scale your instances, then you can use **manual scaling**. With the use of manual scaling, you would set the minimum, maximum, and desired capacity to what you desire for your workload, and your ASG will adjust accordingly.

There are also times when the traffic that is coming to your workload is rather predictable. These may be cases where your customers are more active during normal business hours, or you may have business workloads that you use internally, and your staff that uses the workloads are all in one or two particular time zones. These are perfect cases for the use of **scheduled scaling**. With the use of scheduled scaling, your ASG will scale in and out automatically based on the times and dates that you set.

If you want your ASG to automatically scale out when there is a high demand for the application's services along with scaling back in when that same demand has diminished, then this is a case for **dynamic scaling**. Dynamic scaling allows you to choose a specific metric that matters to your application and set a percentage whereby once that percentage has been reached, the associated CloudWatch alarm on the metric will be triggered and one or more instances will be launched so that they can be added to the ASG. In the same vein, if you have previously scaled up capacity in response to a particular metric and that metric has fallen to a level where it is being underutilized, then the scaling policy will look at which one of the instances it should start to put in a terminating state. Based on your configuration, this could be the oldest instance, the newest instance, or even the instance that is closest to the next billing hour.

Dynamic scaling also allows you to put in **step scaling**. These step adjustments can vary based on the type of metric alarm breach, meaning that you can scale up to multiple instances at once if you have a large spike in traffic if your maximum capacity allows.

Taking the properties of dynamic scaling one step further is **predictive scaling**. Predicative scaling analyses your traffic over time and then increases and decreases the number of EC2 instances in your ASG based on these trends. Predictive scaling is a good choice when you can't decide exactly which metric would be best suited to your application's needs or you have applications that take a long time to initialize. It can also be a wise choice if you have workloads that seem to come up and down, such as batch processing. The predictive scaling analysis can analyze these workloads and decide when it would be best to bring on more capacity.

We have just looked at the different ways in which ASGs can be set and adjusted to accommodate the influx of traffic for our workloads. Now, let's look at the primary use cases for Auto Scaling.

The four primary use cases for AWS Auto Scaling

There are four popular use cases that Auto Scaling solves for customers, outlined as follows:

1. It automates the provisioning of servers.

2. It reduces paging frequency.

3. It makes it easier to use spot instances.

4. It allows you to scale your cloud infrastructure up and down and save costs.

Now that we know some of the main use cases for using Auto Scaling, let's look at how we can put ASGs into action by creating our own launch template.

Deploying EC2 instances with Auto Scaling

The best way to get a full understanding of a service is to jump in hands-on to see how it performs, and the Auto Scaling service is no different. In this hands-on exercise, we will create a launch template for our ASG. We will then create an ASG. Follow these next steps:

1. Log on to the **Amazon Management Console** using your administrative user account. Once logged in, navigate to the EC2 service. Once on the EC2 service, locate and click on the **Launch Templates** sub-menu item on the left-hand menu, which is located under the **Instances** menu heading.

2. Once you are on the EC2 **Launch Templates** main screen, click on the orange **Create launch template** button in the main window.

3. We should now be on a screen labeled **Create launch template**. Use the following values in the first box labeled **Launch template name** and the fields that come after this:

    - **Launch template name**—chapter18
    - **Template version description**—version 1
    - **Auto Scaling guidance**—Check the box, as shown in the following screenshot:

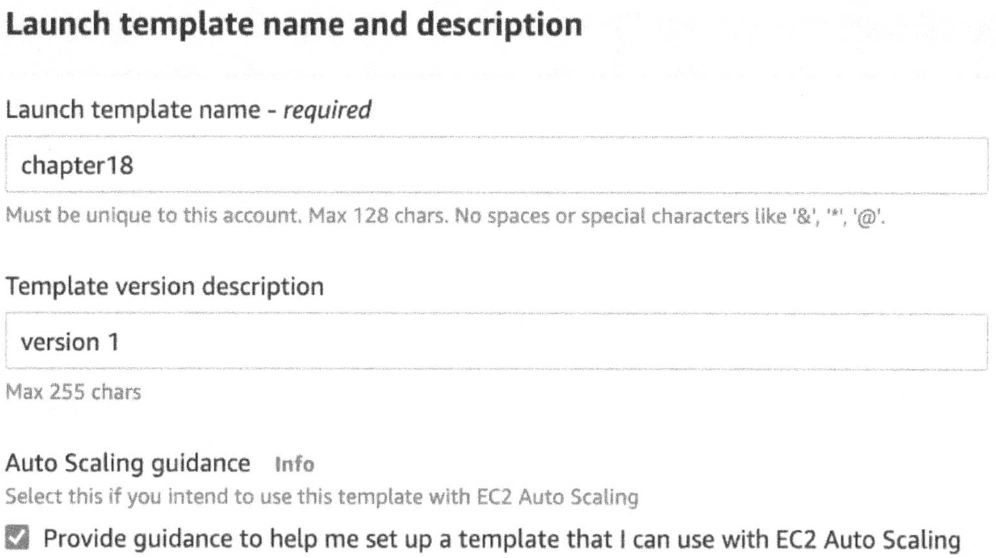

Launch template name and description

Launch template name - *required*

chapter18

Must be unique to this account. Max 128 chars. No spaces or special characters like '&', '*', '@'.

Template version description

version 1

Max 255 chars

Auto Scaling guidance   Info
Select this if you intend to use this template with EC2 Auto Scaling

☑ Provide guidance to help me set up a template that I can use with EC2 Auto Scaling

Figure 18.4 – The first section of the launch template completed

4. Next, scroll down the page for the next box of selection criteria. We will fill out the rest of the fields using the following values:

    - **Amazon machine image**—Select **Amazon Linux 2 AMI (HVM), SSD volume (x86)**.
    - **Instance type**—t2.micro.
    - **Key pair (login)**—Don't include this in the launch template.
    - **Network settings**—**VPC**.
    - **Go to Network settings | Security groups** and choose your default security group.

5. Once you have completed filling out all of the values, we can then click the orange **Create launch template** button. You should see that you have successfully created a launch template at this time.

6. Use the link in the middle of the page under the **Create an Auto Scaling group from your template** heading to be brought to the **Auto Scaling** section where we can create our new ASG, as illustrated in the following screenshot:

Create an Auto Scaling group from your template

Amazon EC2 Auto Scaling helps you maintain application availability and allows you to scale your Amazon EC2 capacity up or down automatically according to conditions you define. You can use Auto Scaling to help ensure that you are running your desired number of Amazon EC2 instances during demand spikes to maintain performance and decrease capacity during lulls to reduce costs.

Create Auto Scaling group

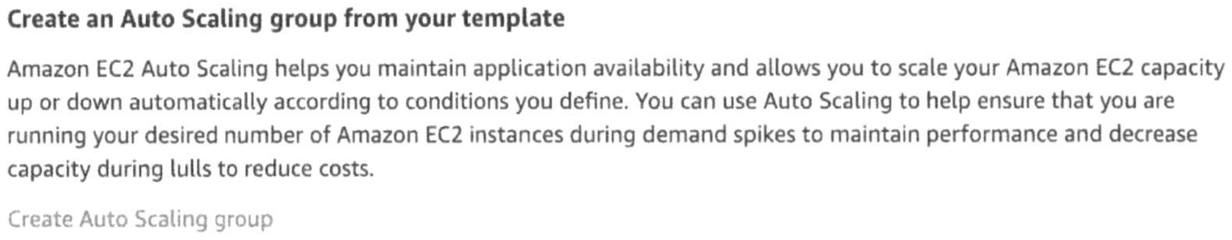

Figure 18.5 – Create an Auto Scaling group section once you have completed your launch template

7. You should now be at the **Create Auto Scaling groups** page. If not, you can find this page by navigating back to the EC2 service page and then using the left-hand menu to choose **Auto Scaling groups**. Once on the **Auto Scaling groups** page, you can choose to create an ASG. We will name our group `eighteen`.

8. In the next box, labeled **Launch template**, choose the launch template that we just created (named `chapter18`) from the drop-down list. Click the orange **Next** button at the bottom of the screen.

9. On the **Configure settings** screen, just leave all the default options, including the default VPC. We will not be using any spot instances or extra VPCs in this exercise. In the **Network** box, add all three subnets in your default VPC by selecting one after the other until all three appear underneath the selection dropdown. After all the subnets have been added, click on the orange **Next** button at the bottom of the page, as illustrated in the following screenshot:

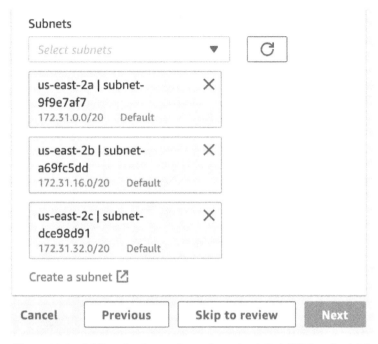

Figure 18.6 – Adding the three subnets from the default VPC to the ASG

10. Now, at the **Configure advanced options** page, under **Load balancer**, leave the radio button selected at **No load balancer**. We do want to make a change under the box labeled **Health checks**. Here, we want to change the health-check grace period from 300 to 30 seconds. Once you have made this change, click the orange **Next** button.

11. On the **Configure group size and scaling policies** page, in the **Group size** box, set **Desired capacity** to 1, **Minimum capacity** to 1, and **Maximum capacity** to 3.

12. Scroll down to the **Scaling policies** box. Choose the radio button for **Target tracking scaling policy**. Use the following values to fill out the boxes:

- **Scaling policy name**—18 ASG Policy
- **Metric type**—Average network in (bytes)
- **Target value**—5000

The following screenshot illustrates this process:

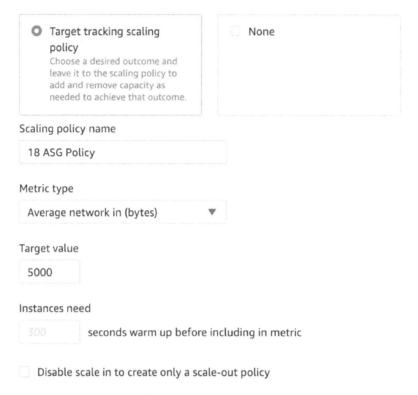

Figure 18.7 – The scaling policy with metric for our ASG

13. Once you have filled out the values for the scaling policies, scroll down to the bottom of the page and click the white button labeled **Skip to review**.

14. On the **Review** page, scroll down to the bottom of the page and click the orange **Create Auto Scaling group** button to create our ASG.

15. Once you click the button, you should be taken to the **EC2 | Auto Scaling groups** page where you will see the status of your group is initially **Updating capacity** as the first instance is being brought online.

16. If you want to see your instance scale out, then you can do so by taking the following steps. In the **Auto Scaling groups** section, click on your group name, which is `eighteen`.

17. In the horizontal menu bar, click on the **Automatic scaling** menu item, as illustrated in the following screenshot. This will bring you to the dynamic scaling policy that we created earlier. Click on the box in the top right-hand corner of our eighteen ASG policy. Once this is selected, then using the **Actions** drop-down, choose **Edit**:

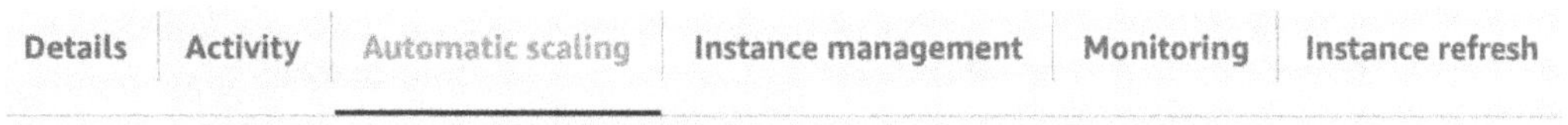

| Details | Activity | Automatic scaling | Instance management | Monitoring | Instance refresh |

Figure 18.8 – The horizontal menu bar on a single ASG highlighting Automatic scaling

18. On the **Edit dynamic scaling policy** page, change the **Target Value** field from 5000 to 100. Once you have changed this value, then click the orange **Update** button.

19. Go back to the main **Auto Scaling groups** page and refresh as you see the instances come online to meet the *demand* of the incoming traffic.

Now that we have seen how to create both a launch configuration and an ASG in the real world, let's move on to gaining a fuller understanding of the Auto Scaling lifecycle and how it differs from just launching an EC2 instance.

The Auto Scaling lifecycle

When you put an **EC2 instance** into an ASG, it follows a particular path that a normal EC2 instance you initiate via the command line or the AWS Management Console does not follow. The instance starts by being launched by the ASG. If this is part of a **scale-out** event, then the instance has an opportunity to have special commands be performed on it via a **lifecycle hook**. Lifecycle hooks allow you to add custom actions when either launching or terminating instances that are part of an ASG. Once the instance becomes healthy, it is then **InService** and is part of the ASG. If that instance fails the set number of health checks, the instance can then go to a **Terminating** state. Moving to a Terminating state can also happen if there is not enough traffic or metric data to support having the number of instances currently running, in which case there could be a **scale-in** event. Once again, just as with a scale-out event, this scale-in event allows us to use a lifecycle hook.

The Auto Scaling lifecycle process is illustrated in the following diagram:

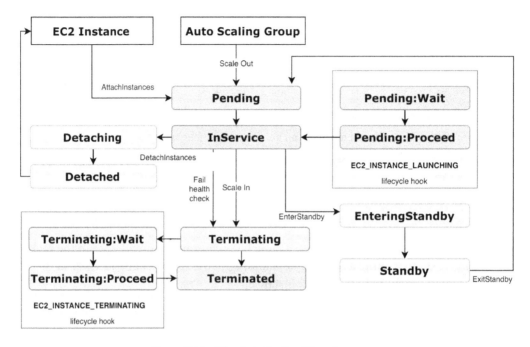

Figure 18.9 – The Auto Scaling lifecycle process

Now that we have an understanding of the lifecycle of ASGs, let's take a deeper look at lifecycle hooks and when it would be most appropriate to use them.

Using Auto Scaling lifecycle hooks

As we just looked at the Auto Scaling lifecycle, there are two states when an instance enters these states that allow for extra actions to occur. These two states are the **Pending** state and the **Terminating** state. When a lifecycle hook is added to an ASG, there is a configurable amount of time set that an instance can wait before moving to the next state. In the case of the `Terminating:Wait` state, you can have the instance pause for up to 30 minutes before moving on to the `Terminating:Proceed` state and then on to `Terminated`.

Use cases for lifecycle hooks

You may want to know some of the use cases for lifecycle hooks. Let's take a quick look at a few.

The first one is using the launching state to invoke a Lambda function. Once our instance has passed the pending state and gone on to the `Pending:Wait` state, we can use this event to call a specific Lambda function for our application. A good use for this would be if we had Windows instances and we needed to join each instance as it came up to a specific **Active Directory** (**AD**) domain and name server. As the instance entered in the initial lifecycle hook, it could run a script to get its **Domain Name System** (**DNS**) name as well as join the domain.

The same is true when we start the termination process of an instance. When you are terminating an instance, there are some cases where to want to be sure to capture all of the data from that set of applications before allowing the instances to proceed to the termination state.

As a word of caution, using complex scripts in lifecycle hooks can cause an extra amount of time between the time we have either requested a new instance to come online and join our ASG or be terminated and allow us to have more elasticity when we need it. Instances in the wait states still take up an instance worth of capacity when calculated against the total maximum size capacity of the ASG.

Cleaning up resources

As with many of our other exercises, it is suggested that you terminate any running instances and clean up the ASG once you are done with this chapter so that you don't incur any unwanted charges on your AWS account.

Summary

In this chapter, we looked at Auto Scaling and how it can automatically manage the demand that is presented to us by both internal and external customers. We took a look at the different ways in which ASGs could be configured using different scaling policies, and even went through the hands-on exercise of deploying an ASG. Lastly, we looked at the Auto Scaling lifecycle and how lifecycle hooks can help perform more complex tasks than just simple scaling.

In the next chapter, we will be starting the first of a few chapters focused on securing your environment and your pipeline by talking about protecting your data both in transit and at rest. This will especially cover the **Key Management Service** (**KMS**) and using **Amazon Certificate Manager** (**ACM**) for server-side certificates.

Review questions

1. You have been brought into a company that is running a business-critical workload using ELB and Auto Scaling. This workload is a two-tier application that consists of an application tier and a database tier. Both tiers are currently deployed across two AZs in the us-east-1 region. The database needs to be replicated in a synchronous fashion from the application. The chief technical officer (CTO) has told you that the application must stay fully available even if a single AZ becomes unavailable and Auto Scaling cannot launch new instances in the remaining AZ. How can you establish this through AWS-specific services and architecture?

   A. Set the configuration in ELB to deploy in three AZs with Auto Scaling set to deploy to handle 33 percent of the load per zone at peak.

   B. Set the configuration in ELB to deploy in three AZs with Auto Scaling set to deploy to handle 50 percent of the load per zone at peak.

   C. Set the configuration in ELB to deploy in two regions using the Round Robin algorithm with Auto Scaling set to deploy to handle 50 percent of the load per zone at peak.

   D. Set the configuration in ELB to deploy in two regions using the Round Robin algorithm with Auto Scaling set to deploy to handle 100 percent of the load per zone at peak.

2. You have created a DevOps pipeline for an application that is running on EC2 instances. Customers interact with this application via an application load balancer that is attached to an ASG. Since releasing the latest version of both the application and the launch template, the application seems to be scaling up and down multiple times each hour of the day. Which fixes should you and your team make in order to stabilize the ASG, preserve elasticity, and optimize costs? (Choose two answers)

   A. Modify the application's ASG so that it uses scheduled scaling actions.

   B. Modify the application's ASG termination policy to terminate the oldest instance first.

   C. Modify the application's ASG cooldown timers so that they are bigger.

   D. Modify the CloudWatch alarm associated with the application's ASG group so that there is a longer alarm period associated with the scale-down policy.

3. The company you work for has an application that consists of EC2 instances that are launched using an ASG. It has come to your attention that the EC2 instances are not scaling up as more demand is needed. How should you check and remedy the situation?

   A. Check to make sure that the ASG is placing instances across multiple regions.

   B. Check to make sure that the ASG is placing instances across multiple AZs.

   C. Check to make sure that the ELB health checks are being utilized.

   D. Check to make sure that the correct metrics are being measured to trigger the scale-out event.

Review answers

1. b
2. c, d
3. d

19
Protecting Data in Flight and at Rest

As you and your developers start to connect to your systems, security is not always front of mind. This is especially the case if you think that encryption keys and certificate handshakes could cause latency. Incorporating encryption both for data in transit and at rest is a must, especially in today's environments.

There are multiple ways to incorporate encryption into your environment. It can start with how to secure the transmissions you are sending back and forth between you and AWS. It then progresses to securing data that you are storing on the Amazon cloud and then moves on to data that you are serving for your customers to access.

In this chapter, we're going to cover the following main topics:

- Understanding KMS keys
- Adding encryption to storage
- Adding encryption to data stores
- Protecting data in transit with AWS Certificate Manager
- Adding a certificate to Amazon CloudFront

Data encryption introduction

There are many reasons why there is a bigger push to encrypt data these days. The requirement could stem from a regulation that your company needs to follow for the industry they are in. It may be that there is internal compliance governance that states that data needs to be encrypted. And it just may come down to taking a proactive stance on security by adding an additional layer of protection to your environment. No matter what the driving reason is, the goal, in the end, is to have a more secure platform for your customers to use and better data protection:

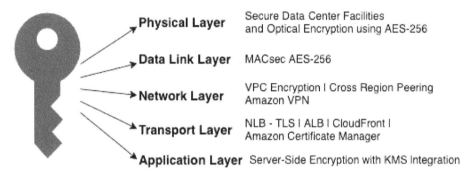

Figure 19.1 – The AWS cryptography stack

In the end, what you and your organization's end goal should be with the use of encryption can be simplified by minimizing the unauthorized physical and logical access to data.

There are three categories to be aware of when looking at data and encryption:

- Data in transit
- Data at rest
- Data in use

If you go by the recommendation from Amazon, then you should encrypt as much as possible in your environment. This includes encrypting data at rest when it's stored and encrypting data in transit, as it's being shipped back and forth between client and server.

Options for encrypting data at rest in AWS

Throughout this chapter, we will be concentrating on two services from AWS, Key Management Service and Certificate Manager. There are other services in the Amazon ecosystem that play a part in the data encryption process, and we wanted to give them a brief mention just in case they happened to appear on the test.

The first is part of Amazon Virtual Private Cloud, and the second is **AWS Virtual Private Network (AWS VPN)**. Using an AWS site-to-site VPN connection to connect your data center to your VPC allows your transmissions of data to happen securely over the IPSec protocol. This is a group of protocols for encrypting internet transmissions.

You get started by setting up a customer gateway, which usually consists of a router, and then create the VPN gateway on your VPC. VPNs have a throughput limit of 1.25 gigabits per second.

Now that we understand the options for encrypting data at rest in our AWS accounts, let's concentrate on using AWS **Key Management Service (KMS)** to protect our data at rest.

Understanding KMS keys

KMS is a managed service from Amazon that makes it easy to produce and manage **Customer-Managed Keys (CMKs)**. CMKs are the encryption keys used to secure and control access data stored on AWS. KMS integrates seamlessly with many other services, especially the IAM service, so that you can control who is allowed access to the keys.

AWS creates the key material for the KMS key. You, as the customer, cannot extract, export, view, or manage this key material. You can delete the key but cannot delete the key material itself:

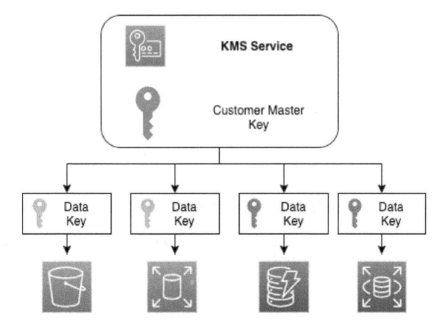

Figure 19.2 – Master and data keys in AWS KMS

One of the key concepts to understand in KMS is **envelope encryption**. When AWS encrypts your data, your data is secured, but your key needs to be protected as well. AWS does this by encrypting your data with a data key and then encrypting the data key with another key. That top-level plaintext key is known as the master key.

Using the envelope encryption process provides a number of benefits:

- **It protects the data keys** – This way, you can store the encrypted data key beside the encrypted data.

- **You can encrypt the same data using multiple keys** – KMS allows you to only re-encrypt the data keys that protect the raw data instead of all raw data that the key encrypts.

- **You can combine the strengths of multiple algorithms** – Symmetric key algorithms are smaller than public-key algorithms generally. However, public-key algorithms allow for inherited separation of roles. Using envelope encryption allows you to use both strategies at the same time.

Using AWS-managed CMKs keys to encrypt storage

Each AWS service that offers encryption has its own AWS-managed KMS key.

One of the key differences between AWS-managed CMKs and user-created CMKs is the rotation schedule. With AWS-managed keys, key rotation is required, and it happens every 1,095 days or every 3 years.

The similarity between the two different types of keys is the fact that both AWS-managed keys and CMKs are used only for your AWS account.

Using KMS with S3 to protect objects stored at rest

KMS integrates seamlessly with S3 object storage. S3 has its own service-managed key that allows you to encrypt objects in your buckets with almost little to no effort on your part. There are also the following security controls available to help incorporate KMS with S3:

- Amazon S3 has the ability to encrypt all new objects placed into a bucket automatically. This is done through a bucket setting in the **Properties** section labeled **default encryption**. Once you have turned this setting on, then any new object that is uploaded into the bucket is automatically encrypted with either the Amazon-managed S3 key (by default) or the specified customer-managed S3 key.

- If you need to retroactively encrypt items in an S3 bucket, you can do so using the S3 batch process. Amazon S3 has its own ability to perform batch processing on objects within the buckets. If your internal guidance has changed, or an audit has found that a user (either internal or external) has been adding items to a bucket without encryption, then you can use S3 Batch Operations to remediate the situation.

- If you want to check which items are or are not encrypted in an S3 bucket, then you can run an S3 inventory report.

Now that we have seen how KMS works smoothly with Amazon S3, especially in the case of Amazon-managed keys, we will now look at how to create and manage our own key, and the important differences between a customer-managed key and an Amazon-managed key.

Creating and managing a customer-managed key in KMS

There may be cases where certain users or teams may need to have access to data and others should not have the right to access it. This is one of the ideal use cases for integrating a CMK along with an IAM policy for who can access the key.

As we will see in this exercise, there is a separation of duties between key administrators and key users. Key administrators are the users who control the key access, including the ability to rotate and delete the key. There are also key users. These are the users and service roles that have the ability to use the key on a day-to-day basis. These are important concepts to understand as you both create and manage the keys in your account. There may be individuals and groups who need the ability to manage the keys but do not need access to data that is protected by the keys.

In the following exercise, we are going to create our own customer-managed key using KMS. We will also designate both key administrators and key users before we are finished creating the key:

1. Log on to the AWS Management Console and search for `Key Management Service` from the top search menu bar:

Figure 19.3 – The Key Management Service icon from the search menu

2. Once at KMS, we can now click on the orange **Create a key** button.

3. Now on the **Configure Key** page, choose the **Symmetric** key type and click on the orange **Next** button at the bottom of the page.

4. This should bring us to the **Alias** page. In the **Alias** box, use `chapter19` as the KMS key alias. We can use `example key` as the **Description**. Scroll down to the bottom of the page and click on the orange **Next** button:

Figure 19.4 – Adding the KMS key alias and description

5. Now it's time to choose key administrators. This is an important step that determines who can control the key access and who can delete the key, but not the people and groups who can use the key. Since you have most likely logged in with your `devops` user that we created at the beginning of our exercises, choose this user, and if you have another administrator account that you either use or would like to have access to the key, check the box next to that name as well. Once you have selected the names for the administrators, click the orange **Next** button at the bottom of the screen.

6. We should now be on the **Define key usage permissions** page. And although this looks extremely similar to the previous page we were on, this is the page where we actually set which users and roles can use and access the key. This time, besides our `devops` user, let's add the developer user that we created previously, `mariel`, to the list of authorized users as well. Once you have made your selections, click the orange **Next** button at the bottom of the screen.

7. On the **Review** page, scroll down and click the orange **Finish** button to create the key.

Now that we have gone through the process of creating our CMK, let's see how we can use this key to keep our data secure via encryption.

Adding encryption to data stores using our custom KMS key

Let's now take a look at how we can use the CMK we just created to encrypt objects in an S3 bucket that we have created previously. For the examples shown, we will be using the bucket that we created back in *Chapter 4, Amazon S3 Blob Storage*:

1. Log on to the AWS Management Console and navigate to **S3 Service**. Find the original bucket that you created back in *Chapter 4, Amazon S3 Blob Storage*. In our case, it was a bucket named `devopspro-beyond`. Once you find it, click on the name of the bucket to enter it.

2. Once inside the bucket, click on the top horizontal menu on the tab named **Properties**:

Figure 19.5 – The Properties menu item in the S3 horizontal menu

3. Now displaying the properties items of the S3 bucket, scroll down until you find the **Default encryption** heading. Click on the white **Edit** button on the top right corner of the **Default encryption** box to enter the settings.

4. You should now be on the page labeled **Edit default encryption**. Under the label of **Server-side encryption**, click on the radio button next to **Enable**. Once you have selected this radio button, another set of selections will appear below for **Encryption key type**. Choose the radio button next to the **AWS Key Management Service key (SSE-KMS)** option:

Server-side encryption

◯ Disable

⦿ Enable

Encryption key type
To upload an object with a customer-provided encryption key (SSE-C), use the AWS CLI, AWS SDK, or Amazon S3 REST API.

◯ Amazon S3 key (SSE-S3)
An encryption key that Amazon S3 creates, manages, and uses for you. Learn more [↗]

⦿ AWS Key Management Service key (SSE-KMS)
An encryption key protected by AWS Key Management Service (AWS KMS). Learn more [↗]

Figure 19.6 – Enabling the KMS key on an S3 bucket

5. Once the **AWS Key Management Service key (SSE-KMS)** option is selected, another set of options will appear below on the **Edit default encryption** page. Under the **AWS KMS key** heading, select the radio button next to the **Choose from your AWS KMS keys** label. This will unveil a dropdown where we can select the key named `chapter19` that we created in the previous hands-on exercise. Once selected, you can then scroll down to the bottom of the page and click the orange **Save changes** button.

6. With our bucket set up for object encryption, click on the **Objects** menu item on the horizontal menu to be brought back to the main bucket page. Click on the orange **Upload** button, and on the **Upload** screen, click on the white **Add files** button. Choose any random file to upload to your bucket. Once your file has been queued to upload, click the orange **Upload** button.

7. When your file completely uploads to the S3 bucket, you can click on the orange **Close** button. This will bring you back to the **Objects** page for the S3 bucket. Click on the name of the object and then scroll down to **Server-side encryption settings** to see that your object has been encrypted with your KMS key.

> Important Note
>
> If you are done with your KMS key, then you might want to go back and schedule its deletion. KMS keys do not get deleted right away and have a minimum of seven days before they can be deleted. Encrypted objects whose key has been deleted will not be able to be accessed.

With an understanding of how to protect our objects at rest using KMS under our belt, we will next move on to protecting data in transit using AWS Certificate Manager.

Protecting data in transit with AWS Certificate Manager

When your external site presents a **Transport Layer Security** (**TLS**) certificate and uses the secure HTTP protocol (HTTPS/443), customers know that you are protecting data they are sending to and from your systems in an encrypted manner:

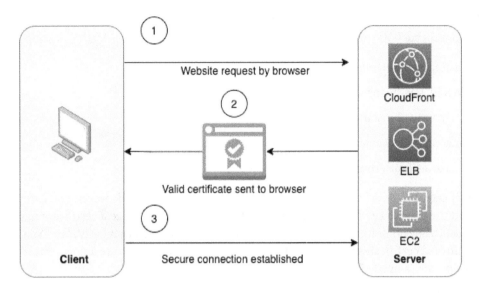

Figure 19.7 – How SSL/TLS works from browser to client

When you or your customers request a website that is presenting as HTTPS secured by an SSL/TLS certificate, the following steps take place:

1. The server attempts to connect to the website over the secure 443 port. That web server then identifies itself.

2. The server then sends a copy of its SSL certificate.

3. The client will then check the certificate to see whether it's been issued from a trusted authority. If it trusts the certificate, it then sends a message of acknowledgment back to the server.

4. The server then issues a digitally signed acknowledgment that will start the SSL session.

5. Data is then shared securely between the client and the server.

AWS Certificate Manager (ACM) is a server that can help you quickly and easily create, manage, and deploy TLS certificates on AWS services. ACM also supports public certificates. ACM handles the hardest and most complicated parts of **Private Key Infrastructure (PKI)**. This includes tasks such as creating the certificate, certificate renewal, and issuing the certificate.

ACM can quickly and easily provision the TLS certificates for use with external-facing AWS resources such as the following:

- Elastic Load Balancing

- AWS Elastic Beanstalk

- Amazon EC2 instances

- Amazon API Gateway

- Amazon CloudFront

Any certificate that ACM creates is valid for 13 months (395 days). However, you cannot download the private key for an ACM certificate that you create. Once a certificate has been issued, you can neither add nor remove domain names from that certificate. You can create wildcard certificates (`*.domain.com`) or you will need to create a new certificate if a new domain is needed.

When we talk about domains in the following paragraph, we are referring to domains registered with a registrar and being hosted on a DNS server or service such as Route 53. This many times can be thought of as the domain name. When first issuing a requested Amazon certificate for a domain, ACM must verify that you either own or control the domain for which you are requesting the certificate. This can be done in two ways. The first is via email verification. ACM will send out email messages to the three email addresses listed on the WHOIS database and you have up to 72 hours to validate by acting on that email. The second way for validation is via DNS validation. ACM creates two CNAME records for your domain and requests that you put those entries into your DNS database for validation. Once it sees those entries, the service knows that you are the owner of the domain name.

The two functions that ACM can serve

We have explained the role of secure certificates mainly as they pertain to public-facing websites. Issuing these certificates is one of the critical roles that ACM plays. However, this is not the only role that it can serve in your AWS environment. ACM can provide two separate services. The first is to manage, provision, and renew enterprise-grade TLS certificates for use with services such as Elastic Load Balancing, API Gateway, and CloudFront, along with other front-facing services. The second service that ACM provides is being a private certificate authority. This allows you to issue secure and trusted private certificates for your applications and infrastructure for your workloads and organization.

Now that we know what functions ACM can perform, we will go through a hands-on exercise of creating and provisioning a certificate using ACM and then using that provisioned certificate in a publicly available CloudFront distribution.

Adding a certificate to Amazon CloudFront

With a solid understanding of the ACM service, we will now move on to the final hands-on exercise of this chapter. We will start off by creating a TLS certificate using ACM. Then, we will create a very simple CloudFront distribution and secure it for anyone who tries to connect to the distribution using our secure certificate.

If you do not have a domain name for use, you can easily register one using the Amazon Route 53 service to complete this exercise at a very nominal cost. Otherwise, you can read through the steps of the following exercises to understand the process:

1. Log on to the AWS Management Console and search for the ACM service. Once it appears from the drop-down menu, click on the icon to be taken to the main service page. For the certificate to work with the CloudFront service, we need to create it in the us-east-1 region. If you create the certificate in any other region, you will not be able to see it in the second part of this exercise.

2. On the ACM main page, find the icon and heading that is labeled **Provision certificates**. Click on the blue **Get started** button under this heading.

3. On the **Request a certificate** page, since we do not have a Private Certificate Authority available, there is only one option available to us. Keep this option selected and press the blue **Request a certificate** button at the bottom of the page.

4. The first step when creating a certificate is to add domain names to the certificate. Unless you only want the certificate to be used for a single subdomain (for example, www), you can use an asterisk to serve as a wildcard for your entire domain. Type your wildcard domain in the textbox in the form of *.devopsandbeyond.com and then click on the blue **Next** button:

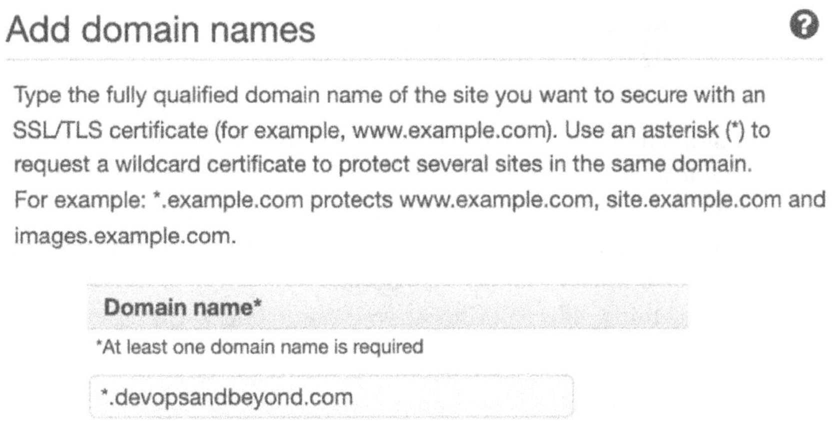

Add domain names   ❓

Type the fully qualified domain name of the site you want to secure with an SSL/TLS certificate (for example, www.example.com). Use an asterisk (*) to request a wildcard certificate to protect several sites in the same domain. For example: *.example.com protects www.example.com, site.example.com and images.example.com.

Domain name*
*At least one domain name is required

*.devopsandbeyond.com

Figure 19.8 – Adding the wildcard domain name to our ACM-created certificate

5. The next step will be validating that you actually own the domain. We are going to assume that you are managing your DNS via Route53 and will go for **DNS validation**. Keep the **DNS validation** option selected and then click the blue **Next** button.

6. This will bring you to the **Add tags** page. Simply click the blue **Review** button at the bottom of the page. When you are brought to the **Review** page, after verifying that your domain was typed in correctly, press the blue **Confirm and request** button.

7. You will then need to add the requested CNAME to your DNS file. Take the value and go to the Route53 service. Add the entry as a new CNAME record in your domain's Route53 hosted zone. You will most likely need to wait a few minutes for the values to propagate before you see **Validation status** turn green and say **Success**. Once this happens, you can click the blue **Continue** button at the bottom of the screen:

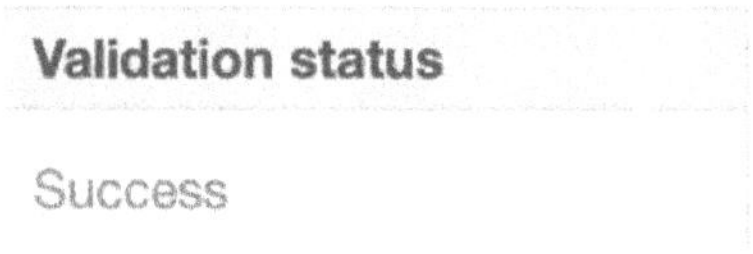

Validation status

Success

Figure 19.9 – Validation status showing Success after the DNS entry has been added

With our certificate now created, we can now move on to creating a CloudFront distribution. Amazon CloudFront is the AWS native **Content Distribution Network (CDN)**. CloudFront allows you to serve content using a single origin to many customers all over the world simultaneously with much lower latency than if you were trying to serve the content from a single point of presence, such as a server or Auto Scaling group.

8. We will now need to navigate to the CloudFront service from the top search bar on the AWS Management Console. Once you arrive on the service main page, click on the orange **Create a CloudFront distribution** button to start creating a new CloudFront distribution.

9. The page heading should now be titled **Create distribution**. The first option under the **Origin** section allows you to choose an origin – this is where your content will reside. We will use the S3 bucket that we added our KMS key to in the earlier hands-on lesson in the chapter as our origin. The example bucket we have been using throughout this book has been named `devopspro-beyond`; however, your bucket will be named something different. Choose your origin bucket from the drop-down list. The origin will be suffixed by `.s3.us-east-2.amazonaws.com`.

10. In the next textbox, **Origin path**, we will add a specific directory where we will upload the content for our origin. Enter `/pages` in the **Origin path** textbox.

11. Allow the name created by CloudFront to remain and move on to the **S3 bucket access** section. Choose the **Yes Use OAI** option. This option will allow us to keep our bucket private and make users go through the CloudFront distribution, rather than trying to bypass the CDN and access the assets in the origin directly. Click on the white **Create a new OAI** button; you can keep the name that CloudFront provides. Then click the orange **Create** button.

12. When the dialog window closes, under the **Bucket policy** heading, select the **Yes, update the bucket policy** option.

13. Scroll down until you get to the **Viewer** option and ensure that the Redirect HTTP to HTTPS value is selected.

14. We can keep other options in place until we get down to the **Settings** heading. The first thing we want to do is optimize on costs, so under **Price class**, we want to select **Use only North America and Europe**.

15. Now, we are about to install our custom certificate from ACM. Under the **Custom SSL certificate** – *optional* label, use the drop-down menu option to find the certificate you created in the first part of the exercise under the **ACM certificates** heading.

16. With your certificate selected, scroll down to the bottom of the page and click on the orange **Create distribution** button.

17. While the CloudFront distribution is being created, make a note of the **Distribution domain name**; this is how you will access CloudFront without adding a Route53 alias:

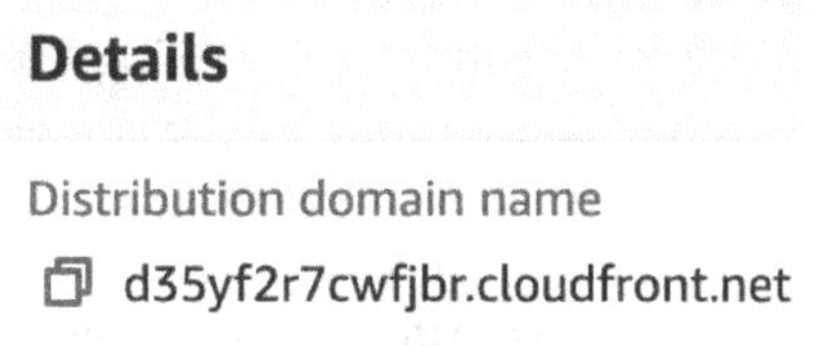

Details

Distribution domain name

☐ d35yf2r7cwfjbr.cloudfront.net

Figure 19.10 – The CloudFront Distribution domain name

While the distribution is deploying, if you would like to test out the certificate, we need to go and perform a few more steps at the origin. Let's go back to S3 and get our origin ready to serve content. One of the first things we need is a simple webpage to serve. Take the following code and add it to a file named `index.html`; we will upload it to our S3 bucket in just a moment:

```
<html>
<head> <title> Chapter 19 - Protecting Data in Flight and
at Rest </title> </head>
```

```
<p> Protecting Transmission of data - <b> VPN </b> </p>
<p> Protecting Data at Rest - <b> KMS </b> </p>
<p> Protecting Data in Flight - <b> SSL / TLS
certificates </b> </p>
</html>
```

18. Navigate back to S3. Once there, find the bucket that you designated as the origin for your CloudFront distribution. Click on the name of this bucket to enter it.

19. Once inside the bucket, we will need to click on the white **Create folder** button in the upper right of the **Objects** screen. When the dialog box comes up, name the folder `pages`, just as we did in the options for the CloudFront distribution. Click the orange **Create folder** button once you have named the folder.

20. Click on the newly created pages directory and then click the orange **Upload** button. Click the white **Add files** button and then add the `index.html` file that we created at the beginning of this section. Once the index page has been added for upload, click the orange **Upload** button.

We have just seen how to create and implement a secure certificate using ACM. Let's now review what we learned in this chapter.

Summary

In this chapter, we looked at protecting your data both at rest and in transit in your Amazon account. We began truly at an account level, looking at how to implement encryption during the transmission of data to your VPC using secure VPN connections over an IPSec tunnel. After that, we took a look at KMS and the differences between an Amazon-managed key and a CMK. Finally, we looked at securing data in transit using ACM. We saw how easy ACM makes it to create and implement SSL and TLS certificates on several AWS services.

In the next chapter, we will look at how to enforce standards and compliance within your organization with two powerful automation tools offered by AWS: Systems Manager and Config.

Review questions

1. A new set of compliance regulations has come down from the security department in the company you are working for. They are now stating that all encryption keys must be rotated every 12 months without exceptions. Which one of the following options would not meet the proposed new guidelines?

    a. Using imported key material with a CMK

    b. Using an AWS-managed key

    c. Using an AWS customer-managed symmetric CMK

    d. Using an AWS customer-managed asymmetric CMK

2. You have been brought into a company that is working with confidential data. However, they are transferring the data unencrypted, both on the Amazon administrative console and using the CLI. What steps can you take to immediately secure the transport of data using encryption?

    a. Create a set of CMKs in KMS. Using envelope encryption, have each user encrypt each transaction before any CLI command.

    b. Use ACM to create a certificate to create a secure login and encrypt transmissions to the Amazon administrative console.

    c. Create a customer gateway and a VPN gateway on the primary VPC. Ensure that any other VPCs that need to be accessed by the development team are either peered or connected via a transit VPC connector.

    d. Have all users add a multi-factor authentication to their account to ensure secure communications via both the CLI and the AWS Management Console.

Review answers

1. b
2. c

20

Enforcing Standards and Compliance with System Manger's Role and AWS Config

Systems Manager is a service comprising multiple individual capabilities grouped into five categories: Operations Management, Application Management, Change Management, Node Management, and Shared Resources. Learning to use this tool effectively can make you, as a DevOps professional, much more efficient, especially when implementing standards and checking for compliance in one or multiple AWS environments or accounts. And as automation and compliance demands increase from organizations, having a tool that can check for compliance, alert against discrepancies, and remediate violations can be essential in preventing environmental drift.

In this chapter, we're going to cover the following main topics:

- The various capabilities of AWS Systems Manager
- Using runbooks in Systems Manager
- AWS Config essentials

The various capabilities of AWS Systems Manager

DevOps is a marriage of two responsibilities: development and operations. AWS **Systems Manager** (**SSM**) focuses on the operations part of those responsibilities, giving you a vast array of tools to use for everyday operations tasks. These tools range from creating predefined runbooks to quickly, easily, and repetitively performing functions on your instances, whether they're Linux or Windows instances. Systems Manager can also provide you with an interface to track your resources or groups of resources from a common place.

As an added benefit, Systems Manager not only helps with your instances in the AWS Cloud, but it can also help manage your servers on-premises by installing the lightweight agent on those servers and allowing that agent to communicate back to the AWS account.

Key features and benefits of Systems Manager

Systems Manager is not just one service; instead, it is a whole set of tools for you to use. You may not use all of its capabilities, but knowing them can help when you are looking to solve a problem quickly:

Figure 20.1 – An overview of many of the critical components that make up Systems Manager

Now that we have looked at some of the capabilities that Systems Manager provides, let's dive deeper into managing our instances and nodes more effectively with it.

Node management with Systems Manager

Systems Manager provides many capabilities to operations team members, but there is a whole category of items that specifically fall into the category of **Node management**.

We will look at some of the specialty services in AWS Config that have been created for the express purpose of node management.

Fleet Manager

Knowing all the different instances that you are controlling at once can be an overwhelming task. SSM Fleet Manager provides you with a user interface to see all the instances in which you are in control. Fleet Manager also allows you to gather data from those instances to relay back without ever having to go into the servers or instances individually.

Inventory

If you need visibility into either your on-premises servers, your Cloud EC2 instances, or both, then the **Inventory** capability of Systems Manager can be extremely useful for you.

Hybrid activations

If you have servers that are not on the AWS cloud or on-premises, you can use SSM **Hybrid Activations** to manage both those devices, along with other EC2 instances that need to be managed in a single management console. The hybrid activations capability helps walk you through the prerequisites required to create an activation on an on-premises server. This includes creating an IAM service role, ensuring that the operating system is compatible, and installing the Systems Manager agent.

Session manager

Managing all the different PEM keys, for instance, especially when developers create their own keys, can be a burdensome task. **Session Manager** helps ease this burden by allowing authorized users to use a web-based console to perform commands without the need to have a copy of the key. Instead, you can find the name of the machine that you need to create a session remotely and then instantly start a session.

We will go through an exercise that covers using the Session Manager capability later in this chapter.

Run command

With the **run command** capability of Systems Manager, you can securely update the configuration of your registered managed instances. **Command documents** are created for the run command to execute. Commands are then run for specified targets of managed instances using either instance IDs or tags.

> **Tip**
> You should always test the run commands that you create on non-production or test instances. This way, you can make sure that the command is going to perform the way you expect before enacting the commands or servers taking on production traffic.

State manager

Systems manager's **State Manager** allows you to perform the following actions:

- You can configure network settings.
- You can have your instances join a Windows domain.
- It allows you to install software at instance startup using bootstrap scripts.
- It lets you run scripts on Windows-, Linux-, or even macOS-managed instances throughout their life cycle.

Patch manager

If you are not running immutable infrastructure in your environment, then you should have a strategy to keep your instances up to date system patches. Patch manager allows you to add these system patches on a maintenance schedule.

Running remote commands on an EC2 instance

Previously, if you wanted to run a remote command on an EC2 instance, then you had to create a PEM-encoded key. SSM Session Manager changes all of this as it allows you to log in to an instance using a web browser using a role that has the correct permissions instead of managing and rotating keys.

> **Note**
> **PEM** stands for **Privacy-Enhanced Mail**, but this is just extra knowledge, and knowing the definition of PEM will not be a requirement for the exam.

We'll see how this works in the following exercise. Before we can go about using Systems Manager to remotely access an instance, we need to set up an IAM role that will allow the SSM service permissions to access any instance that has that role. Luckily for us, AWS already has a managed IAM policy that contains all of the permissions that we need to accomplish this task; however, we still need to create a role that we can attach to our EC2 instance. Let's get started:

1. Open your browser and go to the Amazon web console, navigating to the IAM service: `https://console.aws.amazon.com/iam/`.

2. From the left navigation bar, choose **Roles** and then press the blue **Create Role** button in the mainframe once it appears.

3. On the **Select type of trusted entity** page, make sure that **AWS service** is selected.

Select type of trusted entity

Figure 20.2 – Selecting AWS service as the type of entity when creating a role

4. Then, from the list of common use cases, find **EC2** and click on it to select it. Once selected, you can click the blue **Next: Permissions** button located at the bottom right of the page.

Common use cases

EC2
Allows EC2 instances to call AWS services on your behalf.

Lambda
Allows Lambda functions to call AWS services on your behalf.

Figure 20.3 – Common use cases when creating a role in IAM

5. Now that you're on the **Attach permissions policies** page, in the search box in the middle of the screen, search for the policy named *AmazonEC2RoleforSSM*. Select the box next to the policy once it appears in the search results. Once selected, you can click the blue **Next: Tags** button to continue.

Select type of trusted entity

Figure 20.4 – Adding a managed permission policy to the IAM role

6. On the **Add Tags** page, just select the **Next: Review** button at the bottom of the page since we are not adding tags to this role.

7. On the **Review** page, you are going to need to name your role. I suggest using `Remote2EC2SSM`. You can change the role description if you like to something that is more fitting of what the role allows. Once you have made these updates, click the blue **Create Role** button at the bottom of the page. Great! Your role should now be created.

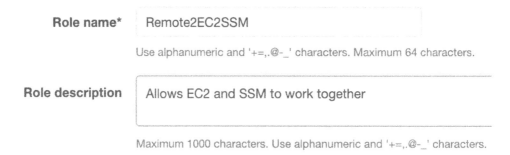

Figure 20.5 – Adding the role name and description

With our role created, we can start creating a temporary EC2 instance that we will use for testing the capabilities of SSM.

8. Find the **Services** menu item on the left-hand side of the top menu. Click on the word **services** to bring up the drop-down menu, which will show all the AWS services. Click on **EC2**, which should be near the top, under the **Compute** heading, to be taken to the **EC2** service page.

Figure 20.6 – EC2 service page

9. Once you're on the **EC2** page, find the **Launch instance** section in the middle of the page. Click on the orange **Launch instance** button inside this section to begin launching your instance. When you click on that button, you will be presented with two options; choose the option labeled **Launch instance**.

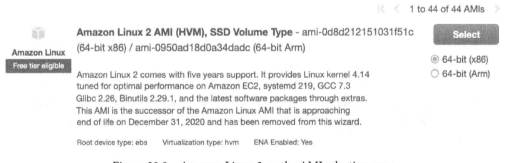

Figure 20.7 – Launching an EC2 instance from the main EC2 page

10. Now you should be on a page where you should be able to select the AMI that you want to use. Select any **Amazon Linux** or **Amazon Linux 2** version dated 2017.09 or later as it will contain the AWS SSM agent already preinstalled. Click the blue **Select** button to choose your image.

Figure 20.8 – Amazon Linux 2 on the AMI selection page

11. We won't need much computing power for our test, so we can use a **t2.small instance**. Click on the gray **Next: Configure Instance Details** button as we need to ensure that we attach our role to our instance.

12. Once you're on the **Configure Instance Details** page, find the line that contains the **IAM role** value – this is where we need to make a change. From the drop-down menu to the right, find the role that we just created, called `Remote2EC2SSM`, and choose this value for the field. Once you have done this, click on the gray **Next: Add Storage** button.

Figure 20.9 – Choosing the Remote2EC2SSM role while configuring our EC2 instance

13. There is nothing to change on the **Add Storage** page, so we will click through to the gray **Next: Tags** button. Once you're on the **Add Tags** page, click on the gray **Add Tag** button in the middle of the screen. Use `name` as the key and `Remote Test` as the value. Make sure that you capitalized the *N* in Name, otherwise, it will not set the value of the name of the instance correctly and only set a tag and value pair. Once you have set the key-value pair, you can click on the blue **Review and Launch** button at the bottom of the page.

Manage Tags   Info

A tag is a custom label that you assign to an AWS resource. You can use tags to help organize and identify your instances.

Key	Value - *optional*	
🔍 Name ✕	🔍 Remote Test ✕	Remove

Add tag

You can add 49 more tags.

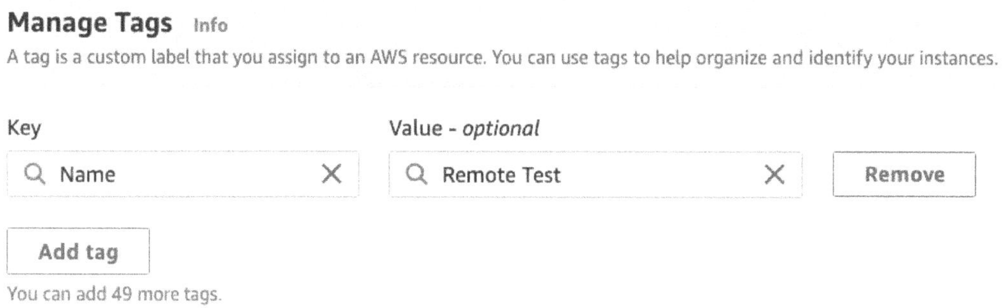

Figure 20.10 – Setting the name for the instance using tags

14. On the **Review Instance Launch** page, click the blue **Launch** button at the bottom of the page. This will bring up a dialog about which key pair you would like to use. However, we are going to clear out the values from the top drop-down menu so that the value that is selected reads **Proceed without a key pair**. Click on the box to acknowledge that you are proceeding without a key pair, and then click the blue **Launch instances** button to start your EC2 instance.

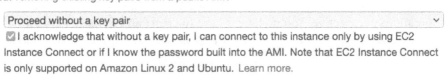

Select an existing key pair or create a new key pair   ✕

A key pair consists of a **public key** that AWS stores, and a **private key file** that you store. Together, they allow you to connect to your instance securely. For Windows AMIs, the private key file is required to obtain the password used to log into your instance. For Linux AMIs, the private key file allows you to securely SSH into your instance.

Note: The selected key pair will be added to the set of keys authorized for this instance. Learn more about removing existing key pairs from a public AMI.

Proceed without a key pair ⌄

☑ I acknowledge that without a key pair, I can connect to this instance only by using EC2 Instance Connect or if I know the password built into the AMI. Note that EC2 Instance Connect is only supported on Amazon Linux 2 and Ubuntu. Learn more.

Cancel   **Launch Instances**

Figure 20.11 – Launching our EC2 instance without a key pair

Now that our instance has started launching, we have finished all the prerequisites for this exercise. We can now move on to the Systems Manager service in the AWS Console to log in without the key and perform the command(s).

15. In the search bar at the top of the AWS Management Console, type in *Systems Manager* and click on the menu item when it appears in the list:

Figure 20.12 – The Systems Manager service as it appears from searching on the AWS Management Console

16. Once you're on the **Systems Manager** service page, find the **Node Management** heading on the left-hand menu. Under this heading, click on the menu item named **Session Manager**.

17. When the main **Sessions Manager** page appears, click on the orange **Start a session** button at the top right-hand side of the main window.

18. Now that we are on the **Start a session** page, we should see our instance, which we named `Remote Test` when we created the instance earlier. Click on the radio button next to the instance's name to select this instance and then click the orange **Start session** button to remotely access the instance without using a PEM key.

19. A new tab or browser window will appear once the **Start session** button has been pressed. Congratulations! You are now inside the instance that you created, without the need to start an SSH command from your terminal. Run the following command to see who you have just logged in as:

```
$whoami
```

It should return a value of `ssm-user`. This is the user that the Systems Manager Sessions Manager connects to.

You just saw how Session Manager can make accessing instances much more accessible from an operations perspective. You no longer have to create and manage keys for your instances. However, let's talk about some of the issues that need to be taken into consideration when using Sessions Manager.

Using runbooks in Systems Manager

Since we already have a managed instance up and running, we can create a runbook in AWS Systems Manager to see how easy it is to execute commands on one or a thousand instances, all with the single click of a mouse remotely.

Before we start, the `clamav.json` scripting file, which we will use for our runbook, is going to be available in the `Chapter-20` folder of this book's GitHub repository. Since the file is not extremely long, we will also be showing it throughout this exercise as well.

This exercise also builds upon the previous exercise, so if you wish to complete this exercise but did not do the *Running remote commands on an EC2 instance* exercise, then you will need to set up your environment by going through steps 1 to 14 in the previous exercise. Let's get started:

1. If you have previously logged out of your AWS Management Console, log back in as your administrative user and navigate to the **Systems Manager** service, either through the top search bar or via the **Services** dropdown in the top left.

2. We will need to create a document for Systems Manager to run. Although several predefined run documents are created by AWS, we will create a custom document. From the left-hand navigation menu of the **Systems Manager** service, under the **Shared Resources** header, click **Documents**.

▼ **Shared Resources**

Documents

Figure 20.13 – The Documents menu item on the Systems Manager service

3. Now that we're on the **Documents** page, we will need to find the orange **Create document** button at the top right-hand side. When you click on it, two options should appear. Choose the **Command or Session** option.

Figure 20.14 – The Create document button showing the two different options

4. Once the **Create document** page appears, under **Document details,** enter in the following information:

    • **Name**: Linux_ClamAV_Installer.

    • **Target type**: Leave this value blank.

    • **Document: Command document.**

Document details

Documents define the actions that AWS Systems Manager performs on your resources.

Name
Enter a unique name for the document.

```
Linux_ClamAV_Installer
```

The name must be between 3 and 128 characters. Valid characters are a-z, A-Z, 0-9, and _, -, and . only

Target type - *optional*
Specify the types of resources the document can run on. For example, "/AWS::EC2::Instance" or "/" for all resource types. **Learn more** ☑

Document type - *optional*
Select a document type based on the service that you want to use.

```
Command document                                                    ▼
```

Figure 20.15 – The Document details page when creating the document

5. At the bottom of the **Create document** page will be the **Content** section. This is where we load the script that we want to run for our run command. You can either cut and paste the script from the **clamav.json** script that you downloaded at the beginning of this exercise, or you can type in the script, as shown here. Once the script is in the content box, press the orange **Create document** button:

```
{
    "description": "Install ClamAV on Amazon Linux, Run
freshclam and clamscan",
    "schemaVersion": "2.2",
    "mainSteps": [
        {
            "inputs": {
                "runCommand": [
                    "#!/bin/bash",
                    "sudo amazon-linux-extras install -y epel",
                    "sudo yum -y install clamav",
                    "sudo touch /var/log/freshclam.log",
                    "sudo chmod 600 /var/log/freshclam.log",
                    "sudo freshclam ",
                    "sudo clamscan -r /var --leave-temps"
                ]
            },
            "name": "ALclamInstall",
            "action": "aws:runShellScript"
        }
    ]
}
```

6. After pressing the **Create document** button, you will be taken back to the main documents page. Here, you should see a green banner at the top of the page, confirming that your document was created successfully. If you would like to quickly and easily find the document that you just created, click on the middle tab of the top menu labeled **Owned by me**. At this point, you should be able to see your document.

Figure 20.16 – The notification showing that your document was created successfully

7. With our document created, we can find the **Run Command** menu item underneath the **Node Management** heading from the left-hand set of menu options. Click on **Run Command** to be taken to the **Run Command** capability.

8. On the **Run Command** capability screen, click on the orange **Run command** button at the top right of the main window to start the process.

9. For the command document to run, we are going to run the document that we just created. You can find it by typing Linux in the search box. Use a radio button to select an instance of the document. We will just leave the default version (1) selected since we only have one version of the document.

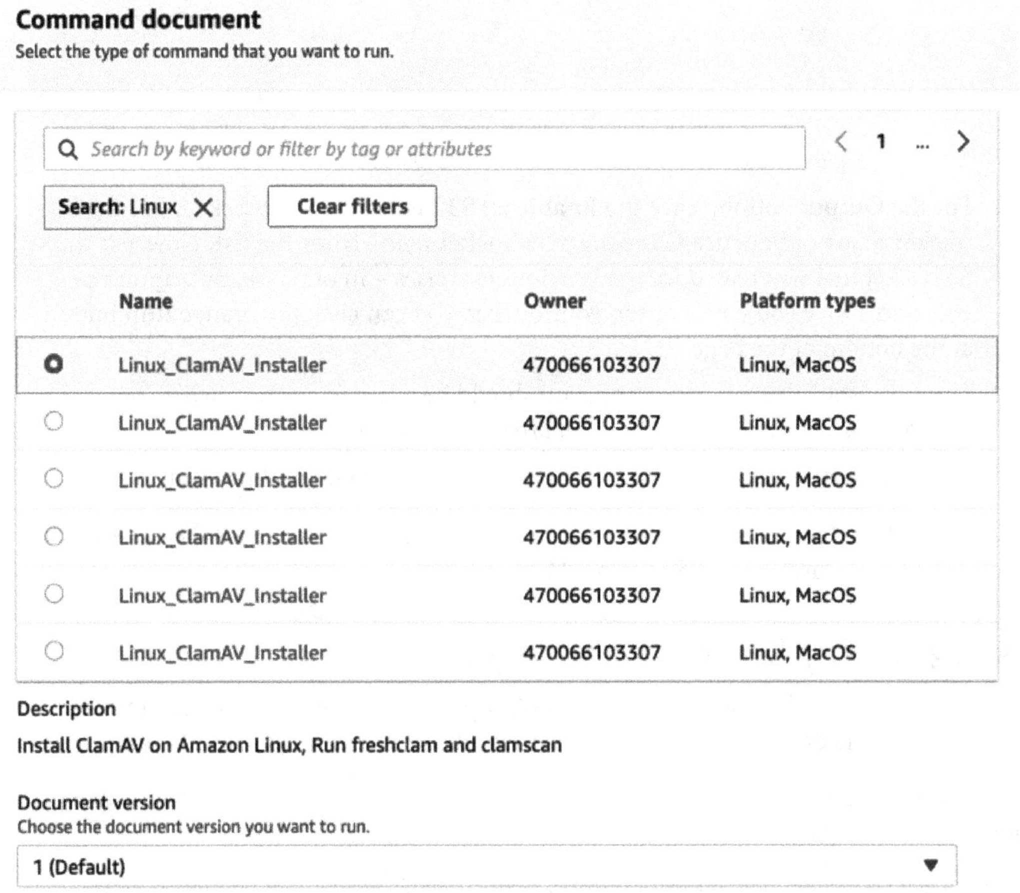

Command document
Select the type of command that you want to run.

	Name	Owner	Platform types
●	Linux_ClamAV_Installer	470066103307	Linux, MacOS
○	Linux_ClamAV_Installer	470066103307	Linux, MacOS
○	Linux_ClamAV_Installer	470066103307	Linux, MacOS
○	Linux_ClamAV_Installer	470066103307	Linux, MacOS
○	Linux_ClamAV_Installer	470066103307	Linux, MacOS
○	Linux_ClamAV_Installer	470066103307	Linux, MacOS

Q Search by keyword or filter by tag or attributes   < 1 ... >

Search: Linux ✕   Clear filters

Description
Install ClamAV on Amazon Linux, Run freshclam and clamscan

Document version
Choose the document version you want to run.

1 (Default) ▼

Figure 20.17 – Finding the document that we created based on a search term

10. In the **Targets** section, we will choose our instances manually since the only tag we used when spinning up our instance was the name tag. When you click on **Choose instances manually**, if you don't see the instance, then type the name of your instance; that is, *Remote Test*. Choose the selection box next to the instance's name.

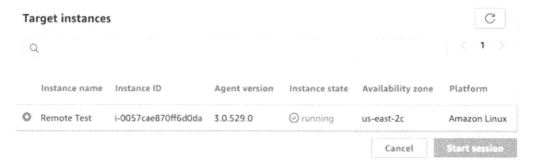

Figure 20.18 – Choosing the target instance to run the command on

11. For the **Output** option, leave the **Enable an S3 bucket** option checked. For the **S3 bucket name** option, use **Choose an S3 bucket name from the list**. Now, use the S3 bucket that we created for the previous exercises – in our case, `devopspro-beyond`. Once you have chosen your bucket, you can click the orange **Run** button at the bottom of the page.

12. Once you start the run command, you should be taken to a screen where you can see your instance and the status of *in progress* as the command runs.

13. After a few minutes, you should see success after the command has run.

Now that we have learned how to run remote commands on multiple instances, let's look at some of the use cases for AWS Systems Manager.

Use cases for Systems Manager

Systems Manager can help with your day-to-day operations management tasks in numerous ways. Let's take a look at some real-world use cases for Systems Manager.

You need to make sure that the EC2 instances running in your AWS account stay up to date with security patches.

If you have a compliance or security requirement that all the security patches that are released must be installed in a specific timeframe, such as 7 days from release, then this can be a daunting task without automation. This is especially the case if you are not using immutable infrastructure and are maintaining long-running instances that need to be updated.

When using the **Maintenance Windows** service, which is part of SSM, you can assign a maintenance window to a group of resources that would be out of their primary task schedule and then combine this with the **Patch Manager** service to install the security patches. This combination works with both Linux and Windows operating systems. It will also work for on-premises servers that have the SSM agent installed.

One of the keys to success in this is making sure that you have the correct tags on your instances so that you can group the resources according to tags. With Maintenance Windows, you can combine tags when making your resource group that only target the specific types of instances that you are planning to patch.

Now that we have learned how we can manage many of our operations tasks and help keep our resources in a compliant state using AWS Systems Manager, we will move on to the AWS Config service. With AWS Config, we will learn how we can constantly record the history of our environment, visually see what has changed, and even proactively remediate items that don't comply with the rules we have set for our account.

AWS Config essentials

In most cases, the resources in an AWS account are constantly changing in one form or another. Instances are being started and stopped, along with being created and destroyed. Security settings are changing as users turn ports on and off to allow the correct protocols for communication.

The AWS Config service allows you to gain a comprehensive view of what happens in your AWS account. It also lets you see how things change over time.

With the AWS Config service, the following capabilities are presented to you:

- You can assess the AWS resource configurations to see whether they conform to the desired settings for the account.

- The ability to save the current configuration settings as a snapshot for supported resources.

- You can retrieve the historical configuration for just one or even multiple supported resources.

- You can set up the ability to receive notifications when resources are created, deleted, or modified.

- Gain a view of the relationships that exist between resources.

Another value that Config provides is that you can use common predefined rules or even write your own rules to check your environment for compliance.

Concepts to understand about AWS Config

Before we dive into the AWS Config service, let's look at some of the key terms and concepts that will be discussed.

AWS Resources

The entities that you create and manage either from the AWS Management Console, the CLI, or any of the SDKs available, are **AWS Resources**. These can include EC2 instances, EBS volumes, DynamoDB tables, RDS instances and security groups, S3 buckets, SNS topics, and much more. AWS Config uses the **Amazon Resource Name** (**ARN**) to identify each of the unique resources. For a complete list of supported resources, visit https://docs.aws.amazon.com/config/latest/developerguide/resource-config-reference.html.

Configuration Recorder

As you run the AWS Config service, the **Configuration Recorder** is what saves and stores the values and changes of the various supported AWS Resources as configuration items. Before it can start recording, however, the Configuration Recorder must be created and then started.

Configuration History

When you want to understand what changes have been made to a particular supported AWS Resource and when they occurred, then you can use the **Configuration History**. The Configuration History is a cumulation of data (configuration items) for a specific resource over any period in which the configuration recorder has been running.

Configuration Snapshot

Unlike the name somewhat implies, a **Configuration Snapshot** is not a graphical picture in time of configuration items. Instead, it is a code-based recording of how your supported resources and their different settings are used at a particular point in time. These snapshots can be saved in a designated S3 bucket so that they can be retrieved or compared with future or past configuration snapshots.

Configuration Items

Point-in-time views for supported AWS resources are captured as **configuration items**. The information that's captured includes attributes, metadata, current configuration, relationships, and related events.

Now that we have an understanding of the concepts and pieces that AWS Config uses, let's take a deeper look at how AWS Config works.

Understanding how Config works

After starting the service, AWS Config starts to scan for supported resources within the account. As it finds these resources, it produces a configuration item for each one.

As those resources change, a new configuration item is then generated and recorded by the config recorder.

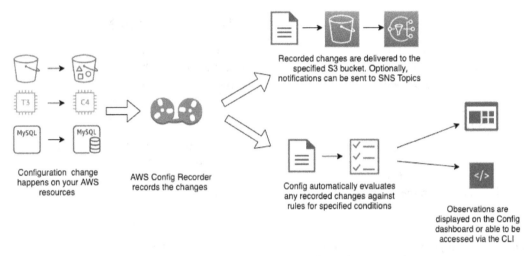

Figure 20.19 – Understanding the workflow of AWS Config

If you have enabled rules for the Config service to evaluate, then the Config service will continuously check the configuration items against the rules. If an item does not meet the criteria of a rule, then the lambda that is associated with that rule will be triggered, and the associated action will be taken.

For example, if a company has enabled a mandate that all EBS data volumes must be encrypted, regardless of whether they are deployed from CloudFormation, the CLI, an SDK, or created by hand, then a simple rule in AWS could be created to send a notification to an SNS topic that contains an email group list once an offending volume has been found.

A more complex rule could be created that not only pushes out the notification to the SNS topic but also uses one of the AWS SDKs programmed in the corresponding lambda function for the rule, which would terminate any created EBS volume that was encrypted upon creation.

As your Config item history is compiled, you can view both the inventory of the resources that AWS has discovered, as well as view the compliance history for any particular AWS resource.

If you are using Config rules, then resources can be constantly checked whenever changes happen. They can also be set to be checked on timed intervals, such as every 12 or 24 hours.

Now that we have a good understanding of how AWS Config works, let's try it out for ourselves with the next exercise so that we can get some hands-on experience.

Standing up AWS Config – a practical example

To see how the AWS Config service works, we will stand up a configuration recorder along with two rules that will check our account. Once the recorder has been stood up and then turned on, we can go back after a while to see its findings.

> **Note**
>
> Since we are using a CloudFormation template to stand up our Config recorder, we can easily take it down once we are done with it. If you were to stand up the Config recorder from the AWS Management console, it would be challenging to take down the default Config recorder and reset everything back to zero if needed.

Before we begin, make sure that you have downloaded the CloudFormation template named `configTemplate.yml` from this book's GitHub repository, under the `Chapter-20` folder:

1. Open your browser and log in to your AWS account using the Administrator user. After logging in, navigate to the CloudFormation service.

2. In the top right-hand corner, click on the **Create Stack** button. Once pressed, choose the **With new resources** option.

3. Choose the following options on the **Create stack** page:

    • **Prerequisite – Prepare Template**: The template is ready.

    • **Specify template**: Upload a template file.

    • **Choose file**: `configTemplate.yml`.

4. After using these options, click on the orange **Next** button.

5. You will be presented with all the parameters on the **Specify stack details** page. There will be several values that need to be filled in. You can use all of the default values that were filled in when you were spinning up the template. However, some values will need to be filled in manually, as follows:

    • **Stack name**: *Chapter20-config*

    • `MaximumExecutionFrequency: One_Hour`

    Once you have filled in these parameters, click the orange **Next** button.

6. After clicking the initial **Next** button when creating the stack, you will be taken to the **Configure stack options** page. Just click the orange **Next** button at the bottom right of the page to continue.

7. Once you're on the **Review** page, at the bottom of the page, you will need to check the box in the **Capabilities** section, acknowledging that the stack may create IAM resources. Once you have done this, click the orange **Create Stack** button to start the creation of the Config recorder and rules via CloudFormation.

8. In less than 5 minutes, the resources should have been created, including the Configuration Recorder, a new S3 bucket to capture the configuration snapshots, and an SNS topic especially for notifications from the AWS Config service.

9. Now, use the search bar in the middle of the AWS Management Console to search for the Config service. Once the icon appears, click on it to be taken to that service:

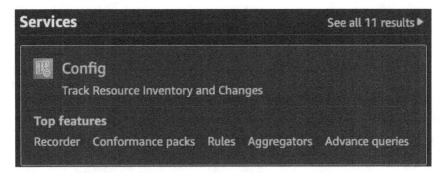

Figure 20.20 – The AWS Config service icon from the search results

10. From the **AWS Config** dashboard, which should be the first screen you see as you are taken to the service, we will see our compliance status right away. From our CloudFormation template, we enabled two rules automatically. These rules on my demo account are already showing to be in compliance. However, if you did not put an MFA on your root account, then it may show that you are out of compliance with one or more of the Config rules.

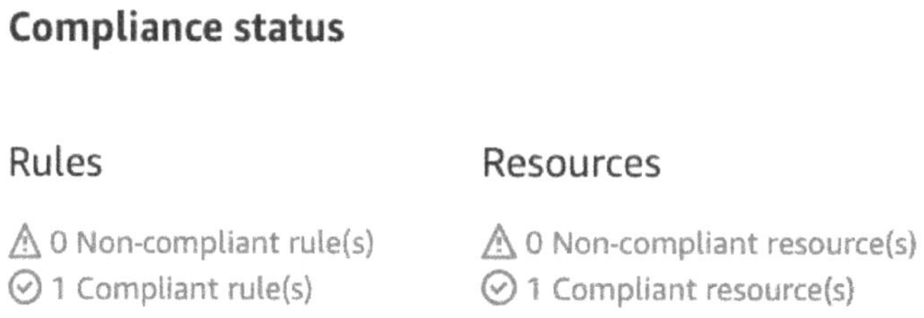

Compliance status

Rules

⚠ 0 Non-compliant rule(s)
⊘ 1 Compliant rule(s)

Resources

⚠ 0 Non-compliant resource(s)
⊘ 1 Compliant resource(s)

Figure 20.21 – The compliance status on the AWS Config dashboard

11. To view the rules that we currently have in AWS Config, simply go to the **Rules** item via the left menu. Once we click on this item, we should see the names of the two rules that were loaded with the template.

12. If we want to look at the resources in our account that AWS Config can check, we can go to the **Resources** menu item. Once **Resource Inventory** has loaded on the main screen, under the **Resource** category dropdown, change the selection to **AWS resources**. This will allow you to see the AWS items in your account that Config can keep track of.

13. On any of these resources, click on the blue **Resource Identifier** button to be taken to the **Resource details** page. If you would like to see the timeline of how the resource has changed over time, go to the top-right corner of the **Resource details** pages. Here, you will see a button labeled **Resource Timeline**. Clicking on this button will bring up the timeline interface, showing how the resource has changed over time.

I would suggest that you leave the Config recorder up for a day or so while you make a few changes to different resources in your account. Remember to take down the Config recorder via the CloudFormation template; otherwise, you will incur the charges of the AWS Config service.

Now that we have learned how to stand up a Config recorder and enable some rules in our AWS account, let's take a deeper look at what different types of rules are available from AWS Config.

The Config rule structure

AWS Config can constantly check the compliance of your AWS resources by using **Config rules**. These rules portray the ideal state of how your resources should be constructed.

You can use Config rules to help determine how your environment and resources conform to industry guidelines, internal best practices, and specified policies you need to enforce.

Rules can be run on a few different types of triggers:

- **Configuration changes**: When the config service detects that a change has been made to the resource.
- **Periodic checks**: The Config service runs on a regular schedule such as every 3 hours or every 24 hours, looking for changes to the resources.

Managed rules versus custom rules

There are lots of predefined rules that have already been designed by Amazon to use on many of the most common use cases. These rules can be checking resources, such as whether EBS volumes are encrypted or whether certain ports have been opened to allow traffic to flow.

These managed rules can be customized with extra items for remediation.

Custom rules can be created for any resource or policy that your organization needs checking and or enforcement on.

Using both managed and custom rules, you can automatically keep your organization in the state of compliance that it desires.

Now that we have learned how to use both Config and SSM to help keep our AWS account in compliance, let's review what we have learned in this chapter.

Summary

In this chapter, we looked at some of the tools that can help make the operations part of our DevOps role easier. This included AWS Systems Manager, which, in and of itself, provides a variety of services to help you accomplish tasks. These range from being able to quickly create a session on a remote instance to creating repeatable processes and installing software or gathering files from an instance. We also learned that Systems Manager works with both compute instances in the cloud as well as servers on-premises.

We also examined the AWS Config service. We saw how it can keep a timeline of the state of supported AWS resources. We also looked at how Config rules work and how they can be used to flag resources that don't meet the standards we have set for our organization.

In the next chapter, we will look at using Amazon Inspector. This is a service that helps you proactively find security vulnerabilities in your account.

Questions

1. You have been asked by a company to configure EC2 instances so that they can be managed by AWS Systems Manager. The company is using a distinct version of Linux for their base Amazon Machine Image. What actions will ensure that once the instance is launched from this AMI, it will be able to be found in the Session Manager console? (Choose two answers)

    A. Install the Systems Manager Agent as part of the base AMI.

    B. Create a key pair named `ssm-user` and use this key pair when launching the instance.

    C. Make sure that any Security Group that the instances are associated with allow traffic from port 22.

    D. Add an instance role to any launched instance that allows Systems Manager permissions.

2. The company you have been working with is asking for a weekly report of the top five operating systems that are being used across the EC2 instances in the two regions where they are currently operating their production systems. What is the quickest and most cost-effective way to get them this information?

    A. Use Amazon Athena to gather the CloudTrail data about all running instances. Create a report in QuickSight that will display a graph and detailed data of the top five operating systems per region. Share this report with the stakeholders who have requested it.

    B. Make sure all the instances have instance profiles that allow access by the Systems Manager service. Use Systems Manager Inventory to create a report that shows the top five operating systems for each region.

    C. Use the EC2 instances console to group all instances by operating systems. Create a report from the sorted data.

D. Create a lambda function that uses the CLI to call the operating system type of all EC2 instances. Have this output sent to an S3 bucket to be downloaded by the stakeholders.

3. The security department of an e-commerce company has put in force a new company policy that no web traffic may travel over non-secure HTTP, or port 80. Any traffic that was previously allowed on port 80 must now be redirected to port 443 using a secure certificate. Any security groups found open to port 80 will be in violation of this new policy and suspended immediately. What is the best way to regularly monitor whether any security group is allowing traffic to enter port 80?

    A. Use Trusted Advisor to scan for insecure security groups that have been created. Set up an SNS topic where Trusted Advisor can send the notifications.

    B. Add a network ACL rule to all VPCs that blocks traffic from entering port 80.

    C. Use CloudTrail logs to determine whether any security group has opened port 80.

    D. Set up a rule in AWS Config that checks for traffic being allowed on port 80 on any security group. If a security group is found in violation, send a notification to the security department's SNS topic.

Review answers

1. a and d
2. b
3. d

21
Using Amazon Inspector to Check your Environment

Amazon Inspector allows you to make security testing a regular part of development and **Information Technology (IT)** operations. As **Development Operations (DevOps)** makes the shift left to **Development Security Operations (DevSecOps)** and security responsibilities fall more to developers, using tools such as Amazon Inspector can help to form a more proactive approach to your and your organization's security posture.

In this chapter, we're going to cover the following main topics:

- Understanding Amazon Inspector
- Configuring the Inspector agent both manually and automatically

Understanding Amazon Inspector

Amazon Inspector enables you to analyze the behavior of your **Amazon Web Services** (**AWS**) resources and helps you identify potential security issues. With Amazon Inspector, you can run automated assessments over any or all applications you are hosting in the AWS cloud, based on hundreds of rules created by AWS security experts. These rules look for vulnerabilities as well as deviations from the best practices. After performing an assessment, Amazon Inspector delivers a detailed list of findings, which are categorized by their level of severity. The process is illustrated in the following diagram:

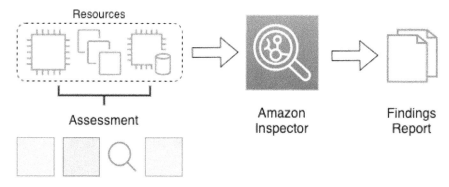

Figure 21.1 – The Amazon Inspector process

At the time of writing this, the assessments that Amazon Inspector can perform are contained to only those on Amazon **Elastic Compute Cloud** (**EC2**) instances.

Since we have just had an overview of Amazon Inspector, let's look at how we can get started with the Amazon Inspector service.

Getting started with Amazon Inspector

To get started with Amazon Inspector, there are three initial steps to take, outlined as follows:

1. Install the **AWS Inspector agent** on the **Amazon EC2** instances that you wish for Inspector to scan.

> **Note**
>
> It's both a good idea and a good practice to tag the instance(s) with a unique tag so that you can add these instances to a specific assessment target for an assessment run.

2. Create an assessment target, which is a collection of the AWS resources that you want Inspector to examine.

3. Create an assessment template that serves as a blueprint for your assessment.

4. Run an assessment on your target.

5. Review your findings and remediate any security issues.

There are ways to incorporate other AWS services into the mix as well. You can configure a **Simple Notification Service** (**SNS**) topic to send out notifications to a particular email address or distribution group once a finding report has been published. There are also ways to have **Lambda** functions automatically kick off Inspector scans either periodically, using **CloudWatch** events, or whenever a particular event happens, such as the creation of a new **Amazon Machine Image** (**AMI**).

We just looked at how to get started with Amazon Inspector and even took a high overview of how some of the other AWS services could be brought in to be incorporated with Inspector assessments. Let's now look at some real-world use cases with Amazon Inspector.

Use cases for Amazon Inspector

Companies are using Amazon Inspector both as a standalone service and by integrating Amazon Inspector into their DevOps pipelines to ensure that instances are free from vulnerabilities, especially in their production environments.

In the case of companies working in regulated industries, such as those that need to comply with **Health Information Portability and Accountability Act (HIPAA)** guidelines or **Payment Card Industry Data Security Standard (PCI DSS)** scanning for vulnerabilities. Amazon Inspector's assessment reports provide not only a trail of which vulnerabilities have been found, but they can also show a timeline of what has been fixed by comparing past reports. The following diagram shows Inspector being used in a **Continuous Integration/Continuous Deployment (CI/CD)** pipeline:

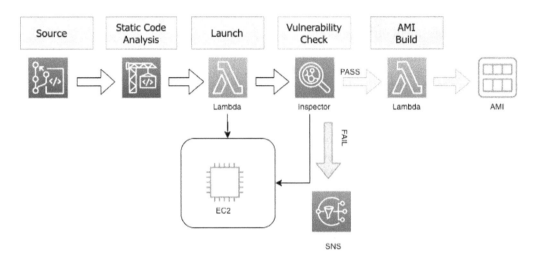

Figure 21.2 – Inspector in a CI/CD pipeline

As the preceding diagram shows, Amazon Inspector can also be added as a vulnerability assessment to a CI/CD pipeline as one of the final checks before creating a final AMI. This final AMI can either be used as a golden image or just a versioned image for an application. If Inspector has findings in this pipeline setup, the build is failed, and any issues need to be remediated before being passed on to the final stage where an image is created.

Now that we see how Amazon Inspector is being used in the real world, let's see how we can install the Amazon Inspector agent so that we can perform our own assessments.

Configuring the Inspector agent both manually and automatically

The **Amazon Inspector agent** can be installed on your target in one of three ways. The first way is to install the agent manually, which includes logging on to the target instance. The second way is to use **SSM** and take advantage of the `run` command feature to automatically install the Inspector agent on the instances in which we want the Inspector service to look for vulnerabilities. The third way is to incorporate a simple script into the user data so that the agent will be installed during the launch of the instance. These three methods are depicted in the following diagram:

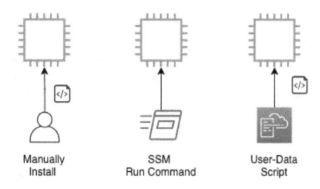

Figure 21.3 – Three ways to install the Inspector agent

Now that we have seen how to install the agent, let's go through the exercise of spinning up some instances and using Systems Manager to install the agent.

Using Amazon Inspector hands-on

In the following hands-on exercise, we are going to launch two instances at the same time and provide them with a tag of `Inspector` and a value of `TRUE`. Since we are going to be using the `run` command option from Systems Manager, we are going to need to have an **Identity and Access Management** (**IAM**) role that allows access to our instances. The good news is that we have previously created this role in our exercise for *Chapter 14, CloudWatch and X-Ray's Role in DevOps*. If you never completed this exercise and you would like to perform it, then you will need to go back and create this role before starting this exercise.

We will start off by creating two instances in our group, one of which will be of the Ubuntu 16.04 **Operating System** (**OS**) and one that will be running the Amazon Linux OS. For the Ubuntu OS instance, we will need to install the AWS **Systems Manager** (**SSM**) Agent using a user-data script. Fortunately, we already have this script as we used it in *Chapter 14, CloudWatch and X-Ray's Role in DevOps*. This same script will be included in the chapter 21 GitHub repository resources. The script to install the SSM agent once again is called agents.sh.

We just went through how to set up the Amazon Inspector agent using a Systems Manager run command. We also conducted an Inspector assessment. Next, we will see how to ingest the findings contained in the assessment report, as follows:

1. Open up your terminal so that we can start our commands. Make sure that in the directory you are working in, you have either made a copy of the agents.sh script from where you originally created it when you did the exercises in *Chapter 14, CloudWatch and X-Ray's Role in DevOps* or that you have downloaded a fresh copy from the Chapter 21 GitHub folder. We will start off by obtaining the AMI **identifier** (**ID**) for the Amazon Linux AMI. Use the command shown next to store the value of the AMI in the IMAGE parameter:

    ```
    IMAGE='aws ssm get-parameters --names \
            /aws/service/ami-amazon-linux-latest/amzn2-ami-
    hvm-x86_64-gp2 \
        --query 'Parameters[0].[Value]' --output text
    --region us-east-2'
    ```

    If you would like to see the actual value of the AMI ID, then you can run the $IMAGE command echo.

2. With the current AMI stored in the IMAGE parameter, we can now start to create our first instance. Use the following command to create an instance:

    ```
    aws ec2 run-instances \
    --image-id $IMAGE \
    --instance-type t2.micro \
    --iam-instance-profile 'Name=CW_SSM' \
    --tag-specifications 'ResourceType=instance,Tags=
    [{Key=Name,Value=AmazonLinux},
    {Key=Inspector,Value=TRUE}]' \
    --region us-east-2
    ```

3. Now, we will get the AMI ID for the Ubuntu instance. The command will look a lot like the command to store the variable for the Amazon Linux AMI, but the `names` value is different. Run the following command so that we can store the Ubuntu AMI ID in a value named `AMI` and then create our second instance:

```
AMI='aws SSM get-parameters --names \
        /aws/service/canonical/ubuntu/server/16.04/stable
/current/amd64/hvm/ebs-gp2/ami-id \
    --query 'Parameters[0].[Value]' --output text --
region us-east-2'
```

4. Now, just as we created our first instance, we will create our second instance. Two values will change in this command—the first value will be the `image-id` value, and the second will be the `name tag` value. You should also notice that we have added a second tag of `Inspector` with a value of `TRUE` to each of these instances. Use the command shown next to create a second Ubuntu instance:

```
    aws ec2 run-instances \
 --image-id $AMI \
 --instance-type t2.micro \
 --user-data file://agents.sh \
 --iam-instance-profile 'Name=CW_SSM' \
 --tag-specifications 'ResourceType=instance,Tags=
[{Key=Name,Value=Ubuntu},{Key=Inspector,Value=TRUE}]' \
 --region us-east-2
```

5. Both instances should be up and running at this point in time. We can now switch over to our browser. Log in to the **Amazon Management Console** and navigate to the **Systems Manager** service. Make sure that you are in the correct region (`Ohio`) in which you spun up the instances. Once at the **Systems Manager** service, under **Node Management** on the left-hand menu, find the sub-menu option named **Run Command** and click on it.

6. With the AWS Systems Manager `run` command now in the main window, click on the orange **Run a command** button so that we can use the `run` command to install the Inspector Agent.

7. We should now be on the **Command Document** screen. In the search box, type
    `Inspector` and hit *Enter*. This will bring up the `run` command document named
    `AmazonInspector-ManageAWSAgent`. Click on the **Radio** button next to the
    document name, as illustrated in the following screenshot:

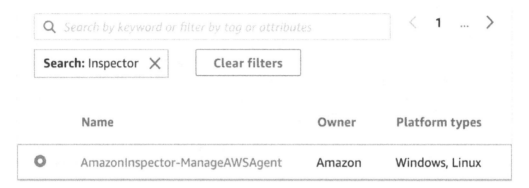

Figure 21.4 – The Systems Manager run document to install the Inspector Agent

8. Once we have the document selected, scroll down on the page until we come to a
    heading of **Targets**. Here, we can specify the instance tags that we want to target. We
    will use a **Tag key** value of `Inspector` and a **Tag value** value of `TRUE`, then click the
    Add button to specify the tag in our search, as illustrated in the following screenshot:

Targets

Targets
Choose a method for selecting targets.

⦿ **Specify instance tags**
Specify one or more tag key-value pairs to select instances that share those tags.

○ **Choose instances manually**
Manually select the instances you want to register as targets.

○ **Choose a resource group**
Choose a resource group that includes the resources you want to target.

Specify instance tags
Specify one or more instance tag key-value pairs to identify the instances where the tasks will run

Tag key	Tag value (optional)	Add

Enter a tag key and optional value applied to the instances you want to target, and then choose **Add**.

Inspector : TRUE ✕

Figure 21.5 – Specifying the instance tags to target with our run command

9. Next, scroll down to **Output options**. If you still have your S3 bucket that you made in *Chapter 4, Amazon S3 Blob Storage*, use this bucket for the output from the `Systems Manager` command. If you do not have an S3 bucket readily available, then simply uncheck the **Enable an S3 bucket** box, as this step is optional.

10. Scroll down to the bottom of the page and click on the orange **Run** button. You will be taken to the **Command Status** page, where you should see **In Progress** in the **Status** section of the targets and outputs. After about a minute, you can refresh the **Command Status** page, and you should see **Success** for both instances, as illustrated in the following screenshot:

○	i-0c445bbd24ceb2e2a	ip-172-31-7-57.us-east-2.compute.internal	⊘ Success	⊘ Success
○	i-09176fafd2c6c2dd3	ip-172-31-36-174.us-east-2.compute.internal	⊘ Success	⊘ Success

Figure 21.6 – The Inspector agent successfully installed on our instances

11. With our agent installed, we can now go to the **Amazon Inspector** service. In the top search box from the **Amazon Management Console**, type in **Inspector** and then click on the service once it appears. When the main **Inspector** page appears, click on the blue **Get started** button.

12. You should now be brought to a page with a heading on the top titled **Welcome to Amazon Inspector**. Under **Assessment Setup**, uncheck the box next to **Network Assessments**, as we are only trying to run assessments on our EC2 instances. This will leave the box next to **Host Assessments** checked. At the bottom of the page, click on the gray button labeled **Run once** and *NOT* the blue button.

13. When the dialog box pops up, click the blue **OK** button.

14. It can take around an hour for the assessment report to become ready.

We have now seen how to set up the Amazon Inspector agent on different OSes. Next, after our assessment is run, we will look at the report and see how to digest the findings of the report.

Comprehending the findings of Inspector assessment reports

Once Amazon Inspector has completed its assessment, it will then return an assessment report. Any findings that the report returns will contain both a detailed description of the security issue along with a recommendation of how to fix the problem.

You have the ability to store the reports as well as share these reports with your team members so that they can perform remediation actions on the findings contained in the report.

Summary

In this chapter, we learned about Amazon Inspector and how it can scan instances running applications for vulnerabilities. We learned about the three ways in which we could install the Inspector agent and even performed a hands-on exercise, installing the Inspector agent using the Systems Manager run command. Finally, we discussed the findings that Amazon Inspector returns.

In the next chapter, we will look at a number of other policies and standard services to know for the certification exam as well as for use in the real world. These include services such as **AWS GuardDuty** and **Macie**, as well as others that can be present in exam questions. Understanding these services will also help you in your current and future positions.

Review questions

1. You have been brought into a company where they are trying to rectify their security and compliance posture. The security and compliance teams are now requiring that all EC2 instances use approved AMIs. As a DevOps engineer, you must find a way to implement a process to find any EC2 instances that have been launched from unapproved AMIs. Which solution will satisfy the requirements?

    a. Use Trusted Advisor checks to identify EC2 instances that have been launched from unapproved AMIs.

    b. Create an AWS Config rule that identifies any non-approved AMIs and then sends a notification to the security and compliance distribution lists.

    c. Have Systems Manager Inventory create a custom report of all the EC2 instances using unapproved AMIs.

d. Have Amazon Inspector run across all the instances in the account. Share the findings of unapproved AMIs with the security and compliance team members.

2. You and your team are running an application in a production environment. The application is built and deployed using Elastic Beanstalk but runs on EC2 instances. You want to make sure that the application is free of any vulnerabilities; which AWS service should you implement to help with this?

   a. AWS Trusted Advisor

   b. AWS **Web Application Firewall (WAF)**

   c. AWS Shield

   d. AWS Inspector

3. There has been a recent investigation at the company you are working for. During this investigation, it was discovered that all application programming interface (API) calls were turned off for the production AWS account. It was also discovered that a new administrative user account was created, along with an access key and a secret access key. This key pair was used several times to create extra-large EC2 instances. How could you have used automation to detect and prevent this incident?

   a. Use Amazon CloudTrail to create a new CloudTrail that looks for the events of disabling any CloudTrails as well as the creation of any API keys with the Amazon administrator managed policy.

   b. Use Amazon Inspector to review all API calls leveraged. Configure the Inspector agent to use an SNS topic if any changes are detected to the CloudTrail. Use an IAM permissions boundary to prevent the spin-up of any extra-large EC2 instances.

   c. Use AWS Config to create a config rule that detects when CloudTrail is disabled. Create a Lambda function to re-enable CloudTrail if the rule is triggered. Use an IAM permissions boundary to prevent the spin-up of any extra-large EC2 instances.

   d. Use the Trusted Advisor API to periodically check if CloudTrail has been disabled. Create a Lambda function to re-enable CloudTrail if the rule is triggered. Use an IAM permissions boundary to prevent the spin-up of any extra-large EC2 instances.

Review answers

1. b
2. d
3. c

22

Other Policy and Standards Services to Know

There are a number of other services that are incorporated into the DevOps professional exam that span the realm of policy and standards. Each of these services has a specific role to play in the AWS ecosystem, and knowing their use case as well as where they are a good fit can help solve problems for a particular organization. Although all of these services are important to know, we will review them only briefly.

In this chapter, we're going to cover the following main topics:

- Detecting threats with **GuardDuty**
- Seeing how to protect data with **Amazon Macie**
- Understanding Server Migration Service

Detecting threats with Amazon GuardDuty

Amazon GuardDuty gives you a new type of threat detection service that was made specifically for the cloud. GuardDuty continuously monitors feeds from one or more accounts. It then continuously analyzes the network and account activity from the sources that are being driven into the GuardDuty service. From the input sources, the GuardDuty service then uses threat intelligence coupled with behavior models and machine learning to intelligently detect threats to your environment:

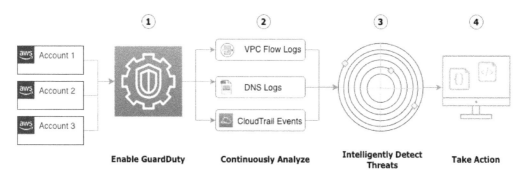

Figure 22.1 – The GuardDuty flow from enablement to taking action

From the preceding diagram, we can see that getting GuardDuty up and running in your account takes a few simple steps:

1. **Enable the GuardDuty service** – Activating the GuardDuty service will then start to analyze multiple types of logs within your account: VPC Flow Logs, DNS log entries, and CloudTrail Events.

2. **Continuously analyze the incoming events** – The GuardDuty service is constantly looking at various sources of data trying to detect either unauthorized or unexpected activity in your AWS environment. These sources include AWS CloudTrail event logs, CloudTrail management events, CloudTrail S3 data events, VPC Flow Logs, and DNS logs.

3. **Detect threats to your account intelligently** – Using a multitude of machine learning algorithms to process the data it collects, GuardDuty then analyzes that data on a consistent basis. These become the different types of findings that GuardDuty reports on, which are categorized into EC2 findings, S3 findings, and IAM findings.

4. **Take action on the threats presented** – Once a finding has been presented to you, GuardDuty will present the details of where the threat was located so that you or a member of your team can take action. This could be a specific EC2 instance, an S3 bucket, or an IAM user or group.

Key information to understand regarding GuardDuty

When studying for the DevOps professional exam, although the GuardDuty service may not be one of the main services showcased on the exam, it may appear either as part of a question or an answer. In these cases, it helps to have an understanding of the service and its capabilities.

GuardDuty has a number of detection categories. These include the following categories:

- **Reconnaissance** – Activity showing a potential attacker or unusual API activity or port scanning, or items such as multiple failed login requests.

- **Instance compromise** – Unusual activity such as cryptocurrency mining, malware activity, and even outbound denial of service attacks are grouped into this category by the GuardDuty service.

- **Account compromise** – If your account has an attempt to turn off the CloudTrail logging, then this would fall into the account compromise category. Also, if there are attempts to try and log into the account from locations around the world that are not normal according to the account baseline, these would fall into this category.

- **S3 bucket compromise** – These include items such as credential misusage, unauthorized access of an S3 bucket from a user or IP address that has no prior history of accessing the bucket, and even access of the bucket from known malicious IP addresses; all this information is gathered via CloudTrail API calls.

GuardDuty allows your security team and other system administrators to use IAM Access Analyzer to help them identify resources inside of your accounts that would be able to be accessed from outside of your accounts.

Within a findings action that the GuardDuty service provides, there are a number of details that will help you and/or your security team help to remedy the situation. These include the following:

- **CreateThreatIntelSet** – This grants permission to create GuardDuty ThreatIntelSets.

- **CreateDetectory** – This grants permission to create a GuardDuty detector.

- **GetFindings** – This allows GuardDuty to retrieve findings.

- **StartMonitoringMembers** – This allows a GuardDuty administrator to monitor member accounts.

Your skills as a DevOps professional are also enhanced as you understand the integrations of security and the security tooling readily available to you.

Next, let's look at some of the use cases for Amazon GuardDuty.

Use cases for Amazon GuardDuty

Now that we have an understanding of the abilities of the GuardDuty service, we can take a look at some of the potential use cases where using GuardDuty would make sense.

You have suspicions that internal employees or outside users may be mining for cryptocurrency on a corporate account

Amazon GuardDuty has the ability to notify you that an EC2 instance on any account in your organization has been contacting an IP address that is associated with the Bitcoin network or other cryptocurrency-related activity. GuardDuty, in the same vein, has the ability to detect when an instance is trying to direct internet traffic with a DNS name that is associated with cryptocurrency activity.

> **Important Note**
> If you or your organization are legitimately trying to mine cryptocurrency or doing legitimate business with this activity, there are ways with the GuardDuty service to suppress these particular rules so that you don't get alerts that do not apply to your organization.

You have a variety of data stored in your S3 buckets or you are using S3 as storage for your data lake

Amazon GuardDuty has a special category of S3 protections, which it performs, and then once enabled, it will monitor all buckets in your account. In monitoring the access patterns of the S3 usage, you do not need to take extra steps such as turning on S3 logging for any or all S3 buckets since GuardDuty monitors actions that are happening to the various S3 buckets at an API level using CloudTrail.

Understanding how Amazon GuardDuty is used in the real world is key to answering the questions about the service in the DevOps professional exam. Next, let's look at how the GuardDuty service unites with another key security service on AWS, Security Hub.

Amazon GuardDuty integrates naturally with AWS Security Hub

When trying to get a comprehensive view of the state of the security of your AWS environment, **AWS Security Hub** is a service that can provide you with this view. Security Hub collects the data from the different accounts that you manage, along with the different services being run inside of those accounts. AWS Security Hub also supports third-party integrations from partners such as Cloud Custodian, Check Point, IBM's QRadar, CrowdStrike Falcon, and others. The full list of partner integrations can be found here: `https://docs.aws.amazon.com/securityhub/latest/userguide/securityhub-partner-providers.html`.

Once initiated, Security Hub begins to ingest, organize, and even prioritize the findings from other AWS security services, such as Amazon GuardDuty, Amazon Inspector, and Amazon Macie. Using Security Hub also allows account holders to prioritize the findings that are discovered for an account so that the most vulnerable items may be addressed first.

It's important to note that on the Security Hub console, detection and consolidation of events are only possible after you have started the Security Hub service. Any events that have occurred prior to starting the service are not collected, and hence findings are not generated for those events.

With the enablement of the AWS Security Hub console in combination with the GuardDuty service, GuardDuty findings are not only sent to the security team but are also visible to anyone who has been given IAM permissions to use the Security Hub service. This helps with the visibility of issues and delegation of responsibility.

Now that we have seen how AWS Security Hub and Amazon Guard Duty work together to help protect your account, we will next look at how the Amazon Macie service can work constantly to help protect your data.

Seeing how to protect data intelligently with Amazon Macie

Organizations that are either moving to the cloud or have established a cloud presence over the past few years start to quickly accumulate data. In *Chapter 17, Advanced and Enterprise Logging Scenarios*, we discussed the importance of having a good system for data hygiene that includes tagging and metadata. And although this is the optimal solution, this is not always possible. The data that you create, acquire, and accumulate can fall into a range of sensitivity and classification levels. Once you get past the public level, access to that data from those who were not authorized to access it could be classified as a breach:

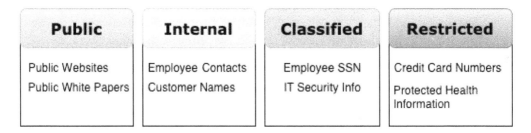

Figure 22.2 – An example of data classification levels

Not every organization takes the time to classify and categorize the information stored in their cloud. This can lead to a number of vulnerabilities, both internal and external. From an internal point of view, allowing internal users access to data that they should not have the ability to reach can leave an organization exposed from the standpoint that users have either not gone through the proper cybersecurity training, have unintentionally accumulated a piece of malware on their system and have now given this item access to data that is at a higher classification level, or even the fact that the internal user has malicious intentions.

Amazon Macie is a fully managed data privacy and security service from AWS that utilizes machine learning along with pattern matching. This helps Macie to discover sensitive data in your account even without any tags in place for assistance. It then helps you to protect that data by presenting you with a list of the findings.

There are also predetermined patterns that utilize Amazon EventBridge so that Macie can send all of the findings to EventBridge and then, using event-driven architecture, rules in EventBridge can be utilized to trigger Lambda functions that would then process the findings automatically.

Amazon Macie provides you with four main features regarding the discovery and protection of data in your accounts:

- You increase visibility and the ability to evaluate your data as Macie will do the following:

  a. It will evaluate the S3 bucket inventory in your accounts.

  b. It will evaluate the S3 bucket policies in your accounts.

  After completion, it will provide a high-level overview of the total number of buckets along with the total number of objects, the storage size of those objects, and will provide an overview of the policies of these objects and buckets. With the Macie service enabled, anytime there is a change to these policies, you will be provided with a real-time notification.

- It allows you to discover sensitive data.

- You can manage all of the data in all of your accounts centrally.

- It has the ability to add automation based on the findings.

Once Amazon Macie has been set up, it allows you to set up **alerts**. These alerts allow you to be notified about a potential security issue that the Macie service has discovered. There are two different levels of alerts that get sent in the Macie service: basic and predictive. *Basic alerts* are the type of alerts that Macie generates from the security checks that it performs. *Predictive alerts* are sent based on the baseline that the Macie service determines for your account and then a determination of a deviation from normal activity in your account.

When an alert is sent by the Amazon Macie service, it is assigned one of four security levels.

It's important to note that Amazon Macie is a regional service. This means that you will need to enable Macie for each region that you would like to utilize its scanning and protection capabilities.

Getting started with Amazon Macie is fairly simple. You start by going to the AWS Management Console and then navigating to the Amazon Macie service.

Amazon Macie use cases

As we continue to look at the Amazon Macie service, exploring different use cases of businesses and enterprises that use it will help us gain a better understanding of the value that the Macie service brings.

Assessing an organization's data privacy and security

If your organization does not currently know the classification levels of the data being stored, or it does not currently have a baseline of data classification levels, then using the Amazon Macie service can see whether there is any sensitive data stored in the specified accounts and lets you take appropriate actions.

Knowing when sensitive data has been placed in your datastore

Even if you have established a baseline for data storage in your accounts, allowing the Macie service to constantly monitor any new data being stored in your account can provide you with alerts if sensitive data has been added. This is especially useful if you are combining Macie with the Security Hub service so that prioritization can be added to any new alerts that get created.

After reviewing some of the other security tools that might appear in the AWS DevOps exam, we will next examine one other service – **Server Migration Service** (**SMS**).

A brief look at the migration tools available from AWS

Many organizations are still in the beginning parts of their journey to the cloud. Yet, as they look at contracts that are up for renewal at their current data centers, there comes a pressing need for an expedited exit. Many of these companies are running their workloads these days not on bare metal servers but on virtualized server farms, using popular software products such as VMware or Microsoft Hyper-V.

Seeing the need to help customers move into the cloud quickly and easily, Amazon created SMS.

This service helps navigate many of the challenges that are presented when trying to move current workloads into the cloud.

Bringing your servers into AWS with SMS

SMS helps simplify the migration of multiple servers into the AWS cloud.

Using SMS helps you achieve the following:

- It helps simplify the cloud migration process.
- It allows you to orchestrate multi-server migrations.

- It minimizes downtime during the migration process.

- SMS supports the most commonly used operating systems, including both the Windows operating system and quite a few of the major Linux distributions.

- It allows you to incrementally test your server migrations.

Although originally crafted for VMware cases, SMS now works for VMware vSphere, Microsoft Hyper-V, and Microsoft Azure migrations. These are performed using an agentless migration.

Now that we have an understanding of AWS SMS, let's take a look at the key features we need to know, both in practice and for the exam.

There are a few different options to move workloads into AWS. There is the CloudEndure product, which requires an agent to be installed. There is also the Import/Export service, which includes the Snow Family of instances such as the Snowball service; however, these services are more for moving and migrating large amounts of data into the AWS cloud.

The main use case for SMS is when you have a number of workloads on either VMware, Microsoft Hyper-V, or in Azure and either cannot install an agent or prefer not to install an agent to perform the migration.

When would you want to use SMS? The following use cases are the optimal times for using SMS:

- If you have a need to migrate to AWS but need to perform an agentless migration

- If you need a migration solution that meets **HIPAA** eligibility standards

- If you need a migration solution that follows **Federal Risk and Management Program** (**FedRAMP**) compliance

- If you need to be able to have the ability to execute an automated launch of your migrated virtual machines

- If you would like to have the option of having either **Amazon Machine Images** (**AMIs**) or CloudFormation templates as the output of your migration

Now that we have an understanding of the benefits of SMS, along with which use cases would be best optimized for it, let's examine the key features to remember about SMS.

Key features of SMS

The following are some of the key features to know about SMS:

- SMS allows you to automate the VMware, Microsoft Hyper-V, or Microsoft Azure virtual machines you have into the AWS Cloud.

- SMS allows you to incrementally replicate virtual machines as AMIs.

- Using SMS, you can migrate a group of servers as a single application that has multiple tiers and intertwined dependencies, such as a web tier, a database tier, and an application tier.

Now that we understand the key features, let's look at how we would go about migrating an application that is composed of a number of layers. This would include a database tier, a file-server tier, a web tier, and an application tier.

Migrating multi-tier applications with SMS

Because SMS is agentless, there is a need to download and install a special image (**Open virtualization Appliance** (**OVA**) in the case of VMware), which is called a connector. With the connector installed on the environment, you then create a vCenter user with the correct permissions who can create and delete snapshots on selected VMs.

Servers that contain multiple tiers of an application can be either replicated as a single AMI or they can be subdivided into smaller subgroups, creating multiple AMIs. If you do subdivide a larger application, then SMS can create a corresponding CloudFormation template to launch those AMIs in a coordinated manner:

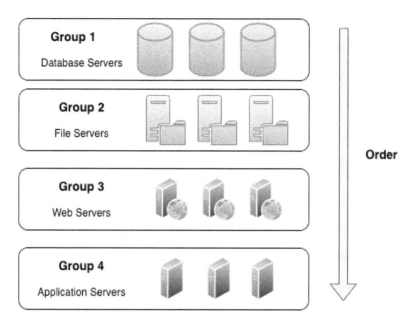

Figure 22.3 – An example of a launch order for an SMS job

It's important to understand that SMS cannot migrate everything to AWS. What does this really mean? This means that although the majority of your workloads can be copied into AWS and turned into AMIs, you will still need to create the underlying infrastructure for those instances to run on.

Summary

In this chapter, we discussed other relevant services that can appear in the DevOps professional exam with a focus on security and policy services. This included looking at the Amazon GuardDuty and Amazon Macie services, and how they can work with AWS organizations on a regional basis to protect your assets and accounts from bad actors looking for vulnerabilities. We also examined SMS. This service is useful to know since as companies are quickening the pace of moving into the cloud from their current data centers, the ability to update those workloads once they are in there starts to fall on DevOps professionals such as yourself.

In the next chapter, we will look at the structure of the DevOps Professional certification exam. We'll discuss the formats available for taking the exam and talk about the exam blueprint. We will also give some pointers for studying for and passing the exam.

Review questions

1. You have been brought into a company that wants to start implementing CI/CD for their builds. This company has recently gone through a security posture analysis assessment and needs to make sure that any solution implemented is secure in nature. If any vulnerability is found after implementation, there then needs to be a plan for remediation, and action must be taken to remedy the vulnerability.

2. After setting up the pipeline builds in AWS CodePipeline, which further steps can you take to ensure that no underlying access on the S3 buckets is available, which you created during the process?

    a. Every 15 days, use the Trusted Advisor service to check and see whether there is a misformed bucket policy for the S3 bucket. If the S3 bucket shows up in the vulnerable access list, then craft a bucket access policy that will only allow for the CodePipeline service account to access that bucket.

    b. Turn on logging for the S3 bucket. Add versioning to the newly created S3 bucket. Create a CloudWatch alarm on the bucket that measures access to the bucket. Have the alarm go off if the bucket access is more than the number of build jobs that the AWS CodePipeline runs. Add object-level encryption to the objects with AWS Key Management Service (KMS).

    c. Create a bucket policy for the S3 bucket that only allows access from the account itself and the service account that is running the pipeline. Add object-level encryption to the objects with KMS.

    d. Turn on logging for the S3 bucket. Enable the GuardDuty service. If something is detected, follow the remediation steps presented by the GuardDuty service. Add object-level encryption to the objects with KMS.

3. Your company has just started to allow partner companies to access subscription data that is being published in a few specific S3 buckets. Access is being provided via cross-account access. Since this is a new service offering and is supposed to be a new revenue source that the company is counting on, how can you ensure that access is only being obtained by those partners that have been granted access via a subscription? Your solution should be easy to maintain with minimal effort:

    a. Turn on CloudTrail for all regions. Create a CloudWatch alarm for access to the bucket, which is greater than the number of customers times 30. Create a bucket access control list (ACL) to block any country ranges that have no current customers.

b. Turn on CloudTrail for all regions. Enable GuardDuty. Add your current customers to the access whitelist. Review the GuardDuty findings on a weekly basis to ensure that no unauthorized access is being granted.

c. Turn on logging for the S3 bucket. Enable advanced Trusted Advisor checks. Review the Trusted Advisor checks on a weekly basis to ensure that no unauthorized access is being granted.

d. Turn on logging for the S3 bucket. Enable advanced Trusted Advisor checks. Create a bucket ACL to block any country ranges that have no current customers.

4. You have just joined a company that is in the midst of moving a large number of workloads out of their current data center into the AWS cloud. Their current workloads in their data center are running on VMware vSphere. These workloads on-premises are a combination of Windows Server, Red Hat Linux, and Ubuntu Server, which need to be moved to AWS. With 400–500 servers and a tight deadline, the company does not want to have to install an agent on each virtual machine. They would also prefer to be able to test the migration incrementally during the process. Which tool would help you accomplish all of the company's objectives?

a. Log on to the AWS Management Console and order a Snowball so that you can download the workloads securely to the Snowball hardware. Once the Snowball has been shipped back to AWS, download the data and create AMIs from the downloaded data.

b. Use AWS CloudEndure to discover the instances which need to be migrated to the AWS cloud. Create a migration job to start the replication of the instances to the specified AWS region.

c. Log on to the AWS Management Console and start a job using SMS. Create a catalog on an on-premises server. Create and configure your migration. Track and test your migration.

d. Make sure that you have the AWS CLI installed on the machine that you plan to perform the migrations. Use the VM Import/Export service to push the VMs to the Amazon S3 service. Create AMIs from the uploaded images.

Review answers

1. d
2. b
3. c

Section 5: Exam Tips and Tricks

In this part, we will look at the structure of the test, its six areas of scoring, study techniques, and study guides.

This part of the book comprises the following chapters:

- *Chapter 23, Overview of the DevOps Professional Certification Test*
- *Chapter 24, Practice Exam 1*

23
Overview of the DevOps Professional Certification Test

Hopefully, by the time you reach this chapter, you will have a solid understanding of the vast amount of material we have covered up to this point, all of which may appear on the **DevOps Professional Certification test**. Once you feel ready to take the test, you must register for the exam and will receive a specified exam time.

In this chapter, we're going to cover the following main topics:

- The DOP-CO1 exam guide
- How the exam is scored
- Understanding the different testing options
- Study tips for preparing for the exam

The DOP-C01 exam guide

Each AWS certification that is currently offered publishes an exam guide. This is the blueprint for success in passing the exam. It explains the expected experience and requirements a testing candidate should have before taking the exam. It also outlines what topics (or domains) will be covered in the exam. If you would like to look at the current exam guide, it is available at the following URL: `https://d1.awsstatic.com/training-and-certification/docs-devops-pro/AWS-Certified-DevOps-Engineer-Professional_Exam-Guide.pdf`.

The exam requirements as per the exam guide

It is recommended that you have 2 or more years of experience using the AWS platform to provision, operate, and manage AWS environments before taking the DevOps Professional Certification exam. You should also have experience automating infrastructures and constructing systems using Infrastructure as Code (preferably with CloudFormation templates). You need to have proficiency in administrating modern operating systems such as Windows and Linux. Finally, it is recommended that you have a solid understanding of the modern development and operations processes, as well as the methodologies that make up these processes.

AWS will also explain what is not in scope for the test. This is helpful as the certification team is telling you what you do not need to concentrate on with your studies. The following topics are outside the scope of the exam:

- Advanced networking (advanced routing algorithms, failover techniques)
- Deep-level security recommendations for developers
- Database query and performance optimization
- Full stack application code development
- Normalization of data schemes

Now that we understand what AWS expects from us, as candidates, as far as requirements go, let's take a look at how the exam is scored.

How the exam is scored

AWS certification exams are broken down into domains. Although the domains have different scoring levels, it is important to have a mastery of all domains since missing just one or two questions can mean the difference in passing or failing the exam. There are two types of questions that are presented in the exam:

- **Multiple choice questions**: These questions have one correct response.

- **Multiple selection questions**: These questions have two or more correct responses.

For multiple selection questions, no partial credit is given. You must select all of the correct answers to gain the points for that question. Also, unanswered questions are counted as incorrect answers. This means you are better off guessing at an answer than leaving it blank.

Domain	% of Exam
Domain 1: SDLC Automation	22%
Domain 2: Configuration Management and Infrastructure as Code	19%
Domain 3: Monitoring and Logging	15%
Domain 4: Policies and Standards Automation	10%
Domain 5: Incident and Event Response	18%
Domain 6: High Availability, Fault Tolerance, and Disaster Recover	16%
TOTAL	**100%**

Figure 23.1 – The six domains on the DevOps Professional Certification exam

The exam itself is 180 minutes or 3 hours in length, with 80 questions in total. This leaves you a little over 2 minutes to read, comprehend, and answer each question. Let's examine each of the six domains in more detail:

- **Domain 1: SDLC Automation – 22%**: The questions in this section of the exam concentrate on CI/CD pipelines, source control, testing, and artifact storage/management. These questions can test your knowledge on how to set up and configure code repositories, along with automating deployment pipelines that have multiple stages. You also need to have an understanding of the workflow for accommodating changes or new commits into a repository, along with incorporating secure practices and requirements with your **code versioning system** (**CVS**).

In domain 1, you will also be tested on your ability to add and run integrated testing in your code process. This includes allowing a process to either pass or fail based on the test results. Once your code has passed, you need to be able to create and store artifacts in a secure manner. You also need to have the capability to determine the correct delivery strategy, such as Blue/Green, Canary, or AB for a particular business based on the needs expressed.

- **Domain 2: Configuration Management and Infrastructure as Code – 19%**: In domain 2, you are tested on your ability to determine the deployment of different services based on the deployment's needs and requirements. You also need to understand how to use life cycle hooks for both optimizing starting up and shutting down instances. Also present in this domain is the need to identify the pros and cons for different configuration management tools, along with demonstrating the ability to run configuration management in an automated fashion.

- **Domain 3: Monitoring and Logging – 15%**: This domain concentrates on storing, aggregating, and analyzing logs and monitoring your systems. This includes distributed log collection and processing using different agents such as the CloudWatch agent, `flumd`, or others. You will also need to understand how to manage log storage, such as transitioning logs from CloudWatch to S3, using S3 life cycles to phase out older logs, and even S3 events to trigger event-driven actions. Also included in this domain is the process of creating custom CloudWatch metrics and log subscription filters.

- **Domain 4: Policies and Standards Automation – 10%**: Cost optimization through the process of automating manual processes is one of the main themes of domain 4. The other main theme of domain 4 is the ability to implement governance strategies. This includes the ability to enforce compliance using self-service products such as Service Catalog and report on the compliance of an organization.

- **Domain 5: Incident and Event Response – 18%**: In this domain, you need to be able to help troubleshoot an issue and determine its root cause as quickly as possible. This also includes setting up and configuring alerts and event-driven actions from those alerts. You are expected to understand how to use the `Route53` service to direct traffic to another region in case of regional failure.

- **Domain 6: High Availability, Fault Tolerance, and Disaster Recover – 16%**: This final domain will assess your ability to determine whether you need to deploy to multiple regions or multiple availability zones. You also need to evaluate the correct services needed for the organization's RTO and RPO requirements. You will also need to be able to evaluate a deployment for points of failure.

> **Note**
> There are 10 questions in the exam that do not affect your score. You will not know which questions these are. These questions are used to collect performance metrics for use in future exams.

Obtaining a passing score

The DevOps Professional Certification exam is a pass or fail exam. Once you press the final submission button, then your test will be scored, and you will be notified if you have passed or failed. The test is graded on a scoring scale of 1,000, and you need at least a score of 750 to pass the exam. The math breaks down to this: with 70 valid questions (remember, 10 questions do not affect your score), you need to correctly answer 53 questions in the exam to pass. This is not the full scoring system since some domains are more prevalent than others. Or, if you think of it the other way around, you can only miss 17 questions throughout the entire exam.

AWS will later email you a detailed report of how you scored on each of the six domains. This breakdown will also be available in your **Certmetrics** account, under the previous exams section.

Now that we know what it takes to pass the exam, let's briefly discuss your options if you do not pass on your first attempt.

If you do not pass the exam on your first attempt

If you have seen that dreaded fail word after you have submitted your exam, do not fret. All is not lost. There is a mandatory 14-day waiting period between attempts, and you will need to pay the necessary fee to retake the exam. There is no longer a limit on the number of times a candidate can attempt the exam.

One of the key things to do is have a piece of paper nearby, either in your vehicle if you are taking the test on location at an actual exam site, or find one after your exam is over if you have chosen the online proctoring route. Write down some of the topics that you feel you were not strongly prepared for. This is an immediate opportunity to capture information regarding what you did not feel you were prepared for while it is still fresh in your mind. If you do not take this immediate step, you may regret it later when you go back to think about areas that you need to concentrate on.

Commit to retaking the test in a short timeframe after restudying the material. This is key. It does not have to be directly after the mandatory 14-day waiting period, but it should not be more than 60 days later. You have already put in the time and effort to study for the initial exam and now have a better understanding of what the actual questions will be like. At this point, you will have received the results back from AWS, showing how you scored on each of the six domains, showing you where your strengths and weakness are in the testing process. Don't make the mistake of failing to review the topics where you scored higher than others. The certification exams contain a large bank of questions, and it is rare to see the same question multiple times in a row.

Understanding the different testing options

AWS has begun to offer multiple choices for taking the exam. Not only do you have the option of taking the exam at a testing center, but there is also the option of having an online proctored exam right from the comfort of your own home, without the need to go anywhere. Let's examine both options as there are some pros and cons to each.

Registering to take the exam

When you feel like you have a solid understanding of the material that will be presented in the exam, then it's time to register for a time slot to take the test. To register, you will need to go to the `https://aws.training` site and either sign in if you already have an account or create a new account.

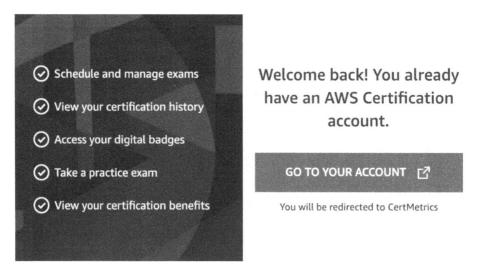

Figure 23.2 – The welcome screen of aws.training to schedule your certification exams

It is a good practice to schedule your exam about 2 weeks out from when you feel that you will be able to take the exam. Using this technique will allow you to push away any final distractions around you, knowing that you have a hard deadline that you have to meet. If you continue to wait until you feel completely confident in scheduling your exam, then this can lead to hesitancy in scheduling.

> **Note**
> If you are a non-native English speaker, then you are allowed an additional 30 minutes for your test. However, before scheduling your exam, you must request this accommodation via the certmetrics.com portion of the AWS training site. This can be found once you click on the **Upcoming Exams** top menu item. From here, you can **Request Exam Accommodations**.

Figure 23.3 – The dropdown showing the selection of adding 30 minutes to the exam for English as a second language

If you cannot make the exam on the date that you set, you can cancel up to 24 hours in advance without any penalties. If you skip your exam date without canceling within this advanced window, then you will forfeit the cost of the exam, but it will not count as a failed exam.

Now that we have covered the basics of taking the exam, let's look at some of the differences between taking the exam at the testing center and a remote location using online proctoring.

Taking the test at a testing center

Testing centers, which allow the certification exams to be taken on site, are operated by either PSI or Pearson VUE. You can search for local testing centers around you and then see what time slots are available to take the tests. Sometimes, you need to schedule a week or more in advance, so as you get closer to being ready to take the test, start looking at the availability in your area and how long in advance you will need to book your test.

After scheduling the date and time for your exam, you will need to arrive at the testing center a few minutes before your scheduled time so that you can check in and be seated. It is a good idea to either have a print-out of the email sent by AWS, which will specify your test confirmation number, or to have the email available on your smartphone, just in case there are any issues locating your name and test in the system. When checking in, you will need to provide two forms of ID, one of which will need to be a government-issued ID card and one of which can be something such as a school ID or credit card.

Before stepping into the exam area, you will be asked to empty your pockets, remove any extra jackets, cellphone, hats, sunglasses, pens, or other items as they are not allowed in the testing room. The testing facility will provide a locker and a lock for you to store your items while you are taking the exam. They will provide you with a pencil and 2-3 sheets of blank scratch paper. You will need to return all of the scratch paper when you are finished with the exam.

Taking the test via online proctoring

If there is not a Pearson VUE or PSI center available for taking the test around you or you feel like you would be more comfortable in your surroundings, then you have the option to take the test via the online proctoring method using one of your computers.

If you choose this method, then you will need to install the OnVUE software onto your computer so that the remote person proctoring your exam can keep an eye on your system while you are taking the exam. This is best done a few days (and at a minimum, a few hours) before your exam's start time to ensure that the computer that you are planning to use meets all of the requirements and that the OnVUE software can perform its system test.

The space in which you will be taking the test will also need to be not only neat but also clean. This means removing all notes, pieces of paper, and other items from the area where you are going to be taking the exam. No food or smoking is allowed in the area where you are going to be taking the exam either, but the Pearson VUE site does say that a beverage in any container is allowed. If you have a multi-monitor setup on your desk, then those extra monitors will need to be unplugged for the duration of the exam. The same goes for any extra computers or smart devices that are in the immediate area.

30 minutes before your scheduled exam time, you will need to use a link provided to you to take pictures of the area that you will be using for the exam to verify that it is a clean environment. You will also need to send a picture of a government-issued ID to ensure that the person on the webcam is you taking the exam.

You cannot use your mobile phone once you have started the exam process. Doing so will get your exam revoked. This also goes for leaving the view of the webcam. When the exam starts, the proctor will be listening for any variations in noise in your environment. This includes reading the questions aloud to yourself or even mumbling, trying to get a better understanding of the question. If they do catch you making noises, then you will get a warning; multiple warnings may lead to your exam being revoked.

Some of these last points are one of two major differences between taking the exam in a testing center and taking the exam via an online proctor. One of the other major differences is that at the online testing center, you are given 2-3 sheets of real paper and a pencil to write notes during your exam. This is in strict contrast to the virtual whiteboard that you have to use with your mouse or trackpad.

There is one other option that is usually not spoken of when taking the certification exams, and that is taking the certification exam at the AWS re:Invent conference.

Taking a certification test at re:Invent – a cautionary tale

If you ever have the opportunity to go to the re:Invent conference, then I suggest that you go at least once for the experience. When you get there, there is an even greater push than from the early days to get you certified and gain access to the certification lounge. Taking the certification test in this type of environment is unlike any other. There are rows of computer screens and people packed side by side in a large conference room, staring quietly at the various tests they are trying to take. As you are trying to concentrate on your exam, you cannot help but see the various other people finishing and getting up and sitting down repeatedly. It becomes distracting, and sometimes, thoughts will wander in your head about if they are taking the same exam as you and, if so, how are they finishing so quickly.

One of the things that you may not think about when signing up for a test at the re:Invent conference is that there are a lot of extra-curricular activities going on, especially at night, many times with one or more adult beverages. This can draw you away from the quiet time that you need to do the final preparations for your exam. It can also lead to leaving you feeling not as fresh as you need to be if you have scheduled your exam for the morning.

Given that we have gone over the different options available for taking the actual certification test, let's focus on study tips for the exam to optimize your chances for success.

Study tips for preparing for the exam

Although we have covered a lot of material throughout this book, passing the certification exam takes study and discipline. After taking and both passing and failing multiple AWS exams, some tips can be passed on to help you try and have the best chance of success.

One of the first things to note about the exam is that it is an exercise in reading and comprehension as much as it is on AWS services and knowledge. Skimming through the questions and answers can lead to missing a critical piece of information and therefore missing the question altogether. Even though you only have around 2 and a half minutes to read, comprehend, and answer each of the questions, each of those questions must be carefully read for the correct details and components. Practicing reading the large question format is an important key to success, I have found. The ability to focus on this long format exam can help, especially when you look up at the timer and see that the minutes are running down and there are still several questions that you need to answer.

AWS whitepapers to read

Many of the questions in the exam are based on the whitepapers that have been written through the years at AWS. I would strongly suggest reading these whitepapers to gain a better understanding of the topics presented in this book. Some should just help to reinforce the content that was used to craft this chapter and its hands-on exercises. Others will go deeper into some of the topics that will be covered in the exam. There are also a few questions that will be derived directly from the whitepaper's context:

- **Infrastructure as Code**: `https://d1.awsstatic.com/whitepapers/DevOps/infrastructure-as-code.pdf`

- **Introduction to DevOps on AWS**: `https://docs.aws.amazon.com/whitepapers/latest/introduction-devops-aws/introduction-devops-aws.pdf`

- **Build a Log Analytics Solution on AWS**: `https://d1.awsstatic.com/Projects/P4113850/aws-projects_build-log-analytics-solution-on-aws.pdf`

- **Blue/Green Deployments on AWS**: `https://docs.aws.amazon.com/whitepapers/latest/blue-green-deployments/blue-green-deployments.pdf`

- **Serverless Architectures with AWS Lambda**: `https://docs.aws.amazon.com/whitepapers/latest/serverless-architectures-lambda/serverless-architectures-lambda.pdf`

- **Running Containerized Microservices on AWS**: `https://d1.awsstatic.com/whitepapers/DevOps/running-containerized-microservices-on-aws.pdf`

- **Practicing Continuous Integration and Continuous Delivery on AWS**: `https://d1.awsstatic.com/whitepapers/DevOps/practicing-continuous-integration-continuous-delivery-on-AWS.pdf`

- **AWS Governance at Scale**: `https://d1.awsstatic.com/whitepapers/Security/AWS_Governance_at_Scale.pdf`

- **Infrastructure Event Readiness**: `https://d1.awsstatic.com/whitepapers/aws-infrastructure-event-readiness.pdf`

- **Operational Excellence Pillar (AWS Well Architected Framework)**: `https://docs.aws.amazon.com/wellarchitected/latest/operational-excellence-pillar/wellarchitected-operational-excellence-pillar.pdf`

While this seems like a large amount of extra reading, it needs to be emphasized once again that many of the questions come from the whitepapers. Gaining the necessary knowledge and having an understanding of its context will help you feel more comfortable while reading through the complex questions presented in the exam.

Final test tips

The test itself is long. It is 3 hours in length. Especially if you are taking the test in the morning, I suggest that you do not fill up on coffee. Remember that, if you are doing an online proctoring session, you are not allowed to leave the view of the webcam. Before you start the test, it's a good idea to head nature's call so that you can concentrate fully on the exam.

Don't second guess yourself too much by marking lots of questions to review. Most of the time, your initial choice will be your best choice. I have had some handwringing instances in the past with taking certification tests where I marked 5-7 questions for review and changed my initial answer on at least four questions only to fail the test by 1-2 questions when the scoring finally came out. Did I miss the questions where I changed the answer at the end of the test? I cannot say definitively since AWS does not give you feedback on those types of metrics, but I have since changed the way that I use the review button. If I am having a hard time understanding the question and/or the answers, I will select something and then go on to the next question.

Even though they have taken out the lower associate test prerequisite before you take the professional level exam, I would not suggest just jumping right to the professional-level test. Sitting the exam for either the AWS SysOps Associate or the Developer Associate certification, which was once part of the necessary prerequisites to taking the DevOps professional exam, gives you a proper warmup to taking the professional-level exam. Even in my recertification, I usually take the lower test so that I can judge what new services and questions they have added to the exam in a simpler format, before going after the professional-level recertification a few weeks later.

Code Commit	Code Deploy	Code Pipeline	Life Cycle Hooks
ssh	AppSpec - YAML	Transitions	pending:wait
https	deployment group		
Triggers		Approval	pending:proceed
Lambda	In Place		
SNS		Job Worker	terminating:wait
KMS	Blue / Green		
			terminating:proceed

Figure 23.4 – Example of the brain dump format

I recommend going to the testing center to take the test. One of the main reasons for this is for the actual pieces of paper that you get. The night before taking the exam, practice with a blank sheet of paper and write some quick notes on the main heading topics that can just jog your memory, as shown in the preceding figure. I would practice this about 3-5 times the night before the test so that I could do this in less than 5 minutes once I was seated at my terminal at the testing center, as I was typing in my username and testing ID, but before I would press the start button. I used to teach this technique in the AWS meetup group I ran for 6 years, where many of the members used to challenge each other on accomplishing multiple levels of AWS certifications. You can't bring any type of cheat sheet into the testing center. You are only quickly dumping out information from your brain onto the scratch piece of paper provided and then handing that piece of paper back to the testing center once you are finished with the test. You can refer to these quick notes as the test is taking place; however, since you created them, it allows you to take a breath and relax while concentrating on reading and comprehending the questions.

Summary

Becoming a certified (or even recertified) AWS professional is a task that takes dedication, study, as well as the ability to successfully pass a strenuous testing format. Believe that your hard work and studying will pay off. The AWS DevOps Professional Certification is a badge of honor that is recognized by colleges and employers alike.

In the next chapter, we will present you with a mock exam so that you can both test your knowledge and continue to become familiar with the question-and-answer format that is shown on the professional exam.

24
Practice Exam 1

This section will have sample questions that are similar to the test so that you can become familiar with how to read and answer questions in the allotted time.

In this chapter, we're going to cover the following main topics:

- SDLC automation
- Configuration management and infrastructure as code
- Monitoring and logging
- Policies and standards automation
- Incident and event response
- High availability, fault tolerance, and disaster recovery

Before you start to answer the following questions, it is suggested that you use a timer of some kind to keep track of how long it is taking you to answer each question. None of the questions are actual AWS test questions. They are, however, made to be in the format that the test appears in. This includes long scenario-type questions along with multiple-choice answers that are very similar. Small details can mean the difference between the correct answer and one of the incorrect answers.

Both the questions and answers are long. Many of the answers have the same components in them as well, which makes this different than some of the associate tests where you can just eliminate a few of the answers that you know are wrong and then pick the answer from the one or two that are left.

Test questions

1. You have developed a set of CloudFormation templates that can be used by your current company to deploy its middleware solution for processing files. This application is deployed to a number of EC2 spot instances, and there is a user data component of CloudFormation that downloads a number of scripts, including an initialization shell script from an S3 bucket to help configure those instances. The S3 bucket that holds the scripts has versioning enabled. The only known copy of the scripts at the company is located in that S3 bucket. A new intern accidentally deletes the script when trying to upload a different static asset to the bucket. What is the quickest way to restore the script so that the servers may be redeployed using the CloudFormation templates?

   a. You will need to recreate the script and upload it with the same name to the S3 bucket. With file versioning enabled, you can only see the differences between files, not recovered deleted files.

   b. You will need to make a modification to the CloudFormation scripts so that they use the previous version ID of the script, which got deleted in the S3 bucket.

   c. You will need to go into the AWS Management Console, navigate to the S3 bucket that holds your deployment scripts, and then choose the List versions option. You can then find the script that was deleted and remove the delete marker.

   d. You will need to go into the AWS Management Console, navigate to the S3 bucket that holds your deployment script, and then choose the List versions option. You can then download the previous version of the script, which was deleted. After the script has been downloaded, you can then upload it back up to the bucket and rename it the original script name so that the EC2 instances can find it from the CloudFormation template.

2. You are part of a team that has recently developed a Spring Boot application and is ready to deploy it to the AWS cloud. Because the traffic to the application varies, you have configured the application to be launched in an Auto Scaling group. It is important to make sure that the application is running, so you have created a Bash script that will run on each EC2 instance and check the application's health periodically. If the instance is unhealthy, then it should be marked as such, and that instance should be replaced by a new, healthy one. What is the best way to construct the Bash script to perform this task?

   a. Construct the script to reboot the instance if the health check fails. Run the script as the root user. Once the Auto Scaling group detects the reboot, it will terminate the instance and then create a new instance in its place.

b. Construct the script to use the AWS CLI. Have the instance use the `autoscaling set-instance-health` command, letting the Auto Scaling group know that the instance is unhealthy. The Auto Scaling group will then terminate the instance and create a new one in its place.

c. Construct the script to use the AWS CLI. Have the instance use the `autoscaling put-notification-configuration` command to notify the Auto Scaling group that the instance is unhealthy. The Auto Scaling group will then terminate the instance and create a new one in its place.

d. Construct the script to use the AWS CLI. Have the instance use the `autoscaling enter-standby` command, letting the Auto Scaling group know that the instance is unhealthy. The Auto Scaling group will then terminate the instance and create a new one in its place.

3. You have been brought into a company that has recently established configuration and tagging standards for the infrastructure resources that it is running on AWS. You have been asked to design and build a near-real-time dashboard showing the compliance standpoint that will emphasize any violations.

a. Define the tagging and resource requirements in Amazon Inspector. Create a Simple Notification Service (SNS) topic to be notified if anomalies are found. Have Inspector check periodically for the compliance requirements and then send a notification to the SNS topic if found. Use AWS Security Hub to quickly visualize the violations.

b. Create a customized CloudWatch metric to track all the resource and tagging standards. Use the CloudWatch service to create a dashboard to visually track any resources that do not meet the tagging standards.

c. Enable the AWS Config service and use the configuration recorder to record all resources that are created and deleted. Have the configuration changes sent to an S3 bucket. Use Amazon QuickSight to create turn the data in the S3 bucket into visual information and dashboards for analysis so non-compliant resources can be easily spotted.

D. Define multiple resource configurations in AWS Service Catalog. Use Amazon CloudWatch to monitor any compliance violations from Service Catalog. Use the CloudWatch service to create a dashboard to visually track any resources that do not meet the tagging standards.

4. You have joined an application team that is moving their MySQL database from on-premises to the AWS cloud. This is a critical application, and therefore it needs more stability than the single on-premises server can give it. The application is read-intensive and has a 10/1 read-to-write ratio. Cost is the main objective, yet they do need to be aware of the project's budget. None of the current members of the team are skilled as DBAs, and all would rather focus on application development. What setup would you suggest to keep their database running should an Availability Zone fail in their current Region?

    a. Create a MySQL RDS instance. When setting up the instance, choose the Multi-AZ feature. Once the data has been imported, then create a read replica and program the application to utilized the read replica for the heavy reads.

    b. Create a MySQL RDS instance. Once the data has been imported, create a read replica in the alternate Region and program the application to utilized the read replica for the heavy reads.

    c. Create a CloudFormation template that stands up an Auto Scaling group and, in the launch template, configures the MySQL server on each of the two EC2 instances in different Regions. Have the configuration make the MySQL service run in a Master-Master setup so that any of the servers could be pointed to in case of failure.

    d. Create a CloudFormation template that stands up an Auto Scaling group and, in the launch template, configures the MySQL server on each of the two EC2 instances in different regions. Have a third smaller EC2 instance stood up as part of the template as a MySQL proxy so that the application servers can point to the proxy. The proxy will switch automatically between the current master and the slave.

5. You have previously built a CI/CD pipeline in AWS CodePipeline for your team's application. With the current pipeline, there were automated stages for code checkout, build, test, deploy to dev, and deploy to test. You are now updating CodePipeline to add a new stage to Deploy to Prod and adding a manual approval for the product owner before the code is released into the production environment. On the initial test of the new stage, the product owner informed you that after logging into the AWS Management Console to the CodePipeline pipeline with their IAM user, they found that she did not have permission to approve the build. How can you remedy this issue?

    a. Create a new SNS topic for the Deploy to Prod stage. Add the product owner to the topic. Have the product owner click the approval link in the SNS message when it is sent.

    b. From the AWS Management Console, go to the specific CodePipeline and add the product owner's IAM user as an approver for that pipeline.

c. From the AWS Management Console, go to the IAM service. Create a new group called `CodePipeline Approvers`. Attach the `AWSCodePipelineFullAccess` managed IAM policy to the group. Add the product owner to the group.

d. From the AWS Management Console, go to the IAM service. Create a new group called `CodePipeline Approvers`. Attach the `AWSCodePipelineApproverAccess` managed IAM policy to the group. Add the product owner to the group.

6. An organization has moved its code versioning system for its developers to AWS CodeCommit. Teams of developers create and work on feature branches while testing and then create a pull request when ready to merge into the main branch. The organization has set guidelines that no one should be able to directly commit to the main branch. Any team member who is a developer is part of the IAM group developers. This group was recently modified to add the `AWSCodeCommitPowerUser` managed policy and now all members of this group are able to commit to the main branch of any repository in the organization's AWS CodeCommit. What steps need to be taken to prevent this and enforce the organization's guidelines?

a. Create an added IAM policy that would allow the `codecommit:GitPush` action. Add a condition to the policy that specifies the CodeCommit repositories in the resource statement.

b. Create an added IAM policy that would deny the `codecommit:GitPush` action. Add a condition to the policy that specifies the CodeCommit repositories in the resource statement.

c. Alter the IAM policy and remove the `codecommit:GitPush` action.

d. Alter the IAM policy and add a deny rule for `codecommit:GitPush` that specifies the specific repositories where push is not allowed.

7. You have configured AWS CodeDeploy to automate deployments to both EC2 instances in your development and test environments located in your AWS account, along with a few RHEL servers that are still located on-premises. There is a deployment group configured that defines each of the specific instances that are included for deployment. There has been an announcement stating that there will be a hardware refresh on-premises for one of the instances in the deployment group, which will take 2 weeks to complete. During this period, no new deployments should be pushed to these instances. Which method is the most suitable for enacting this 2-week freeze for the specific on-premises server?

a. Create a new deployment group with a tag that is only used by the user.

b. Verify the tags used by both the deployment groups. Use the AWS CLI to remove the specific instance that will be in service with the `aws deploy remove-tags-from-on-premises-instances` command.

c. Use the AWS CLI to deregister the on-premises instance from the CodeDeploy deployment using `deploy deregister`.

d. Use the AWS CLI to uninstall the CodeDeploy agent from the on-premises instance using `deploy uninstall`.

8. A DevOps engineer on your team has submitted the following `buildspec.yaml` file for security review. The security review failed. You have been tasked with helping the junior engineer to review the file:

```
version: 0.2
env:
  variables:
    AWS_ACCESS_KEY: AS1A17LTH411HPQ11FRA
    AWS_SECRET_KEY: 0+a42An16Aa00qAaaALvLsl4CF+Nt2AaZ1Aaws67
    AWS_DEFAULT_REGION: us-east-2
    DB_PASSWORD: pil6mdr+sUBq
phases:
  build:
    commands:
      -aws s3 cp s3://deploy-bucket/my.cnf /tmp/my.cnf
      -sed -i '' s/DB_PW/${DB_PASSWORD}/ /tmp/my.cnf
      -aws s3 cp s3://deploy-bucket/instance.key /tmp/my.key
      -chmod 0600 /tmp/my.key
      -scp-i /tmp/my.cnf root@10.10.125.126:/etc/my.cnf
      -ssh-i /tmp/my.key root@10.10.125.126 /etc/init.d/mysqld restart
```

Figure 24.1 – The security review file

What changes would you recommend that the junior DevOps engineer make so that the `buildspec` file complies with security best practices (choose three)?

a. Add permissions to the CodeBuild role so that the necessary actions can be performed during the build process. Remove the access key and secret key from the file.

b. Use KMS encryption for the environment variables so that they don't appear in plaintext on the file.

c. Write all environment variables to a file. Store the file in S3 and pull the file down at execution time so that the variables don't appear in plaintext on the file.

d. Use AWS Secrets Manager to store the DB_PASSWORD value. Remove the DB value from the environment values once stored and then retrieve them when needed.

e. Create a run command in Systems Manager that would perform the commands. Use System Manager instead of SSH and SCP directly from the instance.

9. You have been brought into a company that is expanding its presence on the AWS cloud. They want to build out their footprint using CloudFormation. However, they would like to use common components and patterns throughout their various applications. Many of the underlying components such as the infrastructure and networking will not be modified frequently after being generated. The company would like to manage all of the common component items independently and allow other application stacks to reuse the components when they need to. How can you achieve this objective?

    a. Create a CloudFormation stack to hold all of the common resources. Other CloudFormation stacks will be able to use its resources by importing the resources from the AWS Management Console.

    b. Create a CloudFormation stack to generate all of the common resources. Export the output values so that other CloudFormation stacks can import the values using the GetAttribute function.

    c. Construct a CloudFormation stack to generate all of the common resources. Export the output values so that other CloudFormation stacks can import the values using the ImportValue function.

    d. Create a CloudFormation stack to generate all of the common resources. Any application stack can be created as a nested stack from this stack to use all of the common resources.

10. A company has just launched a new booking service that has both a website and a mobile application. Thanks in part to the marketing team's effort, the service has been a smash hit with customers that have tried it and it keeps growing in popularity. The CTO has implemented a new directive for the upcoming quarter to make the application as efficient as possible and make any necessary tweaks to the performance. In order to achieve this, the development team will need to monitor all the different details of the application to see the root causes of any issues, errors, and latency issues. How can they achieve this using native AWS tools and services?

    a. Configure Amazon Inspector to view the application. Periodically read the Inspector assessment reports for any latency issues found along with errors. Use timestamps to trace logs in CloudWatch Logs.

b. Configure Amazon Elasticsearch to subscribe to the CloudWatch log groups for the application. Use Kibana to graph out the latency times from user click to application response. Create a custom Kibana visualization to count the number of errors.

c. Configure a custom metric in Amazon CloudWatch to track latency. Create a dashboard on CloudWatch to track the metric.

d. Configure the AWS X-Ray SDK for the application. Send the segments to X-Ray for processing. View the service graphs and traces in the X-Ray service console.

11.   You have set up AWS CodeCommit for your company to use as the code versioning service. The main application team is developing a mobile phone app and submits the source code in one CodeCommit repository (repository A) in the AWS account (account A). The company has just acquired another company that had its own AWS account (account B) and some of the developers have been tasked to help the development of a new feature of the mobile phone app. What actions should you take to configure cross-account access so that the new developers who have accounts and IAM users in account B have the ability to access repository A in account A (choose two)?

a. Go to the AWS CodeCommit service in the AWS Management Console in account A. Share repository A with account B so that users in account B will have access to the repository.

b. Go to the AWS CodeCommit service in the AWS Management Console in account A. Add the IAM users of account B to the repository as users.

c. In AWS account A, go to the IAM console and create an IAM policy that allows access to CodeCommit's repository A. Then create an IAM role that can be assumed by another account and attach this policy. Allow users in account B to assume this role.

d. In AWS account B, go to the IAM console and create an IAM policy that allows full access to the CodeCommit service and is connected to the repository A ARN resource. Attach this new policy to all users that need access to repository A in AWS account A.

e. In AWS account B, create an IAM policy that allows for Security Token Service to assume role action so that a cross-account role can be assumed. Attach this new policy to all users that need access to repository A in AWS account A.

12. The company you are working for has just undergone an audit. The corrections that came back included the need to retain and store all system logs for 6 years. The development and operations teams need 30–60 days' worth of logs for troubleshooting purposes. The marketing department needs at least 6 months of web traffic logs for their analytics analysis. The management wants to make sure that the solution you come up with is cost-effective as well as meeting the auditor's requirements. How can you satisfy both management and the auditor's needs?

a. Put the logs onto an EBS volume. Create monthly EBS snapshots for long-term storage of the logs after 60 days.

b. Put the logs into Amazon S3 Glacier.

c. Put the logs into the Amazon CloudWatch Logs service and set the retention policy on the log groups to 6 years.

d. Put the logs into an Amazon S3 bucket. Create a bucket policy that moves the logs to infrequent access after 60 days and then to Amazon Glacier after 1 year.

13. Your company runs a .NET application in AWS that relies on around 50 Windows servers for the underlying infrastructure. The company has a policy that all of the servers in the development, test, and production environments must be kept up to date with the latest security patches. These Window servers have all been built from a master AMI image. The DevOps team that is responsible for the patching of the instances only consists of yourself and one other team member, so creating an automated way to perform this process is imperative; otherwise, you will be pressed to complete all of the updates in the allotted time window of 1 A.M.–4 A.M. on Saturday morning when there is very little customer traffic. How can you automate this process using native AWS services?

a. Create a Lambda function that can download and run the updates in PowerShell. Schedule the Lambda to run every week at 1 A.M. using CloudWatch Events.

b. Apply AWS Systems Manager Patch Manager to the Windows instance fleet. Use System Manager run commands to install the updates.

c. Apply AWS System Manager Patch Manager to the Windows instance fleet. Use System Manager Maintenance Windows to schedule the updates to run every week at 1 A.M.

d. Create custom Chef scripts in OpsWorks to download and install the updates in PowerShell. Create a task in OpsWorks to schedule and run the updates every week at 1 A.M.

14. The CTO has recently approached you, concerned about the security of the company's AWS account. They would like you to implement monitoring for any possible attacks that could be coming in against the company's AWS resources. They specifically emphasized monitoring against port scans, brute force attacks, or any SSH attacks. If an attack was detected, they would like it posted to the company's Microsoft Teams security channel. How can you go about achieving this?

    a. Set up Amazon GuardDuty. If suspicious activity is detected, trigger a Lambda function that will post to the Microsoft Teams channel.

    b. Create a Lambda function that will scan the CloudTrail logs for suspicious activity. If suspicious activity is found, it will post it to the Microsoft Teams channel.

    c. Set up Amazon Inspector. If suspicious activity is detected, trigger a Lambda function that will post to the Microsoft Teams channel.

    d. Create a Lambda function that will scan the VPC flow logs. If suspicious activity is found, it will post it to the Microsoft Teams channel.

15. The CEO has visited the DevOps team personally, stating that the company is promising six 9s of uptime to customers or the company would be giving large refunds. This leaves you and your team only 31.56 % of downtime per year. The main workload comprises multiple EC2 instances that are configured in Auto Scaling groups, running behind Application Load Balancers. How can you configure the workload to ensure that the company maintains that uptime promise even in the case of regional failure?

    a. Configure Amazon CloudFront in front of the load balancers and instances. CloudFront will cache the workload for customers in case of failure.

    b. Set up a Route 53 geoproximity routing record. Make sure that the Auto Scaling groups are set to utilize two Availability Zones. Have the Route 53 routing record point to the Application Load Balancers.

    c. Set up a Route 53 weighted routing record. Make sure that the Auto Scaling groups are set to utilize two Availability Zones. Have the Route 53 routing record point to the Application Load Balancers.

    d. Set up a Route 53 latency routing record. Implement your workload EC2 instances, Auto Scaling groups, and load balancers in two different regions. Have the Route 53 routing record point to the Application Load Balancers.

16. Your company wants to implement the Apache Cassandra NoSQL database on AWS. There is no managed service for Cassandra, so you will have to build this on EC2 instances. Your team is looking for you to choose the correct type of EBS volume so that they get the optimum performance for this high-performance NoSQL database. Which type of volume should you choose when building the EC2 instances in the cluster?

    a. Use IO1 EBS volumes when creating the instances.

    b. Use standard EBS volumes when creating the instances.

    c. Use GPL EBS volumes when creating the instances.

    d. Use GP2 EBS volumes when creating the instances.

17. You have been brought into an organization to help create a new AWS CodePipeline pipeline so that the team can implement continuous integration. The pipeline needs to pull the source code and then have a test stage run by AWS CodeBuild. The test includes extracting data from a database that requires a username and password. You will need to put these in the test stage using environment variables. How can you securely configure these variables in the CodeBuild stage of the pipeline?

    a. Use the CodeBuild environment variable options to store the secrets. Select the `Plaintext` type when storing the values and use a KMS key to encrypt the values.

    b. Use the CodeBuild environment variable options to store the secrets. Select the `SecureString` type when storing the values.

    c. Use AWS Secrets Manager to store the values. Update your CodeBuild environment values for the variables and use the names of the secrets as the values in `Plaintext`.

    d. Use AWS Systems Manager Parameter Store to store the values. Update your CodeBuild environment values for the variables and use the parameter names as the values as `Parameter` types.

18. You have been asked to help a team that is using OpsWorks to enhance the monitoring of their stack. This team is rather skilled in developing automations using Chef but just seems to know the basics to run their services in AWS. Which of the following will not help them enhance the monitoring of their application being deployed in OpsWorks?

    a. Utilize Amazon CloudWatch metrics and create a custom metric to track the application.

b. Utilize Amazon Cloud Trail to make sure that only authorized calls are being made to the application.

c. Utilize Amazon CloudWatch Logs to gather the logs from the application.

d. Utilize AWS Config to gather the application's configuration changes.

19. You have been brought into a small start-up whose marketing website is nothing but static content. The marketing department has been complaining that load times are currently too long, and this is affecting their search engine rating. They have only a limited budget for upgrades and would like to make the most effective use of their money. The start-up is also in the process of moving all digital assets to their new AWS account. They also want to make their site as fast as possible. Which of the following suggestions would best meet their needs?

a. Serve their website using an EC2 server. Add Amazon CloudFront in front as the content delivery network.

b. Move all static assets to S3. Serve the website on EC2 spot instances. Add Amazon CloudFront in front as the content delivery network.

c. Move the entire website to S3. Add Amazon CloudFront in front as the content delivery network.

d. Serve their website on EC2 spot instances. Add Amazon ElastiCache for content caching to speed up page load times.

20. An organization is running a successful room booking mobile application on AWS. They are using DynamoDB to store all the records of the transactions and confirmation codes obtained from credit card companies once the reservations have been made. They chose DynamoDB for its ability to quickly autoscale and handle any type of capacity that is needed without much management. These transactional records are of vital importance to the company and must not be lost lost due to any server failures. The organization has a policy that the financial transactions must be stored for 3 years in case of any customer disputes. What is the most cost-effective and reliable way to accomplish this?

a. Use CloudWatch Logs to capture the records from DynamoDB. Set the retention period in CloudWatch Logs to 3 years.

b. Use DynamoDB Streams to stream transition records to a Lambda function. Have the Lambda function write the record to an S3 bucket. Use a life cycle policy to move objects to S3 Glacier storage for cost savings.

c. Use DynamoDB global tables to replicate the data to a secondary region. Create a Lambda function that trims records based on the creation date of the record minus 3 years and runs on a nightly basis.

d. Use DynamoDB Streams to stream transition records to an S3 bucket. Use a life cycle policy to move objects to S3 Glacier storage for cost savings.

21. A company has set up multiple accounts using AWS organizations. They have just started implementing event-driven automation and are taking the first steps to have notifications being sent from the CloudWatch event bus located in the master account via SNS to topics. How can they set up their master account and grant access to all child accounts so that the events can be sent to the event bus located in the master account?

a. Create an IAM policy that allows for sending CloudWatch events. Attach that policy to a role in the master account that can be assumed by all child accounts in the organization.

b. Create an IAM policy in each of the child accounts that allows for sending CloudWatch events and specifies the ARN of the event bus in the master account. Attach this policy for any service-based role that needs to send events to the master account.

c. In the master account, go to the CloudWatch Events console and then choose your event bus. Go to add permission and then choose to add the entire organization by entering the ID of the organization.

d. In the master account, go to the CloudWatch Events console and then choose your event bus. Go to add permission and then add the ID for each of the child accounts that need permission to send events to the event bus.

22. You are working with a development team that is trying to track the performance of an application that they build and are running on a group of EC2 instances. This team is especially interested in any error messages that are being generated from their Java code and would like the full team to be notified if more than 5 error messages occur in a 5-minute time period. Which of the following solutions could you implement to fulfill the team's requirements?

a. Configure the instances to have all the Java logs write to the syslog on the EC2 instances in the user-data script. Use Kinesis Data Firehose to pull the syslog data for the instances and count the number of error messages. Create an SNS topic for the group of developers. Have Kinesis send a notification to the topic if there are more than 5 error logs in a 5-minute period.

b. Configure the instances to write to a single log group in Amazon CloudWatch Logs. Use Amazon Elasticsearch to subscribe to the log group. Build a Kibana dashboard so that the developers can see how many error logs are being generated on a minute-by-minute basis.

c. Configure the instance to install AWS Systems Manager Agent. Have the agent pull the logs to Amazon EventBridge. Create an SNS topic for the group of developers. Create a Lambda function that will send a notification to the SNS topic if there are more than 5 error logs in a 5-minute period.

d. Configure the instance to install the unified CloudWatch agent. Create a custom metric to count the number of errors in the Java logs. Create an SNS topic for the group of developers. Push the logs and the custom metric to Amazon CloudWatch. Create a CloudWatch alarm that will send a notification to the SNS topic if the custom metric reaches a value greater than 5 in a 5-minute period.

23. All of the developers in your organization currently have the ability to start and stop any of the EC2 instances that are currently running in the development account simply by logging onto the AWS Management Console and choosing to stop the instance under the Instance State settings. Some teams have been complaining that their workloads have been disrupted by other developers mistakenly stopping the wrong EC2 instances. How can you implement security measures so that only members from a particular team can start and stop their own EC2 instances using native AWS features?

a. For each development team in the company, create a policy that restricts the starting and stopping of instances to the $\{aws:Principal/Team\}$ tag as the resource.

b. Add a `Team` tag to all the EC2 instances that can help restrict access by comparing the $\{aws:Principal/Team\}$ tag attached by the individual developer to the `ec2:ResourceTag/Team` tag on the instance and seeing whether they match.

c. Add a `Team` tag to all the EC2 instances. Restrict access to each team in the developer policy by seeing whether the EC2 instance matches the team tag.

d. In the IAM developer role, remove the ability to start and stop instances. Create a CodePipeline job for each team that will allow them to see, start, and stop all of the instances for their development team.

24. You have an Extract, Transform, and Load (ETL) application that is sending its logs to CloudWatch Logs. The logs are landing in a CloudWatch log group and are formatted in JSON. The following is a sample of the log file:

```
{
    "eventType": "Process",
    "eventObject": "File",
    "errorCode": "CorruptFile"
    }
```

25. How can you create a metric filter so you can find all events where the error code was "CorruptFile"?

a. Filter Pattern: { $.errorCode = "CorruptFile" }

b. Filter Pattern: { $.errorCode == "CorruptFile" }

c. Filter Pattern { errorCode = "CorruptFile" }

d. Filter Pattern { errorCode == "CorruptFile" }

26. Your company has requested that you create a reliable and durable logging solution for their three AWS accounts so that they can track the changes being made to their AWS resources. Which of the following options would help you successfully do this?

a. Create a new CloudTrail with an existing bucket to store the logs. Select the global services option when creating the trail. Use IAM roles on the S3 bucket and S3 encryption to secure the bucket.

b. Create three new CloudTrails with a single new bucket to store the logs. One trail will be for the AWS Management Console, one will be for SDK commands, and one will be for the AWS CLI. Use multi-factor authentication for any delete actions and S3 encryption on the buckets.

c. Create a new CloudTrail with a new bucket to store the logs. Select the global services option when creating the trail. Use multi-factor authentication for any delete actions and S3 encryption on the bucket.

d. Create three new CloudTrails with three new buckets to store the logs. One trail will be for the AWS Management Console, one will be for SDK commands, and one will be for the AWS CLI. Use IAM roles on the S3 bucket and S3 encryption to secure the buckets.

27. Your company has used AWS Organizations to both create and manage their AWS account. There are multiple accounts, which include child accounts that contain organizational units that have been created using AWS Organizations. As the company grows, there is now a need to add uniform roles to each of the accounts. What is the most effective way to add all of the roles throughout the organization?

a. In the master account, use CloudFormation to deploy a template with which to create the new roles. Use CloudFormation StackSets to replicate the changes across the whole organization's child accounts.

b. In the master account of the organization, create a Service Control Policy (SCP), which will then add the roles to all the child accounts.

c. In the master account, use CloudFormation to deploy a template with which to create the new roles. Use CloudFormation change sets to replicate the changes across the whole organization's child accounts.

d. In the master account, create a run command in Systems Manager to create the new IAM roles. Have Systems Manager perform the command on all of the child accounts to create the new role.

28. Members of your DevOps team have come to you because they have noticed a problem with one of the Auto Scaling groups that has just been updated. Instead of reaching a steady state and serving traffic, the application is constantly scaling up and down numerous times an hour. Using the native features of CloudFormation, which settings could you help your team tune in order to stabilize the application?

a. Examine the current Auto Scaling group termination policy and change the value to terminate the oldest instance first so that newer instances stay online.

b. Examine the current Auto Scaling group termination policy and change the value to terminate to `ClosestToNextInstanceHour` so that the instances become more stabilized.

c. Find the previous version of the Auto Scaling launch template and deploy that version to stabilize the application.

d. Examine the current Auto Scaling group Health Check grace period and expand the time currently allocated for the instances to come online and become healthy.

29. A medium-sized software company has hired you as a DevOps consultant to help set up their deployment pipeline. They want to be able to push their tested code into their production environment in a quick manner but do not want the possibility of dealing with any downtime for their customers. You have worked with the application team to configure their application to run on containers and be deployed to Amazon Elastic Container Service (ECS). Their DNS is hosted on a third-party service and changes for the DNS would require a change ticket. What deployment method would you implement?

a. In the service settings of ECS, set the `minimumHealthyPercent` and `maximumHealthyPercent` values of tasks before you begin your rolling update to the service.

b. Create a CodeDeploy job for your updates. Use a blue/green deployment type. Set the configuration of the blue/green deployment to all-at-once.

c. Create a CodeDeploy job for your updates. Use a blue/green deployment type. Set the configuration of the blue/green deployment to linear.

d. Create a CodeDeploy job for your updates. Use a blue/green deployment type. Set the configuration of the blue/green deployment to canary.

30. Your team has recently been subject to a code audit and there were multiple findings of plaintext database usernames and passwords in the application code. This has been flagged by the company as unacceptable and the team has been given 30 days to remedy the problem. According to the company guidelines, the team needs to be able to store the secrets securely in an encrypted manner using a native AWS service that also has the ability to rotate the secret automatically every 60 days. How can you and your team remedy this issue?

a. Remove the database values that were previously set in the code base. Add environment variables to the deployment process. Insert the username and password as the corresponding deployment variables.

b. Remove the database values that were previously set in the code base. Store the username and values in AWS Systems Manager Parameter Store. Update your IAM role to allow access to retrieve the secret from Parameter Store.

c. Remove the database values that were previously set in the code base. Store the username and password in AWS Secrets Manager. Update your IAM role to allow access to retrieve the secret from Secrets Manager.

d. Use KMS to encrypt the values for the username and password for the database. Replace the previous values in the code with the newly encrypted values.

31. Your developers have created a DynamoDB table and seem to find that the performance always slows down after 20–25 minutes of their testing process. They can see from the basic monitoring on the AWS Management Console that their requests are being throttled. What can you do to help pinpoint the issue?

a. Increase the Read Capacity Units (RCUs) on the table so that the queries are no longer throttled.

b. Add enhanced CloudWatch monitoring with alarms whenever throttling occurs.

c. Enable Contributor Insights on the table so that the keys that are being throttled the most are shown.

d. Add adaptive capacity to the table so that the extra RCUs are spread evenly across partitions that are becoming hot.

Test answers

1. c

   Since the bucket had versioning turned on, removing the delete marker restores the object, and any current or future deployments using that script will be able to find it.

2. b

   You can use the AWS CLI and the `autoscaling set-instance-health` command along with the `--health-status Unhealthy` flag to have the instance be out of service.

   More information can be found on the documentation page for the AWS CLI at the following link: `https://docs.aws.amazon.com/cli/latest/reference/autoscaling/set-instance-health.html`.

3. c

   With the configuration recorder, AWS Config can evaluate new resources being created in an account. Items are recorded as JSON snapshots to an S3 bucket declared in the setup.

4. a

   RDS uses DNS to switch over to the standby replica for a seamless transition in a Multi-AZ implementation.

5. d

    Although the `AWSCodePipeline_FullAccess` policy would give
    approval access, it doesn't follow the AWS principle of least privilege. This
    policy would give the product owner more privileges than they need. Hence
    `AWSCodePipelineApproverAccess` would add the access that they
    were missing.

6. b

    You cannot modify a managed AWS policy, and hence this disqualifies both answers
    c and d. You are trying to prevent the action of users directly pushing to the master
    branch.

7. b

    You remove an on-premises instance tag from an on-premises instance when that
    tag is no longer being used, or if you want to remove the on-premises instance from
    any deployment groups that rely on that tag.

8. a, d, and e

    Writing all the values to a file in S3 would not guarantee their integrity. It is a much
    more secure practice to use roles and to have database values stored in a credential
    store, such as Secrets Manager or Systems Manager Parameter Store.

9. c

    The `Fn::ImportValue` intrinsic function returns the value of an exported value
    of a previously created stack. This is used to create cross-stack references.

10. d

    Amazon Inspector is a security service and would not find latencies in an
    application. The X-Ray service helps developers identify the root cause of
    performance issues and errors.

11. c and e

    In order to gain access to resources in another AWS account, a cross-account
    IAM role needs to be created so that it can be assumed by the other AWS account.
    Likewise, the other account's users must have a policy attached to them that will
    allow for the assumption of the role.

12. d

The use of Amazon S3 life cycle policies will allow you and your team to have both immediate access to the current logs and the ability to store them on the low-cost option in Amazon Glacier as the auditor requires.

13. c

Using AWS Systems Manager, you can use the combination of Patch Manager and Maintenance Windows to successfully automate this task in the recommended way by AWS.

14. a

Amazon GuardDuty can detect all of the different types of events that the CTO was concerned about. Adding a Lambda function that will post to the company's Microsoft Teams channel will satisfy the request.

15. d

Only by deploying to multiple regions can you make sure that you are protected against a regional failure. Using a latency-based record in Route 53 will automatically point to the set of servers that is responding the quickest to a request in case of a failure.

16. a

io1 volumes are crafted for workloads that require sustained IOPS performance and I/O-intensive database workloads.

17. d

Sensitive values should be stored in either Systems Manager Parameter Store or Secrets Manager. If Secrets Manager is used for CodeBuild, then the variable type should be selected as Secrets Manager and not `Plaintext`.

18. d

Amazon Config is not a service used for monitoring and metrics.

19. c

20. c

Since the full site is made of static content, and S3 is the least expensive and reliable solution, this is the most optimum choice. Using S3 as the origin and being fronted by Amazon CloudFront will allow for assets to be served faster to the end user via edge locations.

21. b

DynamoDB Streams cannot directly stream to S3 as a source, so a Lambda Function would need to first GetRecords and then put them to the specified S3 bucket with the life cycle policy.

22. c

A CloudWatch event bus allows you to add permissions on an organizational level. This also helps if your organization grows, as you don't have to keep track of which accounts you have added to the event bus or remember to take the extra step of adding the account to the event bus permissions when it is created.

23. d

24. b

Using tags on EC2 instances can be the first part of differentiating the ownership of EC2 instances between teams. ec2:ResourceTag is a tag that exists on an EC2 resource and can be verified against an IAM policy.

25. a

The syntax of the metric filter would be { $.errorCode = "CorruptFile" }.

26. c

The use of the global option will send all of the API actions recorded to a single S3 bucket. Adding in MFA will prevent any unauthorized deletions of the logs.

27. a

AWS CloudFormation StackSets extends the functionality of CloudFormation stacks, allowing you to create, update, or delete stacks across multiple accounts and regions with a single operation.

28. d

Although you may be able to roll back with a code versioning system, adjusting the current health check of the Auto Scaling group will allow your instances to come online and become healthy.

29. a

A rolling type of deployment would be the most optimal type of deployment, especially when the DNS is hosted on a third-party provider.

30. c

    Although both Systems Manager Parameter Store and AWS Secrets Manager will
    safely secure secrets according to the new guidelines in this scenario, only Secrets
    Manager will automatically rotate the database secrets.

31. c

    Amazon CloudWatch Contributor Insights integrates with DynamoDB to provide
    information about the most accessed and throttled items in a table or global
    secondary index.

Question breakdown

If you are interested in how you are performing in a particular domain, then we have how
the sample test questions would be grouped based on the test domains, as follows:

Domain 1 – SDLC automation:

- Question 6
- Question 11
- Question 17
- Question 23
- Question 26
- Question 27
- Question 28

Domain 2 – Configuration management and infrastructure as code:

- Question 5
- Question 7
- Question 9
- Question 16

Domain 3 – Monitoring and logging:

- Question 10
- Question 18
- Question 21

- Question 22
- Question 24
- Question 30

Domain 4 – Policies and standards automation:

- Question 3
- Question 8
- Question 12
- Question 13

Domain 5 – Incident and event response:

- Question 1
- Question 2
- Question 14
- Question 19

Domain 6 – High availability, fault tolerance, and disaster recovery:

- Question 4
- Question 15
- Question 20
- Question 25

If you are missing questions particularly in a specific area, then go back and reread those chapters, look to the end of *Chapter 23*, *Overview of the DevOps Professional Certification Test*, for one of the AWS Whitepapers that could give you more insight into that topic, or even watch some of the past re:Invent talks or AWS TechTalk videos to gain a better understanding of the domain.

Now, let's have one final summary of our journey to certification.

Summary

In this chapter, you have been presented with a number of sample DevOps professional exam questions so that you can practice all of the items that you have learned up to this point, as well as reading and comprehending the question and answer format that will be on the test.

Hopefully, at this point, you feel confident to take and pass the DevOps Professional certification exam. Once you pass, you will join the small subset of individuals who can be quickly recognized for their skill in not only DevOps but also AWS and cloud technologies.

Why subscribe?

Other Books You May Enjoy

If you enjoyed this book, you may be interested in these other books by Packt:

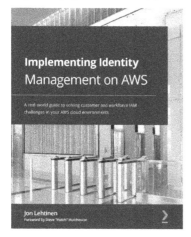

Implementing Identity Management on AWS

Jon Lehtinen

ISBN: 9781800562288

- Understand AWS IAM concepts, terminology, and services

- Explore AWS IAM, Amazon Cognito, AWS SSO, and AWS Directory Service to solve customer and workforce identity problems

- Apply the concepts you learn about to solve business, process, and compliance challenges when expanding into AWS

- Navigate the AWS CLI to unlock the programmatic administration of AWS

- Explore how AWS IAM, its policy objects, and notational language can be applied to solve security and access management use cases

- Relate concepts easily to your own environment through IAM patterns and best practices

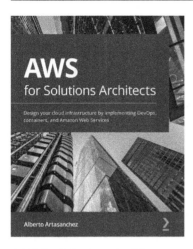

AWS for Solutions Architects

Alberto Artasanchez

ISBN: 9781789539233

- Rationalize the selection of AWS as the right cloud provider for your organization
- Choose the most appropriate service from AWS for a particular use case or project
- Implement change and operations management
- Find out the right resource type and size to balance performance and efficiency
- Discover how to mitigate risk and enforce security, authentication, and authorization
- Identify common business scenarios and select the right reference architectures for them

Packt is searching for authors like you

If you're interested in becoming an author for Packt, please visit authors. packtpub.com and apply today. We have worked with thousands of developers and tech professionals, just like you, to help them share their insight with the global tech community. You can make a general application, apply for a specific hot topic that we are recruiting an author for, or submit your own idea.

Share your thoughts

Now you've finished *AWS Certified DevOps Engineer - Professional Certification and Beyond*, we'd love to hear your thoughts! Scan the QR code below to go straight to the Amazon review page for this book and share your feedback or leave a review on the site that you purchased it from.

https://packt.link/r/1801074453

Your review is important to us and the tech community and will help us make sure we're delivering excellent quality content.

Index

W

X

Y

www.ingramcontent.com/pod-product-compliance
Lightning Source LLC
Chambersburg PA
CBHW060920060326
40690CB00041B/2728